ETHNOPHARMACOLOGY AND BIODIVERSITY OF MEDICINAL PLANTS

ETHNOPHARMACOLOGY AND BIODIVERSITY OF MEDICINAL PLANTS

Edited by

Jayanta Kumar Patra, PhD, PDF
Gitishree Das, PhD, PDF
Sanjeet Kumar, PhD, DELF
Hrudayanath Thatoi, PhD

APPLE ACADEMIC PRESS

Apple Academic Press Inc.
3333 Mistwell Crescent
Oakville, ON L6L 0A2
Canada

Apple Academic Press Inc.
1265 Goldenrod Circle NE
Palm Bay, Florida 32905
USA

Exclusive worldwide distribution by CRC Press, a member of Taylor & Francis Group

International Standard Book Number-13: 978-1-77188-773-1 (Hardcover)
International Standard Book Number-13: 978-0-42939-819-3 (eBook)

Library and Archives Canada Cataloguing in Publication

Title: Ethnopharmacology and biodiversity of medicinal plants / edited by Jayanta Kumar Patra, Gitishree Das, Sanjeet Kumar, Hrudayanath Thatoi.

Names: Patra, Jayanta Kumar, editor.

Description: Includes bibliographical references and index.

Identifiers: Canadiana (print) 20190125144 | Canadiana (ebook) 20190125179 | ISBN 9781771887731 (hardcover) | ISBN 9780429398193 (ebook)

Subjects: LCSH: Medicinal plants—India. | LCSH: Medicinal plants—Variation—India. | LCSH: Ethnopharmacology—India. | LCSH: Ethnobotany—India. | LCSH: Traditional medicine—India.

Classification: LCC QK99.I4 E84 2019 | DDC 581.6/340954—dc23

Library of Congress Cataloging-in-Publication Data

Names: Patra, Jayanta Kumar, editor. | Das, Gitishree, editor. | Kumar, Sanjeet, 1982- editor. | Thatoi, H. N., editor.

Title: Ethnopharmacology and biodiversity of medicinal plants / edited by Jayanta Kumar Patra, Gitishree Das, Sanjeet Kumar, Hrudayanath Thatoi.

Description: 1st edition. | Palm Bay, Florida : Apple Academic Press, 2019. | Includes bibliographical references and index. | Summary: "This book will be especially useful for undergraduate and postgraduate courses focusing on diversity of medicinal plants, ethnobiology, and traditional therapeutic systems. It is also a valuable reference for research students and professionals seeking to find more information about various biological applications. The book focuses on the real-life research aspects of advances in unexplored plants and ethnopharmacology, which provides a new perspective for all interested readers.

Key Features: Includes real-life application for unexplored plants with medicinal properties Incorporates findings from researchers in the field to discuss the biodiversity in India Proposes medicinal usages for plants and compares them to current medical practices Discusses conservation methods for traditional medicine as well as for endangered plants Provides a comprehensive guide to ethnopharmacology and other ethnomedical practices"-- Provided by publisher.

Identifiers: LCCN 2019026715 (print) | LCCN 2019026716 (ebook) | ISBN 9781771887731 (hardcover) | ISBN 9780429398193 (ebook)

Subjects: LCSH: Medicinal plants--Research--India. | Ethnopharmacology--Research--India. | Biodiversity--Research--India. | Materia medica, Vegetable.

Classification: LCC QK99.I4 .E86 2019 (print) | LCC QK99.I4 (ebook) | DDC 615.3/210954--dc23

LC record available at https://lccn.loc.gov/2019026715

LC ebook record available at https://lccn.loc.gov/2019026716

Apple Academic Press also publishes its books in a variety of electronic formats. Some content that appears in print may not be available in electronic format. For information about Apple Academic Press products, visit our website at **www.appleacademicpress.com** and the CRC Press website at **www.crcpress.com**

About the Editors

Jayanta Kumar Patra, PhD, PDF

Jayanta Kumar Patra, MSc, PhD, PDF, is currently working as an Assistant Professor at the Research Institute of Biotechnology and Medical Converged Science, Dongguk University, Gyeonggi-do, Republic of Korea. His current research is focused on nanoparticle synthesis by green technology methods and their potential application in biomedical and agricultural fields. He has about 12 years of research and teaching experience in the field of food, pharmacology, and nanobiotechnology. Since 2007, he has published more than 120 papers in various national and international peer-reviewed journals and around 25 book chapters in different edited books. Dr. Patra has also authored 9 books in various publications like STUDIUM Press (India); STUDIUM Press LLC USA; Springer Nature publisher; CRC Press and Apple Academic Press, Inc., Canada, CRC Press, a Taylor & Francis group. In addition, he is editorial board member of several national and international journals.

Gitishree Das, PhD, PDF

Gitishree Das, PhD, PDF, is currently working as an Assistant Professor at the Research Institute of Biotechnology and Medical Converged Science, Dongguk University, Ilsandong, Gyeonggi-do, Republic of Korea. She has 10 years of research experience in the field of rice molecular biology, breeding, and endophytic bacteria, green nanotechnology and three year of teaching experience. Dr. Das has completed her PhD (Life Sciences) at North Orissa University, India on rice gene pyramiding (Central Rice Research Institute, Cuttack, India) and PDF (Biotechnology) from Yeungnam University, South Korea. She has undergone professional training in the field of biotechnology at various international and national organizations. Her current research is focused on the biosynthesis of nanoparticles using food wastes and plant materials and their application in the biomedical and agricultural fields. To her credit, she has published around 60 research articles in international and national reputed journals and ten book chapters. Dr. Das has also authored four books for Springer Nature and LAMBERT Academic publishers. She is an editorial board member of some national and international journals.

Sanjeet Kumar, PhD, DELF

Sanjeet Kumar, PhD, DELF, is currently working as President at the Ambika Prasad Research Foundation, Bhubaneswar, India, and Zonal Head of the Sustainable Biodiversity Committee, Odisha Wildlife Hub, Odisha, India. His current research is focused on status, taxonomy, diversity, phytochemistry, and antimicrobial activity of medicinal plants; population analysis and restoration of rare, endangered, and threatened plant species; wetlands ecosystems and their biowealth; various management practices in protected areas particularly in biosphere reserve and breeding habitat of tigers; establishment of relationship between flora and fauna; and restoration of medicinal plants found in riverine ecology. He has eight years of research experiences on medicinal plants and on rare, endangered, and threatened taxa. Dr. Kumar has completed his PhD in Biotechnology at Ravenshaw University, India, on Ethnobotanical, Nutritional, and Antimicrobial values of selected wild Dioscorea species of Similipal Biosphere Reserve, India. He holds a DELF from the Alliance Francaise du Bengal, Republique Francaise. He has about 50 publications in journals of national and international repute and many popular articles on medicinal plants.

Hrudayanath Thatoi, PhD

Hrudayanath Thatoi, PhD, is currently working as a Professor and Head at the Department of Biotechnology, North Orissa University, Odisha, India. His has around 30 years of teaching and research experience. His research activities are primarily based on medicinal plants, bioremediation, biodiversity, ethnopharmacology, mangrove biology, etc. Professor Thatoi obtained his MPhil and PhD from Utkal University, Odisha, India, and his research work was based on N2 fixation in legume plants under dual inoculation of Rhizobium and VAM fungi and contributed significantly towards the development of technology for mine waste reclamation. He has handled many research projects from state government and central government organizations, such as the Department of Science and Technology, Government of Odisha; UGC-DAE Consortium For Scientific Research, Government of India; Department of Forest, Government of Odisha, etc. Around 20 students have obtained a PhD under his guidance and supervision. In addition, several MSc, MTech, and BTech students have received his guidance for their dissertation works. Professor Thatoi has published more than 200 research papers in various national and international reputed journals and around 30 book chapters. Professor Thatoi has also authored around 10 books/manuals for different notable publishers including STUDIUM Press LLC (USA), IK International, Narosa Publishers, Biotech Books, APH Publication, etc. He has also authored a textbook on microbiology and immunology, published by India Tech Publication, New Delhi, for MSc and BSc students. Professor Thatoi has contributed immensely to the field of microbiology and biotechnology throughout his research and teaching career.

Contents

Contributors

Ovaid Akhtar
Department of Botany, Kamla Nehru Institute of Physical and Social Sciences, Sultanpur–228118, U.P., India

Asma Khatoon
National Institute of Technology, Rourkela–769008, Odisha, India

Amulya Sai Bakshi
National Institute of Technology, Rourkela–769008, Odisha, India

Gobinda Bal
Department of Biodiversity & Conservation of Natural Resources, Central University of Orissa, Landiguda, Koraput–764021, Odisha, India

Bighneswar Baliyarsingh
Department of Biotechnology, College of Engineering and Technology, Techno Campus, Ghatikia, Bhubaneswar, Odisha, India

Kakoli Banerjee
Department of Biodiversity & Conservation of Natural Resources, Central University of Orissa, Landiguda, Koraput–764021, Odisha, India

Laldingngheti Bawitlung
Department of Horticulture, Aromatic and Medicinal Plants, School of Earth Sciences and Natural Resources Management, Mizoram University, Aizawl–796004, India

Archita Behera
Department of Botany, Ravenshaw University, Cuttack, Odisha, India

Arpita Behera
National Institute of Technology, Rourkela–769008, Odisha, India

Bikash Bhattarai
Regional Centre, Ambika Prasad Research Foundation, Khamdong, Singtam, Sikkim, India

Dipankar Borah
Department of Botany, Rajiv Gandhi University, Arunachal Pradesh, India

Subhendu Chakroborty
Department of Biomedical Sciences and Engineering, National Central University, Jhongli City–320, Taiwan

Anand Kumar Chaudhari
Laboratory of Herbal Pesticides, Centre of Advanced Study in Botany, Institute of Science, Banaras Hindu University, Varanasi–221005, India

Shraddha Chauhan
Department of Biotechnology, National Institute of Technology, Raipur Chattisgarh–492010, India

Srinivas Chowdappa
Fungal Metabolite Research Laboratory, Department of Microbiology and Biotechnology, Bangalore University, Jnana Bharathi Campus, Bangalore–560056, Karnataka, India

Gitishree Das
Research Institute of Biotechnology & Medical Converged Science, Dongguk University-Seoul, Gyeonggi-do–10326, Republic of Korea

Somenath Das
Laboratory of Herbal Pesticides, Centre of Advanced Study in Botany, Institute of Science, Banaras Hindu University, Varanasi–221005, India

Saswati Dash
Department of Biotechnology, College of Engineering and Technology, Techno Campus, Ghatikia, Bhubaneswar, Odisha, India

B. K. Datta
Plant Taxonomy and Biodiversity Laboratory Department of Botany, Tripura University Suryamaninagar–799022, Tripura, India

Chitrangada Debsarma
Department of Biodiversity & Conservation of Natural Resources, Central University of Orissa, Landiguda, Koraput–764021, Odisha, India

Rajkumari Supriya Devi
Ambika Prasad Research Foundation, Regional Centre, Imphal, Manipur, India

Nabin Kumar Dhal
CSIR- Institute of Mineral and Material Technology, Bhubaneswar, India

Nawal Kishore Dubey
Laboratory of Herbal Pesticides, Centre of Advanced Study in Botany, Institute of Science, Banaras Hindu University, Varanasi–221005, India

Abhishek Kumar Dwivedy
Laboratory of Herbal Pesticides, Centre of Advanced Study in Botany, Institute of Science, Banaras Hindu University, Varanasi–221005, India

Kadakasseril Varghese George
Department of Botany, St. Berchmans' College, Changanassery, Kottayam–686101, Kerala, India

Padan Kumar Jena
Department of Botany, Ravenshaw University, Cuttack, Odisha, India

Harbans Kaur Kehri
Sadasivan Mycopathology Laboratory, Department of Botany, University of Allahabad–211002, UP, India

Gopal Raj Khemendu
Department of Biodiversity & Conservation of Natural Resources, Central University of Orissa, Landiguda, Koraput–764021, Odisha, India

Sanjeet Kumar
Ambika Prasad Research Foundation, Bhubaneswar–751006, Odisha, India

Sanjeet Kumar
Bioresource Database and Bioinformatics Division, Institute of Bioresources and Sustainable Development, Govt. of India, Takyelpat, Imphal–795001, Manipur, India

Sanjeet Kumar
School of Life Sciences, Ravenshaw University, Cuttack, India

Siddharth Kumar
National Institute of Technology, Rourkela–769008, Odisha, India

Rajndra K. Labala
Distributed Information Sub-Centre, Institute of Bioresources and Sustainable Development, Govt. of India, Takyelpat, Imphal–795001, Manipur, India

Padma Mahanti
Department of Forest and Wildlife, Trivandrum, India

Madhusmita Mahapatra
Institute of Bioresources and Sustainable Development, Sikkim Centre, Tadong, Gangtok, Sikkim, India

Debashis Mandal
Department of Horticulture, Aromatic and Medicinal Plants, School of Earth Sciences and Natural Resources Management, Mizoram University, Aizawl–796004, India

Jose Mathew
Department of Botany, Kerala University, Kariavattom, Thiruvananthapuram–695581, Kerala, India

Gopa Mishra
Department of Biodiversity & Conservation of Natural Resources, Central University of Orissa, Landiguda, Koraput–764021, Odisha, India

Bernadette Montanari
International Institute of Social Studies, Erasmus University Rotterdam, The Hague, Netherlands

Raghvendra Pratap Narayan
Netaji Subhash Government Girls P.G. College Lucknow–201010, UP, India

Suraja Kumar Nayak
Department of Biotechnology, College of Engineering and Technology, Techno Campus, Ghatikia, Bhubaneswar, Odisha, India

Dheeraj Pandey
Sadasivan Mycopathology Laboratory, Department of Botany, University of Allahabad–211002, UP, India

Jayanta Kumar Patra
Research Institute of Biotechnology & Medical Converged Science, Dongguk University-Seoul, Gyeonggi-do–10326, Republic of Korea

Rakesh Paul
Department of Biodiversity & Conservation of Natural Resources, Central University of Orissa, Landiguda, Koraput–764021, Odisha, India

Ichhamati Pradhan
Department of Biotechnology, College of Engineering and Technology, Techno Campus, Ghatikia, Bhubaneswar, Odisha, India

Rahul Pradhan
National Institute of Technology, Rourkela–769008, Odisha, India

Sushant Prajapati
National Institute of Technology, Rourkela–769008, Odisha, India

P. M. Radhamany
Department of Botany, Kerala University, Kariavattom, Thiruvananthapuram–695581, Kerala, India

Moumita Saha
Plant Taxonomy and Biodiversity Laboratory Department of Botany, Tripura University Suryamaninagar–799022, Tripura, India

Nihar Ranjan Sahoo
Department of Biodiversity & Conservation of Natural Resources, Central University of Orissa, Landiguda, Koraput–764021, Odisha, India

Reecha Sahu
Department of Biotechnology, National Institute of Technology, Raipur Chattisgarh–492010, India

S. C. Sahu
Department of Botany, North Orissa University, Baripada–757003, Odisha, India

P. M. Salim
MS Swaminathan Research Foundation, Puthoorvayal, Kalpetta, Wayanad–673577, Kerala, India

C. Sareena
School of Biotechnology, National Institute of Technology Calicut (NITC), P.O. NIT Campus, Kozhikode–673601, Kerala, India

Angana Sarkar
National Institute of Technology, Rourkela–769008, Odisha, India

Amritesh C. Shukla
Department of Botany, University of Lucknow, Lucknow–226007, India

Akanksha Singh
Laboratory of Herbal Pesticides, Centre of Advanced Study in Botany, Institute of Science, Banaras Hindu University, Varanasi–221005, India

Archana Singh
Laboratory of Herbal Pesticides, Centre of Advanced Study in Botany, Institute of Science, Banaras Hindu University, Varanasi–221005, India

L. Amitkumar Singh
Distributed Information Sub-Centre, Institute of Bioresources and Sustainable Development, Govt. of India, Takyelpat, Imphal–795001, Manipur, India

P. Devanda Singh
Bioresource Database and Bioinformatics Division, Institute of Bioresources and Sustainable Development, Govt. of India, Takyelpat, Imphal–795001, Manipur, India

Puyam Devanda Singh
Ambika Prasad Research Foundation, Regional Centre, Imphal, Manipur, India

Vipin Kumar Singh
Laboratory of Herbal Pesticides, Centre of Advanced Study in Botany, Institute of Science, Banaras Hindu University, Varanasi–221005, India

Pragya Srivastava
Sadasivan Mycopathology Laboratory, Department of Botany, University of Allahabad–211002, UP, India

T. V. Suchithra
School of Biotechnology, National Institute of Technology Calicut (NITC), P.O. NIT Campus, Kozhikode–673601, Kerala, India

Swetha Sunil
School of Biotechnology, National Institute of Technology Calicut (NITC), P.O. NIT Campus, Kozhikode–673601, Kerala, India

A. Anju Suresh
School of Biotechnology, National Institute of Technology Calicut (NITC), P.O. NIT Campus, Kozhikode–673601, Kerala, India

Raghavarapu Swathi
National Institute of Technology, Rourkela–769008, Odisha, India

Sumpam Tangjang
Department of Botany, Rajiv Gandhi University, Arunachal Pradesh, India

Nurpen M. Thangjam
Department of Horticulture, Aromatic and Medicinal Plants, School of Earth Sciences and Natural Resources Management, Mizoram University, Aizawl–796004, India

Sunil S. Thorat
Distributed Information Sub-Centre, Institute of Bioresources and Sustainable Development, Govt. of India, Takyelpat, Imphal–795001, Manipur, India

Sabeela Beevi Ummalyma
Institute of Bioresources and Sustainable Development (IBSD), A National Institute under Department of Biotechnology Govt. of India, Takyelpat, Imphal–795001 Manipur, India

Lata S. B. Upadhyay
Department of Biotechnology, National Institute of Technology, Raipur Chattisgarh–492010, India

Neha Upadhyay
Laboratory of Herbal Pesticides, Centre of Advanced Study in Botany, Institute of Science, Banaras Hindu University, Varanasi–221005, India

Fazilath Uzma
Fungal Metabolite Research Laboratory, Department of Microbiology and Biotechnology, Bangalore University, Jnana Bharathi Campus, Bangalore–560056, Karnataka, India

Ifra Zoomi
Sadasivan Mycopathology Laboratory, Department of Botany, University of Allahabad–211002, UP, India

Abbreviations

2AA	2-amino-anthracene
ABR	Agasthyamala Biosphere Reserve
AGB	above ground biomass
AICRPE	All India Coordinated Research Project on Ethnobiology
AMR	antimicrobial resistance
AMR	Antimicrobial resistance
BF	Bhitarkanika mangrove forest
CBD	convention on biological diversity
CF	colonization frequency
CNH	Central National Herbarium
CP	cyclophosphamide
CR	colonization rates
DBH	diameter at breast height
DHFR	dihydrofolate reductase
DOE	Department of Energy
DPPH	2,2-diphenyl-1-picrylhydrazyl
ECDC	European Centre for Disease Prevention and Control
EIA	Environmental Impact Assessment
ENNG	Ethyl-N-nitro-N-nitrosoguanidine
FDA	Food and Drug Administration
GRAS	generally recognized as safe
HAL	Hindustan Aeronautics Limited
HELA	human cervical epithelioid carcinoma
IMTECH	Institute of Microbial Technology
IPNI	International Plant Naming Index
IR	isolation rates
ISE	International Society of Ethnobiology
IUCN	International Union for Conservation of Nature and Natural Resources
MAPs	medicinal and aromatic plants
MIC	minimum inhibitory concentrations
MNNG	Methyl-N′nitro-N-nitrosoguanidine
MNP	marine natural products
MOEFCC	Ministry of the Environment, Forest and Climate Change

MSSRF	M.S. Swaminathan Research Foundation
NAD	Naval Armament Depot
NFP	National Forest Policy
NPs	nanoparticles
NRSA	National Remote Sensing Agency
NRSC	National Remote Sensing Center
PDA	potato dextrose agar
PDF	passport data form
QSAR	quantitative structure-activity relationships
RF	relative frequency
SBR	Similipal Biosphere Reserve
SF	Sundarban mangrove forest
TB	tuberculosis
TM	traditional medicinal
USGS	United States Geological Survey
WHO	World Health Organization

Preface

The traditional therapeutic systems and biodiversity of the world are under varying degrees of threats due to rapid modernization and urbanization. During the last two decades, there has been a greater consciousness about the need for the conservation of biodiversity and ethnobotanical practices. It is widely recognized that the loss of bioresources and traditional practices have great health impacts on rural and tribal areas as well as in the formulation of new drugs as ethnobotany is the base of pharmacological industries. It is needless to mention here that the forested tribal areas rich with ethnobotanical knowledge which offers the broadest array of options for sustainable health care activities for human welfare and for adapting contemporary changes. Considering the pivotal importance of the great diverse practices and diversity of medicinal plants, there is an urgent need to document and study the ethnobotanical agents from our bioresources for the formulation of new drugs to fight against diseases and disorders as well as contemporary clinical and biodiversity issues. The other exigency steps are the implementation of appropriate conservation strategies for the protection of the traditional therapeutic skills and medicinal bioresources of the country where about 8,000 plants species are reported having potent medicinal and pharmacological values.

The present book, *Ethnopharmacology and Biodiversity of Medicinal Plants,* highlights the importance of medicinal plant wealth to the formulation of new drugs and brings attention towards the conservation of our bioresources along with traditional therapeutic values. Also the present documentation provides transparent and informative ideas covering a wide array of topics like biodiversity bioprospecting, importance of mangroves wealth, urban biodiversity and their floral wealth with uses, medicinal microflora, role of protected areas in the conservation of medicinal plants and traditional therapeutic skills, and role of common nutraceutical in our day-to-day life. The information pertaining to the topics of the documentation is based on fieldwork and literature survey made by the different esteemed researchers of diverse landscapes of India. It is hoped that this edited book will not only provide the knowledge on ethnobotany and biodiversity of medicinal plants but also will give a glimpse on rich traditional therapeutic systems and biodiversity of India from Kanyakumari to Arunachal Pradesh and biodiversity hotspot from Indo-Burma to Western and Eastern Ghats of India. This edited book will be useful particularly for the healthcare sector, and it will also provide a baseline data for the study of ethnopharmacology and reverse pharmacology.

—Jayanta Kumar Patra, PhD, PDF
Gitishree Das, PhD, PDF
Sanjeet Kumar, PhD, DELF
Hrudayanath Thatoi, PhD

Foreword

At this time of resource crunch, sustainable development is the buzzword for the whole world. This is the only solution for the effective and efficient use of all natural resources. Proper information about the quality and quantity of these resources is a must for ensuring sustainable usage and conservation.

The present edited volume, titled *Ethnopharmacology and Biodiversity of Medicinal Plants* by Dr. Jayanta Kumar Patra, Dr. Gitishree Das, Dr. Sanjeet Kumar, and Prof. H. N. Thatoi is an important addition to the knowledge repository in this direction.

The papers in the volume are selected carefully covering varied aspects of the titled subject, providing the reader with an overview and in-depth knowledge on some of the specific aspects, such as bioprospecting, value-added biomolecules of Indian mangroves, and ethnobotanical studies of major medicinal plants.

I congratulate the authors and editors for this fruitful effort and hope that the book will be useful to the scientific community and the entire humanity.

—Padma Mahanti, IFS
Director
Department of Environment and Climate Change
Government of Kerala

PART I
Biodiversity and Conservation

CHAPTER 1

Biodiversity Bioprospection with Respect to Medicinal Plants

ABHISHEK KUMAR DWIVEDY, VIPIN KUMAR SINGH, SOMENATH DAS, ANAND KUMAR CHAUDHARI, NEHA UPADHYAY, AKANKSHA SINGH, ARCHANA SINGH, and NAWAL KISHORE DUBEY*

*Laboratory of Herbal Pesticides, Center of Advanced Study in Botany, Institute of Science, Banaras Hindu University, Varanasi, 221005, India, Mobile: +91-9415295765; *E-mail: nkdubeybhu@gmail.com*

ABSTRACT

Medicinal plants have been in use for the treatment of a range of diseases from the beginning of civilization. Around 65–80% of the world population living in developing countries believes in traditional herbal medicines for their primary health care. Even in the modern allopathic system, medicinal plants are playing a key role in public health care. According to an estimate, approximately 25% of currently used medications are obtained from different higher plants. In addition to this, the interest in plant products has increased exponentially especially the phytotherapeutic supplements (nutraceuticals) and cosmetics over the past decade. Apart from direct clinical use, these plants and their products are also utilized for agriculture in pest control in biodiversity-rich countries like India, China, Sri Lanka, Brazil, and Africa. Every nation has a sovereign right over its biodiversity which is frequently violated by the act of biopiracy or gene robbing. There are many examples of exploitation of traditionally used medicinal plants by the biotechnologically rich but biodiversity poor countries. *Pentadiplandra brazzeana* from tropical Africa, *Vinca rosea* from Madagascar, *Curcuma longa*, *Azadirachta indica*, and *Withania somnifera* from India are some classical examples of biopiracy. Hence, bioprospection would help the native countries in legal exploitation of the bioresources by preventing the act of biopiracy. Hence, bioprospection is a burning issue for biodiversity-rich countries like India, China, and tropical African nations to document their bioresources as well as to identify their useful plants, related phytochemicals and genes controlling them.

1.1 INTRODUCTION

Bioprospecting refers to the process of exploring natural resources for commercialization of new products without perturbation of natural pathways. Traditional bioprospecting includes the huge utilization of birch polypore, *Piptoporus betulinus* fruiting body for recovery of intestinal ailments in Iceman. From pre-historic times, the healing properties of different medicinal plants were recognized and utilized by old world people and other primates. Different animals like monkey, ape, gorilla, and chimpanzee have selected and consumed a particular type of plants for management of their disease, injury and other health problems (Baker, 1996; Glander, 1994). A similar type of style has also been used by humans. The use of traditional therapeutic plants as folk medicines for large-scale production of phytoactive constituents for curative and psychotherapeutic purposes has required a long journey (UNESCO, 1994). Some archeological evidence also indicates the use of medicinal plants from ancient times. The use of medicinal plants for herbal drug preparation was first documented from Sumerian clay slab of Nagpur, India (5000 years old) (Kelly, 2009). The oldest Chinese herbal book "Pen T'Sao" written by Emperor Shen Nung circa (2500 BC), had mentioned about 365 plant products for recovery of different humans ailments (Petrovska, 2012). The Illiad and Odysseys, the great epic of Homer at circa 800 BC, had described 63 plant species for therapeutic purposes (Toplak Galle, 2005). In 'Vedas' the holy mythological books of India, use of various spices like nutmeg, pepper, and clove by Arya Monk had been listed for their health care promises (Agarwal, 2013). More than 500 plants with their therapeutic uses had been classified by Theophrastus (371–287 BC) and were further renowned by an eminent medical writer Celsus (25–50 AD) (Pelagic, 1970). Dioscorides, a military physician, also studied different medicinal plants and their medication during travel with the Roman army. Several important domestic plants *viz.* garlic, nettle, sage, coriander, parsley, sea onion, and chamomile were also described by them (Thorwald, 1991; Katic, 1967). During the 17th century and Middle Ages, physician monks treated health problems by using commonly growing medicinal plants in their monastery. Early in the 18th century "Species Plantarum" by Linnaeus had developed a solid pillar of botanical nomenclature and also supported the proper plant identification and isolation followed by upgrading methods for extraction of secondary metabolites such as saponins, essential oils, hormones and vitamins for the prevention of different diseases (Frodin, 2004). In the 20th century, herbal medication was in great danger due to some poisonous impact of medicinal plants and dose-dependent mode of action (Saad et al., 2006). Over the past years, people are moving towards the allopathic and other surgical treatments for their quick response to human health. However, in modern days, due to the inexpensive price and less chance of side effects, most pharmacopeias of the world focused on isolation and characterization of active plant constituents for real drug development. Recently, different active plant constituent is being encapsulated by different coating materials in order to synthesize nanomedicine with major emphasis on targeted action. The combination of modern technical systems with traditional herbal knowledge could upgrade the old preparatory procedures and help in the discovery of new drugs.

1.2 BIODIVERSITY OF MEDICINAL PLANTS

Since a long time, medicinal plants have served and provided health security to human beings and other life forms. The demand of medicinal plants for health purpose is increasing in developing as well as developed countries. This problem can be resolved by looking into the worldwide diversity of medicinal plants so as to harness as much as resources and knowledge to secure health needs of present generation (Balick et al., 1996). According to an estimate, around $1/10^{th}$ of the world's plant species have been used for the medicinal purpose (Chen et al., 2016). However, their distribution is not uniform over the globe, like China and India have a higher diversity of medicinal plants than others.

The botanical survey has estimated more than 2.5 lakhs plant species on earth, among which more than 70,000s were identified to possess medicinal properties used in folk medicine throughout the world (Alamgir, 2017; Farnsworth and Soejarto, 1991). It has been estimated that about 25% of the drugs prescribed worldwide for the disease treatment are derived from plants (Sahoo et al., 2010). Currently, 17 countries have been identified as a mega-biodiversity center, which are home to the bulk of the world's species including medicinal plants. A few years ago WHO has also made an attempt to identify medicinal plants existing in the world.

The total number of species on earth is unknown, and only a small number have been screened for medicinal properties. However, data available from different countries on medicinal plants is inspiring. India, which is known for one of the most diverse mega biodiversity regions in the world and cultural traditions, uses herbal medicines based on more than 7,000 plants species (Lange, 2002). In China, total numbers of plant species used for the medicinal purpose were around 10,000 (He 1998). In Africa, over 5,000 plant species are used for medicinal purpose (Luyt et al., 1999) followed by USA, Thailand (Phumthum et al., 2018), Mexico (Alonso-Castro et al., 2017) Malaysia, Indonesia (Caniago and Stephen, 1998), Philippines, Nepal and Pakistan (Rajeswara et al., 2012) (Figure 1.1).

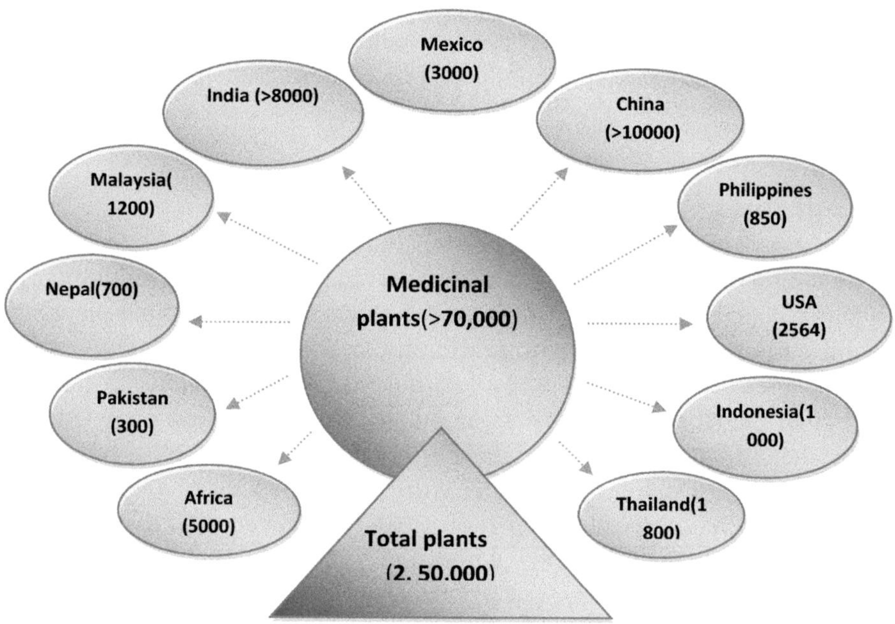

FIGURE 1.1 Schematic status of medicinal plants in some major countries.

These plants bear a diverse array of specific metabolites having one or more therapeutic importance. In the USA, most of the top prescribed drugs and 50% of the drugs all over the world have at least one compound derived from the plant source. Today, many chronic diseases are possible to treat with the help of these herbal remedies. In upcoming few decades, medicinal plants surely will become a new era of the medical system for the management of diseases and exploration of diversity may open up huge opportunities for overcoming such difficult tasks and thus appears promising for exploration of new drugs (Shakya, 2016).

1.3 CONSERVATION OF GENETIC AND CHEMICAL DIVERSITY OF MEDICINAL AND AROMATIC PLANTS

The biological resources of medicinal and aromatic plants (MAPs) of different countries represent a progressive contribution to natural biodiversity, local economies and cultural integrity around the world. Before the development of pharmaceuticals, local mankind of many places used their traditional knowledge and skills to prepare different plant-based formulations to diagnose and treat health associated problems (Traffic International, 2015; WHO, 2013). Aromatic plants have a characteristic fragrance and are known to act as a reservoir of diverse secondary metabolites to protect themselves from natural predators (King, 1996). These metabolites or bioactive constituents have potent therapeutic properties and are being used in novel drug synthesis (Harvey, 2008). In contrast to modern medicine, the traditionally used plant medicines have a holistic approach and less chance of side effects. About 85% of herbal medicines are derived from different medicinal plants based on their habitat, climatic and varied geographical distribution (Farnsworth, 1988; Prasad and Bhattacharya, 2003; Phondani et al., 2014). Although some natural compounds obtained from plant species are not pharmacologically active in their original form while they act as a precursor of several bioactive compounds (Atanasov et al., 2015). Aromatic medicinal plants especially include traditional herbs and spices for their delicious flavor and fragrance. US-FDA recognized spices of key medicinal significance due to the presence of significant amount of flavoring principles (Kaefer and Milner, 2008). Some aromatic plants of Apiaceae have chemoprotective and therapeutic potential up to a certain doses. The leaves of *Melaleuca alternifolia* and *Eucalyptus globulus* were traditionally used for the treatment of chest infection, and decongestion of coughs (Vail and Vail, 2006). India has contributed a major role in Ayurveda using medicinal and aromatic plants. Hamilton (2004) reported that 44% of Indian flora is being used medicinally throughout the world. Due to the enormous natural distribution of different plant species, India is considered as "Herbarium of World" and one of the 12 megadiversity countries. Today, medicinal plants and their bioactive compounds are increasingly recognized not only for its local therapeutic use but also for improvement in the economy of the nations. Iqbal (1993) has estimated that about 4,000–6,000 botanicals are of commercial importance throughout the world. The increasing demand for plant-based medicine has resulted in overexploitation of several high-value medicinal plant populations. Robbins (2000) reported that 29% of total US plant species are facing a major threat due to excessive land use operation and overexploitation. If the ruthless and unscientific collection of medicinal plants from Himalayan regions continues, it will lead to the documented extinction of 19 plants species in the wild, 41 species suspected to become extinct in the recent future, 152 species in great danger of extinction, movement of 102 species towards the higher category of rarity and 251 plant species at greater risk and localized to a particular area with limited access (Walter and Gillett, 1998). Medicinal sector of IUCN has moved towards the identification of threatened plant species due to non-sustainable exploitation or other reasons. Walter and Gillett (1998) have reported about 34,000 out of 49,000 species to be globally threatened with extinction. Some of the medicinal plant species have slow growth rate, localized distribution and low population density which make them more prone to extinction (Kala, 2005; Nautiyal et al., 2002; Jablonski, 2004). Moreover, increasing biotic interference, habitat fragmentation and changing climatic conditions also lead to low regeneration potential of medicinal plants in the natural habitat (Kala, 2005). The weakening of legal, customary rules has resulted in increased exploitation of medicinal plant species (Kala et al., 2006). Destructive harvesting from temperate and alpine zones of Himalaya, India has caused alarming scarcity of genetic stocks of many high-value medicinal species. Ministry of the Environment, Forest and Climate Change (MOEFCC), India has recognized 29 plant species such as *Rauwolfia serpentina*, *Pterocarpus santalinus*, *Picrorhiza kurroa*, *Swertia chirata,* and *Nardostachys jatamansi* which are localized to a particular habitat and banned for export of these plants (Nishteswar, 2014). International Union for Conservation of Nature and Natural Resources (IUCN) has listed 121 plant species in red data book from the Himalayan regions, among these 17 are medicinal plants (Nayar and Sastry, 1987).

Currently, one of the major difficulties to study the traded and commercial details of medicinal plant products is due to its fragmentary reports on a national and international level. Therefore, different plant species with marked medicinal properties are being harvested at a local and regional scale for supply in international markets (Schippmann et al., 2002). Now, scientists are moving toward cultivation approach for the supply of excessive products at the commercial level, but active component/secondary metabolites are often very low in the monoculture conditions due to certain controlled conditions of cultivations (Palevitch, 1991; Uniyal et al., 2000).

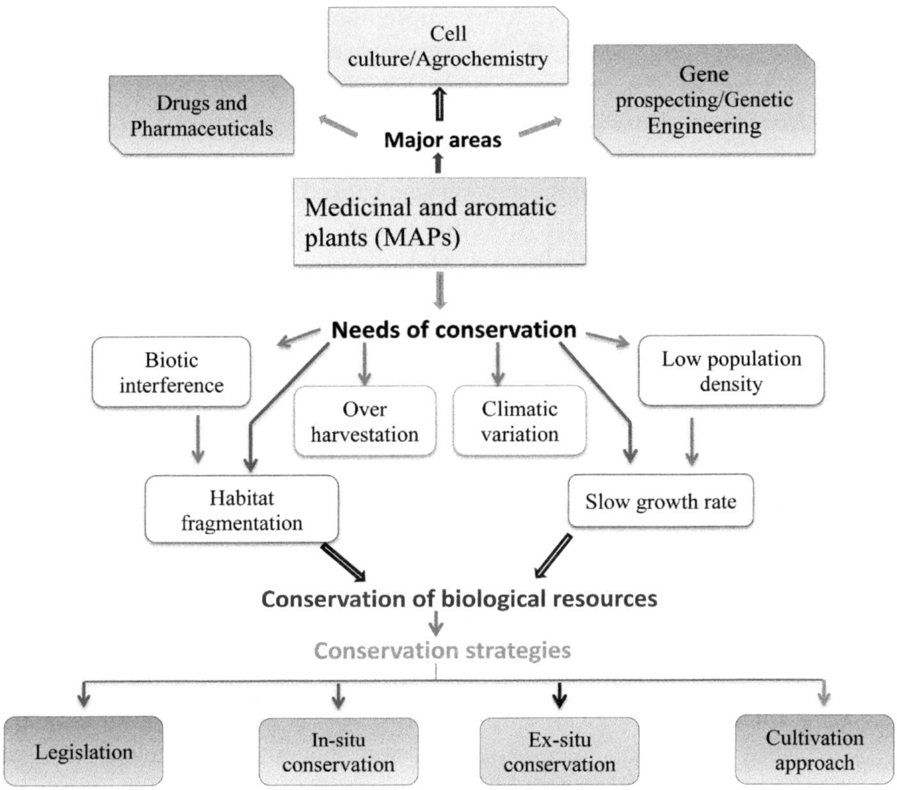

FIGURE 1.2 (See color insert.) An overview of the need for different conservation strategies of MAPs.

1.3.1 CONSERVATION STRATEGIES FOR MEDICINAL AND AROMATIC PLANTS

The progressive increase of developmental activities, gathering, and overexploitation as well as changing environmental situation has caused an alarming decline of medicinal plant wealth throughout the world. Therefore, conservation strategies have envisaged with primary goals of preservation of species, genetic diversity and sustainable use of plant products for human welfare (Chandra, 2016). Basically, there are four scientific techniques applied for the preservation of the diversity of medicinal plants. They are described under as

a) **Legislation:** Rules of IUCN, CBD (Convention on Biodiversity) and environmental laws have been formulated by different countries to protect the herbal flora by implementing forest act (1927), wildlife amendment act (1972), environment protection act (1986), and national biodiversity act (2002).

b) ***In-situ* conservation:** It includes the conservation of specific biogeographic zones with intra- and inter specific genetic variation. National park, biosphere reserves, sacred sites, sacred groves, and wildlife sanctuary are principle policy decision process for in-situ conservation of biological and genetic diversity of medicinal plants at the national, international and global level. MOEFCC (India) has recognized 14 biosphere reserves, 91 national parks and 448 wildlife sanctuary including world heritage site, coral reefs, mangroves and Ramsar convention (for wetlands conservation) for ethnobiological preservation in their respective habitats (Chandra, 2016).

c) ***Ex-situ* conservation:** It includes conservation of medicinal plants outside the natural habitats for long-term preservation such as seed bank, pollen bank and DNA libraries. Seed conservation is an important ex-situ approach, also called as an insurance policy against extinction of medicinal plants and cost has been estimated as little as 1% of in-situ conservation (Hawkes et al., 2012).

d) **Cultivation:** It is an important approach to conserve threatened medicinal plant species to support the ever-increasing market demands. The strategy will be successful only with the immense help of public domestication programmes. There are many threatened plant species such as *Garcinia afzelii*, *Panax quinquefolius*, *Saussurea costus*, and *Warburgia salutaris* that can be marketed at a high price due to its cultivation approaches (Cunningham, 1994). However, domestication will also achieve conservation of threatened plant's diversity for selected plant species. Sometimes, cultivated plants have been found qualitatively inferior than wild plants. Chinese people prefer wild ginseng roots instead of cultivated ones because the cultivated plants did not possess the characteristic features as collected from the wild (Cunningham, 1994).

1.4 MEDICINAL BIOPROSPECTING

The development of infectious diseases and the evolution of multi-drug resistance during the long course of human history has compelled the scientist of the current generation to search the new medicines for disease treatment. In this scenario, medicinal bioprospecting provides a boom in the field of medicine and pharmacognosy (McClatchey, 2005). In general sense, at any time if a person is searching for food or other biological value in their environment, they are bioprospecting. Searching substances for medicines in the natural milieu are known as medicinal bioprospecting, which is a multidisciplinary venture involving various researchers and logical subjects (Juan, 2017). Medicinal bioprospecting may provide new leads against pharmacological targets such as cancer, HIV-AIDS, Alzheimer's disease, malaria, etc. Currently, there is a need for broad revise of natural products derived from plants (herbal remedies) (Tyler, 1999; Cardellina, 2002), which needs further refinement and research on medicines approved by Food and Drug Administration (FDA).

A current literature survey has shown that 60% of the drugs used for oncotherapy and more than 70% for the management of anti-infectious diseases recommended up to 2002 are from natural sources (Newman et al., 2003). Plant-derived drug molecules help in identifying the association between drugs and their target site and enhances the chances to develop more effective drugs by laboratory synthesis.

In modern medicine based disease therapy too, plants contribute significantly in novel drug discovery and development. Some of the important lead bioactive molecules of plant origin having a major development in medical science are under thorough clinical investigations as well as recommended by Food and Drug Administration (FDA) are listed in Table 1.1.

These compounds obtained from different parts of the plants like seeds of *Mucuna pruriens* are an important source of L-dopa which helps in the treatment of most important neuro-degenerative Parkinson's disease. Similarly, the bark of a yew tree (*Taxus brevifolia*), leaf of *Catharanthus roseus* and stem/bark of *Camptotheca acuminata* bears important photochemical compound having antitumor potential. Furthermore, some plants like *Galanthus woronowii* and *Gingko biloba* bear compound having anticholinesterase activity and helps in the treatment of the so called neuro-degenerative disorder, Alzheimer's disease.

Considering these facts, there is a need for bioprospection in order to provide the available natural source for treatment of health threats. In many instances, it has been found that the compounds isolated from the medicinal plants may not serve as the final drug, but they can comprise lead compounds for the development of potential drugs.

TABLE 1.1 Some Important Lead Compounds Obtained from Plants As a Source of Medicine

S. No.	Compounds	Medicinal plant	Treatment	References
1.	L-Dopa	*Mucuna pruriens*	Parkinson's disease	Cotzias (1969)
2.	Taxol	*Taxus brevifolia*	Cancer	Wani et al. (1971)
3.	Vinblastine & Vincristine	*Catharanthus roseus*	Cancer	Cragg and Newman (2005)
4.	Camptothecin	*Camptotheca acuminata*	Cancer	Dancey and Eisenhauer (1996)
5.	Combretastatin	*Combretum caffrum*	Cancer	Dark et al. (1997)
6.	Podophyllotoxin	*Podophyllum peltatum*	Antitumor	Canel et al. (2000)
7.	Galantamine	*Galanthus woronowii*	Alzheimer's disease	Sahoo et al. (2017)
8.	Gingkolides	*Gingko biloba*	Alzheimer's disease	Yoo and Park (2012)
9.	Artemisinin	*Artemisia annua*	Malaria	Klayman (1985)
10.	Tiotropium	*Atropa belladona*	Pulmonary disease	Van Noord et al. (2000)
11.	Reserpine	*Rauwolfia serpentina*	Antipsychotic, and antihypertensive	Singh (1955)

1.4.1 ECOLOGY AND BIOPROSPECTING OF MEDICINAL PLANTS

Over the course of evolution through thousands of years, the plants have developed an array of phytochemicals that had managed their survivability by fighting against, infectious diseases, predator attack, and facilitated food availability. Most of the today's pharmaceuticals constitute these plant-derived chemicals. The World Conservation Union (IUCN) in a recent survey found that 72,000 species of higher plants are used for extraction of medicines worldwide, constituting 17% of the higher plants of the world. The World Health Organization (WHO) has reported earlier that 80% of the world's population relies on traditional medicines for their fundamental health care and the scenario remains the same even today (IUCN, 2006). And for this, the pharmaceutical, agricultural and cosmetic industries depend on plant biodiversity thriving in the wild for their raw materials. *Cinchona officinalis* (the Cinchona tree) from South America is the source of vital quinine and quinidine which are the milestones in the herbal drug industry for their effectiveness against malaria and cardiac arrhythmias, respectively (Efferth et al., 2007). Digitalis has provided the synthetic drugs digoxin and digitoxin which has proved high efficacy against cardiovascular disorders such as atrial fibrillation and even heart failure (Ahmed et al., 2006). Willow tree provides aspirin which is still the most commonly used remedy for pain and inflammation, all around the world (Vlachojannis et al., 2011). There are ample examples of plant metabolites that have been searched for beneficial bioactivity for the welfare of human beings and society.

Progress in biotechnological and molecular biology research has revolutionized the natural product research by the introduction of the concept of 'Bioprospection.' Bioprospection deals with the search for novel biological resources through exploration of biodiversity which may impart some social and economic value. The pharmaceutical industry, agriculture sector, engineering, manufacturing, and many other industries are the basic stakeholders of bioprospection (Beattie et al., 2005).

Ecological studies on medicinal plants deal with the aspects of floristic composition, forest stand structure, distribution of species (Raunkier's frequency classes), species diversity, soil physicochemical properties, etc. (Szafer, 2013). The studies on species ecology hold much importance as they provide new hints/leads for novel drug research as well as the management and conservation of source plant species. There has been increasing attention of the scientific community for the ecological studies on medicinal plants for resource management worldwide.

The Rio de Janeiro Earth Summit on 5th June 1992 presented the Convention on Biological Diversity (CBD) with major emphasis on sustainable development, biodiversity conservation and equal benefit sharing resulting from utilization of biological resources (Dutfield and Suthersanen, 2008). A number of pharmacological projects are undertaken, and several phytochemical studies are performed every year to investigate the potential medicinal value of plant species. However, very few of them actually resulted in the development of novel drugs. Literally, the bioprospection initiatives are rightly phrased to be 'in their infancy.' Bioprospecting is widely broadcasted as a way to discover new phytochemicals to serve mankind; however, the shortcomings of the approach have largely made it farther from its goal (Buenz et al., 2004).

The major causes for such failure may be the methodology related to experimental flaws, the absence of repetition of experiments, non-replicable results, findings based on superficial studies, etc. The technologies utilized during the process are advanced and reliable, but the efforts for drug discovery projects are more or less confined in one direction only. The need is to coordinate multiple disciplines in the same direction to have some vital and significant outcomes (Siqueira et al., 2012). The ecological studies relating the medicinal plants can greatly increase the probability of success in such large bioprospection projects which can save money, time and labor. They provide new clues to predict the potential medicinal properties of certain plants (Albuquerque, 2010).

Chemical ecology is one of the very recent approaches in this context which has been utilized to predict the secondary metabolite allocation in higher plants. Such predictions are based on the widely accepted ecological theories of plant defense. The defense mechanism developed amongst plants against herbivores during the course of evolution led to the formation of a big pool of secondary plant bioactives (Donaldson and Cates, 2004). The trends and patterns related to the habitat of the plant can boost bioprospection efforts (Coley et al., 2003) which can be described as

i) High molecular weight metabolites such as polyphenols get allocated in plants of dry forest (Ecogeographical pattern). For example, Caatinga plants dwelling in semi-arid regions are metabolically rich in phenolic compounds.
ii) Medicinal plant species of humidity rich areas such as Atlantic forest allocate a higher amount of toxic compounds such as alkaloids.
iii) Young leaves contain more active secondary metabolites than mature leaves. Even, they harbor several unique bioactive molecules.

iv) The mature leaves of slow growing plants dwelling in the shade are proposed to have a greater level of chemical and physical defense in comparison to the mature leaves of fast growing plants of light conditions.
v) 'Metabolic specialization' has been reported to be highest in ecotones (The transition zone between two different ecosystems).

The allocation of secondary metabolites has also been interpreted from the phylogenetic patterns (Leonti et al., 2013). The basal angiosperms *viz.* Magnoliids and Chloranthales have been reported to show more activity than the more evolved Asterids and Rosids. On the other hand, Monocots have been demonstrated to harbor a lesser amount of bioactives. The probable cause for such pattern has been attributed to the development of parallel leaf venation in monocots permitting lesser attack by herbivores and thus, fewer chances of the synthesis of secondary metabolites for chemical defense (Coley et al., 2003).

1.4.2 BIOPROSPECTION OF PLANT ESSENTIAL OILS FOR MEDICINAL USES

Diseases are considered as one of the major cause of mortality throughout the globe. Several laboratory studies have validated the diverse role of traditionally used plants and their parts for the treatment of human diseases. The disease curing ability is attributed to the presence of different phytochemicals including alkaloids, flavonoids, and terpenoids. Traditionally used medicinal plants are one of the chief sources of disease treatment among poor people who are unable to afford the cost of modern drugs. Traditional knowledge offers the source of new drugs from plants. Efficacy of traditional medicinal plants has very much influenced to pharmaceutical industries leading to theft of traditional knowledge for their economic benefit which is known as biopiracy. Global climate change and overexploitation are the major factors affecting the existence of numerous traditional medicinal plants used for therapeutic purposes. Under such conditions, searching novel sources of bioactive molecules from the very diverse pool of plant diversity, e.g., bioprospection, can relieve the pressure of overexploitation. Bioprospection is more important for developing countries because it not only increases the economy but also conserves biodiversity. The importance of traditional medicinal plants, few cases of biopiracy along with the need of bioprospection to fill the increasing demand of drugs is addressed briefly

From the early ages of civilization, people are using essential oils for food preservation, fragrance, taste, and disease treatment. Use of medicinal plant-based essential oils by many traditional systems of disease prevention such as Siddha, Unani, Ayurveda, and Chinese are known. Essential oils are defined as low molecular weight, a volatile and complex mixture of various compounds which possess several properties including antimicrobial and antioxidant potential (Ribeiro-Santos et al., 2017). They can be extracted from the whole plant or different plant parts like root, stem, leaf, flower, seed, and bark. The amount of extracted essential oil depends significantly on the type of plant material, the methodology used (Nakatsu et al., 2000), and geographical location. The chief constituents of essential oil can be grouped as terpenes and terpenoids while minor fractions are contributed by aliphatic and aromatic components (Bakkali et al., 2008). Due to their natural origin, and inclusion under generally recognized as safe (GRAS) category by Food and Drug Administration (FDA) (FDA, 2016), there is growing significantly in the application of EOs in food as well as pharmaceutical industry.

Indiscriminate utilization of synthetic antimicrobials by current generation has given rise to drug-resistant bacteria and fungi (Reichling et al., 2009). In order to alleviate the chances of resistance development in microbes, medicinal plants should be investigated for novel bioactive principles. Medicinal plants have been recognized as an important source of drug molecule for the treatment of many diseases. Many plants have found their place in traditional medicine like Ayurveda, Siddha, and Unani because of their disease-curing properties. Enormous evidences are available indicating the significance of plant extract in disease curing. Thus, bioprospecting medicinal plants would be helpful in chemical synthesis and commercial utilization of natural bioactive compounds. Some of the important activities of the essential oils are described below.

1.4.2.1 ANTI-INFLAMMATORY POTENTIAL OF ESSENTIAL OILS

Essential oils obtained from cones of gymnosperm plants have been reported to possess the anti-inflammatory activities (Tumen et al., 2011). Wound repairing activity was assessed by linear incision and circular excision followed by histopathology. Anti-inflammatory actions of essential oils were determined against capillary permeability provoked through acetic acid. Essential oils extracted from *Abies cilicica* subsp. *Cilicica* and *Cedrus libani* were found good in wound models. The activities were suggested to be due to synergistic actions of individual bioactive molecules.

Essential oils form rest of the selected plant did not exhibit good potential. Öztürk and Özbek (2015) have described the anti-inflammatory actions of *Eugenia carryophylla* essential oil. The activity was comparable to that of etodolac (0.025–0.10 mL/Kg) and endomethacin (0.05–0.2 mL/Kg). The major components responsible for observed activities were identified as β-caryophyllen (44.70%) and eugenol (44.20%). The *Eugenia carryophylla* EO should be analyzed and tested further for large scale application. Anti-inflammatory, as well as antioxidant potentials of *Mentha piperata* essential oil, was evaluated against croton oil-induced edema in the mouse. The activities associated with essential oils were found to be due to the presence of major components like methyl acetate, menthone and menthol (Sun et al., 2014). The anti-inflammatory property of *Citrus aurantium* L. essential oil has been described by Khodabakhsh et al. (2015). The chronic and acute anti-inflammatory effects were the result of essential oil components linalool, linalyl acetate, nerolidol, *E,E*-farnesol, a-terpineol, and limonene. The essential oil at 40 mg/kg dose demonstrated anti-inflammatory activity, which was comparable to sodium diclofenac (50 mg/kg). The observed effect supports the ethnomedicinal value of *Citrus aurantium* L. essential oil.

1.4.2.2 ANTIMUTAGENIC POTENTIAL OF ESSENTIAL OILS

Perturbations in metabolic pathways are associated with the development of cancer and health-threatening diseases (De Flora et al., 1996). Use of plant-based products in the human diet as long practiced in Indian food system may be considered as one of the most effective approaches to avoid cancer. Essential oil of ginger (GEO) has been reported to possess potent antimutagenic activity (Jeena et al., 2014). Essential oil inhibited the mutation induced by sodium azide, extract of tobacco and 4-nitro-o-phenylenediamine in a dose-dependent manner. The study established the stimulatory action of GEO over phase II carcinogen-metabolizing enzymes and suggested its use in chemoprevention. The antimutagenic potential of essential oils from *Citrus sinensis* and *Citrus latifolia* was determined against mutation induced by Methyl-N′nitro-N-nitrosoguanidine (MNNG) and Ethyl-N-nitro-N-nitrosoguanidine (ENNG) in *Salmonella typhimurium* TA100 (Toscano-Garibay et al., 2017). Essential oil from both species represented antimutagenic potential against MNNG as well as 2-amino-anthracene (2AA). However, antimutagenic effect against ENNG was shown by *Citrus latifolia* only. The observed antimutagenic activity was attributed to the presence of components like β-thujene, α-myrcene, R-(+)-limonene, and γ-terpinene. Antimutagenic efficacy of *Foeniculum vulgare* essential oil against cyclophosphamide (CP) induced genotoxicity in mice was documented by Tripathi et al. (2013). *Foeniculum vulgare* essential oil was effective in ameliorating the changes induced by CP such as decreased activities of enzymes like superoxide dismutase and catalase. Pretreatment with essential oil resulted in reduced chromosomal abnormality in bone marrow cells suggesting the antimutagenic activity.

1.4.2.3 ANTIDIABETIC POTENTIAL OF ESSENTIAL OILS

Diabetes has long been recognized as disabling disease resulting mainly due to genetic and environmental factors as well as oxidative stress (Akram et al., 2011). Antidiabetic activity of essential oil obtained from *Syzygium aromaticum* and *Cuminum cyminum* has been presented by Tahir et al. (2016). The bioactivity assay of selected essential oil was based on its inhibitory action over the alpha-amylase enzyme. Concentrations of EO used for antidiabetic activity were taken in the range of 1 to 100 ppm. Highest antidiabetic activity was recorded at the highest dose. Five different emulsions of each essential oil were formulated by mixing in tween 80, ethanol and water. One of the emulsion preparations from *Syzygium aromaticum* (A5) and another from *Cuminum cyminum* (B5) demonstrated the highest alpha-amylase inhibitory activity as 95.30 and 83.09%, respectively. The antidiabetic responses were expected to be due to the presence of eugenol (main compound of *Syzygium aromaticum*) and α-pinene (the main component of *Cuminum cyminum*). Antidiabetic activity of essential oils from *Citrus* peels based on inhibition of α-glucosidase activity was demonstrated by Dang et al. (2016). Among six selected *Citrus* essential oils, Buddha's hand citrus essential oil showed remarkable inhibitory action on alpha-glucosidase with IC_{50} value 412.2 ppm. The major components of essential oil were identified as limonene and γ-terpinene. The synergistic action of this essential oil with antidiabetic drug acarbose enhanced the overall efficacy suggesting the use of plant-based essential oil for the treatment of diabetes. Essential oils from two species of *Acacia*, e.g., *Acacia mollissima* and *Acacia cyclops* have been assessed for their antidiabetic potency by determining the α-glucosidase inhibitory activity (Jelassi et al., 2017). The IC_{50} value was found around 89 ppm which is equivalent to acarbose, the commonly used antidiabetic drug. A major component of *A. mollissima* was reported as (E, E)-α-Farnesene (51.5%) and (E)-cinnamyl alcohol (10.7%) whereas the active principle of *A. cyclops* was nonadecane (29.6%) and caryophyllene oxide (15.9%). Adefegha et al. (2017) have illustrated the antidiabetic efficiency of essential oil from *Aframomum melegueta* and *Aframomum danielli*. The selected essential oils significantly minimized the activity of α-amylase and α-glucosidase. EC_{50} value of *A. melegueta* essential oil recorded for α-amylase and α-glucosidase was found to be 139 μl/ml and 91.83 μl/ml, respectively which were better as compared to *A. danielli.*

1.4.2.4 ANTICANCER ACTIVITY OF ESSENTIAL OILS

Cancer is recognized as a group of many diseases. This life-threatening disease has emerged as the first reason for large mortality worldwide. Cancer cells are characterized by uncontrolled cell growth and invasion to newer locations through metastasis (Hanahan and Weinberg, 2011). Therefore, disease treatment should be based upon the inhibition of cell division. Essential oil from *Elsholtzia ciliata* has been reported to possess anticancer activity (Pudziuvelyte et al., 2017). The main components of essential oil obtained through hydro-distillation were dehydroelsholtzia ketone (78.28%) and elsholtzia ketone (14.58%). Anticancer assay of essential oil was performed against human glioblastoma (U87), breast and pancreatic (Panc-1) cancer cell lines. EC_{50} value recorded for selected EO ranged from 0.017–0.021%. Viability test indicated considerably higher survivability of normal human fibroblast in comparison to cancer cell lines at the same essential oil concentration. Yang et al. (2017) have elucidated the anticancer properties of citrus by MTT assay. The essential oil exhibited a pronounced inhibitory effect against the proliferation of lung (A549) and prostate (22RV-1) cancer cell lines. Limonene (74.60%) was the major component of essential oil as determined by GC-MS analysis. The effectiveness of seed essential oil obtained from *Foeniculum vulgare* has been reported to possess anticancer activities against human cervical epithelioid carcinoma (HELA) and breast cancer (MDA-Mb) cell lines as revealed by MTT assay (Akhbari et al., 2018). Essential oil showed remarkable activity against both cell lines with IC_{50} value corresponding to less than 10 ppm. The major active components of essential oil responsible for anticancer activity were identified as Trans-Anethole (80.63%), followed by L-Fenchone (11.57%), Estragole (3.67%) and Limonene (2.68%) through GC-MS analysis. Recently, Han and Parker (2017) have reported the anticancer properties of commercially available essential oil derived from *Foeniculum vulgare*. The essential oil displayed anti-proliferative behavior against dermal fibroblasts.

1.4.3 CONSERVATION THROUGH BIOPROSPECTION AND ECONOMIC BENEFITS

The recent surge in the consumer preference for fascination towards herbal products has resulted in increased demands for plant-based consumables across the world which has definitely elevated the threat for their over-exploitation. In this context, statistical data/reports of habitat destruction and unsustainable collection are piling up day by day. Nearly, 150 plant species have got extinct in the wild due to the prevalent illegal trade and destructive harvesting in a situation where 90% species of plants utilized in the herbal industries are collected from the wild (Lange, 2002). Nearly, 3.5% of medicinal plant species of the Indian Himalayas have been categorized under different categories of threats as a result of the unethical collection, habitat loss and ecosystem disruption (Kala, 2005).

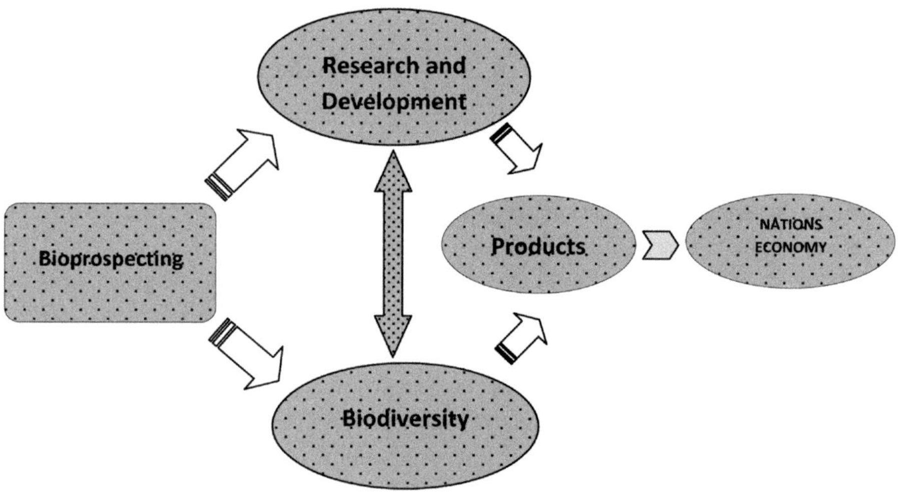

FIGURE 1.3 Graphical relationships between bioprospection and conservation.

1.4.4 BIOPROSPECTING AS GREEN DEVELOPMENTALISM

Bioprospecting natural resources have frequently been cited as a sustainable use of biodiversity, which can serve as a global reserve with anticipated fitting into rational property systems and world markets. Numerous research groups have presented the importance of biodiversity in global development (Gari, 1999). Biological diversity and the 'green' World Bank, follow attempts to normalize global flows of 'natural capital' by means of so-called approach '*green developmentalism*' which comprises the combination of conservation, development and equal benefit sharing. According to this hypothesis, nature is constructed as world cash (McAfee, 1999). Green developmentalism abstracts nature from its spatial and social contexts and strengthens the claims of global leaders to the greatest share of the earth's biomass and all it possesses. Across the range of countries and institutions, there is now a prevalent approval that the ways to protect the environment is to value nature's services, assign property rights, and trade these services within a global market (Liverman, 2004). Green developmentalism provides apparent insights into the current as well as emerging challenges of environmental sustainability, social and financial progress and looms to nature management that proposes to endorse a notable revolution to twist an ecologically caustic market into eco-savior.

1.4.5 NEGATIVE ASPECTS OF BIOPROSPECTING

Bioprospection includes the search for traditional knowledge or screening of biological diversity or indigenous information about plants, with an objective to develop commercial products which are beneficial for our society. The information collected through bioprospecting process is also called "bio-discovery," with the aim to report natural products that can be utilized for beneficial results.

The developing countries don't have much-developed techniques in comparison to developed countries, which can be used in the identification of valuable chemicals from plants. Thus, in order to bioprospect the chemical compound, bioprospector requires from developed countries. This way, bioprospecting is fruitful for our society, but this may lead to patent some biomaterials by multinational companies; as there are no guidelines available to reward the contribution of local people whose knowledge has now become the platform of drug designing and discovery. The illegal collection or patent of indigenous plant chemical compounds by the multinational companies for their own profit comes under biopiracy. From the 90s onwards there has been raised in cases related to biopiracy with patents and trademarks being used to acquire domination rights over native resources without simultaneous benefit sharing. Some examples of biopiracy are mentioned in Table 1.2.

TABLE 1.2 List of Some Plant Resources Patented by Different Countries and Their Biopiracy Status

S. No.	Native country	Native resource	Biopirate country	Patent numbers	References
1.	China	*Momordica charantia*	USA	US Patent No. 5484889	Chang et al. (1996)
2.	China	*Camptotheca lowreyana*	USA	US Patent No. PP11959	
3.	India	*Basmati Rice*	USA	US Patent No. 6274183 and 5663484	Sahai et al. (2007)
4.	Philippines	*Cananga odorata*	France	N/A but used by several perfumeries in France	
5.	Philippines	*Lagerstroemia spp.*	Japan & USA	US Patent No. 5980904	
6.	India	*Curcuma longa*	USA	US Patent 5,401,504	Avantika et al. (2015)
7.	India	*Azadirachta Indica*	USA	US Patents Nos. 5420318, 5391779 and 5371254	Adhikari (2005)
8.	India	*Commiphora mukul*	USA	US Patent No. 6,113,949	Grain (2006)
9	Samoa	*Homalanthus nutans*	USA	US Patent No. 5,599,839	
10.	Thailand	*Croton sublyratus*	Japan	N/A	Sahai et al. (2007)
11.	Andean countries	*Chenopodium quinoa*	USA	US patent no. 5,304,718	Adhikari (2005)
12.	Amazon	*Banisteriopsis caapi*	USA	US patent PP 05751	
13.	South Africa	*Hoodia spp.*	UK	WO 9846243	
14.	Mexica	*Phaseolus vulgaris*	USA	US patent no. 5,894,079	Rattray (2002)
15.	Asia-Pacific	*Piper mythesticum*	USA	US Patents Nos. 6405948, 6277396, 6080410, 6025363, etc.	Afreen and Abraham (2008)

Thus, bioprospecting can improve the ideal distribution of valuable plant chemical compounds throughout world, but the biopiracy weakens global justice (George, 2011).

1.5 BIOPROSPECTING MEDICINAL PLANTS FOR DRUG DISCOVERY AND THEIR CHALLENGES

Current world generation is more inclined and showing interest toward the alternative treatment especially, therapy based on plant-derived compounds (Simoes et al., 1999). The reason lies into important points such as side effect and inaccessibility of currently used therapeutics to poor people. Natural products from plant sources have long been recognized as a huge source of novel components. Many medicinal plants used are not associated with disease cure only, but several of them have religious values too. However, the industrial revolution has aggravated the formation of synthetic products for disease treatment purpose. Globally, approximately 25% of the drugs recommended for disease therapy are plant-originated, and currently, 121 bioactive molecules are under extensive application (Rates, 2001). World Health Organization has recognized 252 drugs as primary and necessary for human health management among which 11% are entirely plant derived. Some important plant produced bioactive molecules associated with human health are quinine and quinidine from *Cinchona* spp, atropine from *Atropa belladonna*, digoxin from *Digitalis* spp, vincristrine and vinblastine from *Catharanthus roseus*, morphine, taxol from *Taxus baccata*, curcumin from *Curcuma longa*, azadirachtin from *Azadirachta indica*, reserpine from *Rauwolfia serpentina* and codeine from *Papaver somniferum*. According to an estimate (Yue-Zhong Shu, 1998), more than sixty percent of drugs commercially available in market or under clinical investigations, applied for treatment of tumor and infectious diseases are originated from natural sources indicating the significance of biomolecules synthesized by plants. So far, many of these drugs could have not been synthesized for commercial application and are still acquired from either cultivated or wildly growing plants, necessitating the conservation of medicinal plants diversity. Natural compounds from a large pool of plant's diversity can give the opportunity for designing as well as the synthesis of novel active principles with characteristics not yet demonstrated by many present-day used drugs (Hamburger and Hostettmann, 1991).

In spite of many evidences of drug discovery from medicinal plants, searching new drugs for diseases treatment is not so easy task. The process needs the interdisciplinary efforts of expertise from medicinal chemistry, modern medical sciences, biotechnology, and molecular biology to improve the compounds at qualitative as well as quantitative level and also for maintaining harmony with other efforts being performed for drug discovery (Butler, 2004). According to an estimate, continuous effort for drug discovery may involve more than 10 years (Reichert, 2003) and cost greater than 800 million dollars (Dickson and Gagnon, 2004). Often, many of these rigorous, expensive experimental efforts result in no fruitful outcome. Surprisingly, only one out of 5000 lead phytochemicals has been estimated to be successful under clinical trials and approved for therapeutic purposes (Balunas and Kinghorn, 2005). The process of drug discovery is exhaustive and involves the key steps including lead identification, lead optimization, and lead development. Since the new drug development from medicinal plants is time-consuming, appropriate methods for plant collection, techniques of compound extraction as well as purification and equipments for the screening of active phytochemicals should be developed further for better results (Koehn and Carter, 2005). Advanced instrumentation techniques for compound identification such as NMR and MS should be applied for drug discovery purposes. One of the major hindrances in drug discovery is very low availability of bioactive principles that may not be sufficient for important steps of drug discovery. The problem can only be resolved by synthesizing these compounds under laboratory conditions which are again a hurdle for chemists as already has happened with many phytochemicals. Another solution to the problem may involve the synthesis of a compound with similar therapeutic properties.

1.6 FUTURE PROSPECTS

Bioprospection of medicinal plants offers several advantages to human society especially the drug development. Analytical techniques have revealed the enormous diversity of components from medicinal plants. Many of the components have been found suitable for disease prevention in humans, and still, many more are under investigation for commercial application. Development in analytical instrumentation techniques like HPLC, NMR, XRD, FT–IR, and GC–MS will speed up the process of drug designing and synthesis. Essential oil components give the idea for the synthesis of lead molecules used in disease therapy. In view of widespread interest and importance of medicinal plants, research and development in the area should be promoted to get the maximum benefit.

KEYWORDS

- **biodiversity**
- **biopiracy**
- **bioprospection**
- **medicinal plants**
- **nutraceuticals**
- **phytochemicals**

REFERENCES

Adefegha, S. A.; Olasehinde, T. A.; Oboh, G. Essential oil composition, antioxidant, antidiabetic and antihypertensive properties of two Afromomum species. *J. Oleo Sci.* **2017,** *66*(1), 51–63.

Adhikari, R. Emerging issues relating to conflicts between TRIPS and biodiversity: development implications for South Asia. *South Asian Yearbook of Trade and Development.* **2005,** 261–288.

Afreen, S.; Abraham, B. P. WPS No. 629/September, **2008**.

Agarwal, M. K. The Vedic Core of Human History: and Truth will be the Savior. iUniverse, **2013.**

Ahmed, A.; Rich, M. W.; Fleg, J. L.; Zile, M. R.; Young, J. B.; Kitzman, D. W.; Gheorghiade, M. Effects of digoxin on morbidity and mortality in diastolic heart failure: the ancillary digitalis investigation group trial. *Circulation.* **2006,** *114*(5), 397–403.

Akhbari, M.; Kord, R.; JafariNodooshan, S.; Hamedi, S. Analysis, and evaluation of the antimicrobial and anticancer activities of the essential oil isolated from *Foeniculum vulgare* from Hamedan, Iran. *Nat. Prod. Res.* **2018,** 1–4.

Akram, M.; Akhtar, N.; Asif, H. M.; Shah, P. A.; Saeed, T.; Mahmood, A.; Malic, N. S. A review of diabetes mellitus. *J. Med. Plants Res.* **2011,** *5*(22), 5337–5339.

Alamgir, A. N. M. *Therapeutic Use of Medicinal Plants and Their Extracts: Volume 1: Pharmacognosy* (Vol. 73). Springer, 2017.

Albuquerque, U. P. Implications of ethnobotanical studies on bioprospecting strategies of new drugs in semi-arid regions. *The Open Complement. Med. J.* **2010,** *2,* 21–23.

Alonso-Castro, A. J.; Domínguez, F.; Maldonado-Miranda, J. J.; Castillo-Pérez, L. J.; Carranza-Álvarez, C.; Solano, E.; Ruiz-Padilla, A. J. Use of medicinal plants by health professionals in Mexico. *J. Ethnopharmacol.* **2017,** *198,* 81–86.

Atanasov, A. G.; Waltenberger, B.; Pferschy-Wenzig, E. M.; Linder, T.; Wawrosch, C.; Uhrin, P.; Temml, V.; Wang, L.; Schwaiger, S.; Heiss, E. H.; Rollinger, J. M. Discovery and resupply of pharmacologically active plant-derived natural products: a review. *Biotech. Adv.* **2015,** *33*(8), 1582–1614.

Avantika, G.; Vinil, T.; Swati, S. Bio-piracy in India: A decline in cultural values. *Int. Res. J. Env. Sci.* **2015,** *4*(9), 80–82.

Baker, M. Fur rubbing: use of medicinal plants by capuchin monkeys (*Cebus capucinus*). *Am. J. Primatol.* **1996,** *38*(3), 263–270.

Bakkali, F.; Averbeck, S.; Averbeck, D; Idaomar, M. Biological effects of essential oils–a review. *Food Chem. Toxicol.* **2008,** *46*(2), 446–475.

Balick, M. J.; Elisabetsky, E.; Laird, S. A. (Eds.). *Medicinal Resources of the Tropical Forest: Biodiversity and its Importance to Human Health.* Columbia University Press: New York, 1996; p 440.

Balunas, M. J.; Kinghorn, A. D. Drug discovery from medicinal plants. *Life Sciences.* **2005,** *78*(5), 431–441.

Beattie, A. J.; Barthlott, W.; Elisabetsky, E. New products and industries from biodiversity. In *Ecology and Bioprospecting*; Hassan, R.; Scholes, R.; Ash. N., Eds). p 273–95; 2005,

Buenz, E. J.; Schnepple, D. J.; Bauer, B. A.; Elkin, P. L.; Riddle, J. M.; Motley, T. J. Techniques: bioprospecting historical herbal texts by hunting for new leads in old tomes. *Trends Pharmacolog. Sci.* **2004,** *25*(9), 494–498.

Butler, M. S. *The role of natural product chemistry in drug discovery. J. Nat. Prod.* **2004,** *67*(12), 2141–2153.

Canel, C.; Moraes, R. M.; Dayan, F. E.; Ferreira, D. Podophyllotoxin. *Phytochem.* **2000,** *54*(2), 115–120.

Caniago, I.; Stephen, F. S. Medicinal plant ecology, knowledge and conservation in Kalimantan, Indonesia. *Econ. Bot.* **1998,** *52*(3), 229–250.

Cardellina, J. H. Challenges and opportunities confronting the botanical dietary supplement industry. *J. Nat. Prod.* **2002,** *65*(7), 1073–1084.

Chandra, L. D. Bio-Diversity and Conservation of Medicinal and Aromatic Plants. *Adv. Plants Agric. Res.* **2016,** *5*(4), 00186.

Chang, M. K.; Conkerton, E. J.; Chapital, D. C.; Wan, P. J.; Vadhwa, O. P.; Spiers, J. M. Chinese melon (Momordicacharantia L.) seed: Composition and potential use. *J. Am. Oil Chem. Soc.* **1996,** *73*(2), 263–265.

Chen, S. L.; Yu, H.; Luo, H. M.; Wu, Q.; Li, C. F.; Steinmetz, A. Conservation and sustainable use of medicinal plants: problems, progress, and prospects. *Chin. Med.* UK. **2016,** *11*(1), 37.

Coley, P. D.; Heller, M. V.; Aizprua, R.; Araúz, B.; Flores, N.; Correa, M.; Gómez, B. Using ecological criteria to design plant collection strategies for drug discovery. *Front. Ecol. Environ.* **2003,** *1*(8), 421–428.

Cotzias, G. C.L-Dopa in Parkinson's disease. *Hospital Practice.* **1969,** *4*(9), 35–41.

Cragg, G. M.; Newman, D. J. Plants as a source of anti-cancer agents. *J. Ethnopharmacol.* **2005,** *100*(1–2), 72–79.

Cunningham, A. B. Integrating local plant resources and habitat management. *Biod. Conserv.* **1994,** *3*(2), 104–115.

Dancey, J.; Eisenhauer, E. A. Current perspectives on camptothecins in cancer treatment. **1996,** *74*(3), 327–38.

Dang, N. H.; Nhung, P. H.; Anh, M.; Thi, B.; Thuy, T.; Thi, D.; Minh, C. V.; Dat, N. T. Chemical Composition and α-Glucosidase Inhibitory Activity of Vietnamese Citrus Peels Essential Oils. *J. Chem.* **2016,** Article ID 6787952, 1–5.

Dark, G. G.; Hill, S. A.; Prise, V. E.; Tozer, G. M.; Pettit, G. R.; Chaplin, D. J. Combretastatin A-4, an agent that displays potent and selective toxicity toward tumor vasculature. *Cancer Res.* **1997,** *57*(10), 1829–1834.

De Flora, S.; Izzotti, A.; Randerath, K.; Randerath, E.; Bartsch, H.; Nair, J.; Balansky, R.; Van Schooten, F.; Degan, P.; Fronza, G; Walsh, D. DNA adducts and chronic degenerative diseases. Pathogenetic relevance and implications in preventive medicine. Mut. Res./Rev. Gen. Toxicol. **1996,** *366*(3), 197–238.

Dickson, M.; Gagnon, J. P. Key factors in the rising cost of new drug discovery and development. *Nat. Rev. Drug Disc.* **2004,** *3*(5), 417.

Donaldson, J. R.; Cates, R. G. Screening for anticancer agents from Sonoran desert plants: a chemical ecological approach. *Pharmaceut. Biol.* **2004,** *42,* 478–487.

Dutfield G.; Suthersanen, U. *Global Intellectual Property Law*. Edward Elgar Publishing, Cheltenham, 2008.

Efferth, T.; Li, P. C.; Konkimalla, V. S. B.; Kaina, B. From traditional Chinese medicine to rational cancer therapy. *Trends Mol. Med.* **2007,** *13*(8), 353–361.

Farnsworth, N. R. Screening plants for new medicines. *Biod.* **1988,** *15*(3), 81–99.

Farnsworth, N. R.; Soejarto, D. D. Global importance of medicinal plants. *The conservation of medicinal plants*. **1991,** 25–51.

FDA: Code Fed. Regul. (CFR). Title 21 Food Drugs. Chapter I–Food Drug Adm. Dep. Heal. Hum. Serv. Subchapter B–Food Hum. Consum. (Continued), Part 182–Subst. Gen. Recognized as Safe (GRAS), 2016.

Frodin, D. G. History and concepts of big plant genera. *Taxon.* **2004,** *53*(3), 753–776.

Gari, J. A. Biodiversity conservation and use: Local and global considerations. *Sci. Technol. Development Discussion Paper*. **1999,** *7,* 1–19.

George, A.E. Bioprospecting and Biopiracy. Encyclopedia of Global Justice, 2011, 77–80.

Glander, K. E. Nonhuman primate self-medication with wild plant foods. *Eating on the wild side: the pharmacologic, ecologic, and social implications of using noncultigens*. Tucson and London, **1994**.

Grain, K. Traditional Knowledge of Biodiversity in Asia-Pacific: Problems of Piracy and Protection. *http://www.grain.org* (accessed on December 27, 2017).

Hamburger, M.; Hostettmann, K.; Bioactivity in plants: the link between phytochemistry and medicine. *Phytochem.* **1991,** *30*(12), 3864–3874.

Hamilton, A. C. Medicinal plants, conservation, and livelihoods. *Biodivers. Conserv.* **2004,** *13*(8), 1477–1517.

Han, X.; Parker, T. L. Anti-inflammatory, tissue remodeling, immunomodulatory, and anticancer activities of oregano (*Origanum vulgare*) essential oil in a human skin disease model. *Biochim. Open.* **2017,** *4,* 73–77.

Hanahan, D.; Weinberg, R. A. Hallmarks of cancer: the next generation. *Cell.* **2011,** *144*(5), 646–674.

Harvey, A. L. Natural products in drug discovery. *Drug Discov. Today*. **2008,** *13*(19–20), 894–901.

Hawkes, J. G.; Maxted, N.; Ford-Lloyd, B. V. *The ex situ conservation of plant genetic resources*. Springer Science and Business Media, **2012**.

He, S. A. Medicinal Plants in China with Special Reference to *Atractylodeslancea. Medicinal Plants: Their Role in Health and Biodiversity*. **1998,** 161.

International Union for Conservation of Nature, Natural Resources. Ecosystems, & Livelihoods Group. *Conserving medicinal species: securing a healthy future*. IUCN. 2006.

Iqbal, M. International trade in non-wood forest products: an overview. FAO, Rome, 1993.

Jablonski, D. Extinction: past and present. *Nature*, **2004,** *427*(6975), 589.

Jeena, K.; Liju, V. B.; Viswanathan, R; Kuttan, R. Antimutagenic potential, and modulation of carcinogen-metabolizing enzymes by ginger essential oil. *Phytother. Res.* **2014,** *28*(6), 849–855.

Jelassi, A.; Hassine, M.; BesbesHlila, M; Ben Jannet, H. Chemical composition, antioxidant properties, α-glucosidase inhibitory and antimicrobial activity of essential oils from *Acacia mollissima* and *Acacia cyclops* cultivated in Tunisia. *Chem. Biod.* **2017,** *14,* e1700252.

Juan, B. Bioprospecting and drug development, parameters for a rational search and validation of biodiversity. *J. Microb. Biochem. Technol.* **2017,** *9,* e128.

Kaefer, C. M.; Milner, J. A. The role of herbs and spices in cancer prevention. *J. Nutr. Biochem.* **2008,** *19*(6), 347–361.

Kala, C. P. Indigenous uses, population density, and conservation of threatened medicinal plants in protected areas of the Indian Himalayas. *Conserv. Biol.* **2005,** *19*(2), 368–378.

Kala, C. P.; Dhyani, P. P.; Sajwan, B. S. Developing the medicinal plants sector in northern India: challenges and opportunities. *J. Ethnobiol. Ethnomed.* **2006,** *2*(1), 32.

Katic, R. The Serbian medicine from 9th to 19th centuries. *Beograd: Scientific work,* **1967,** 22–37.

Kelly, K. History of medicine. New York: Facts on file; **2009,** 29–50.

Khodabakhsh, P.; Shafaroodi, H; Asgarpanah, J. Analgesic and anti-inflammatory activities of *Citrus aurantium* L. blossoms essential oil (neroli): involvement of the nitric oxide/cyclic guanosine monophosphate pathway. *J. Nat. Med.* **2015,** *69*(3), 324–331.

King, S. R. Conservation and tropical medicinal research. In Medicinal resources of the tropical forest: Biodiversity and its importance to human health; Balick, M. J., Elisabetsky, E. Laird, S. A., Eds.; Columbia University Press: New York, 1996; p 63–74.

Klayman, D. L. Qinghaosu (artemisinin): an antimalarial drug from China. *Science.* **1985,** *228*(4703), 1049–1055.

Koehn, F. E.; Carter, G. T. The evolving role of natural products in drug discovery. *Nat. Rev. Drug Disc.* **2005,** *4*(3), 206.

Lange, D. Medicinal and aromatic plants: trade, production, and management of botanical resources. In *XXVI International Horticultural Congress: The Future for Medicinal and Aromatic Plants.* **2002,** *629,* 177–197.

Leonti, M.; Cabras, S.; Castellanos, M. E.; Challenger, A.; Gertsch, J.; Casu, L. Bioprospecting: evolutionary implications from a post-Olmec pharmacopeia and the relevance of widespread taxa. *J. Ethnopharmacol.* **2013,** *147*(1), 92–107.

Liverman, D. Who governs, at what scale and at what price? Geography, environmental governance, and the commodification of nature. *Ann. Assoc. Am. Geog.* **2004,** *94*(4), 734–738.

Luyt, R. P.; Jäger, A. K.; Van Staden, J. The rational usage of Drimiarobusta Bak. in traditional medicine. *South Afr. J. Bot.* **1999,** *65*(4), 291–294.

McAfee, K. Selling nature to save it? Biodiversity and green developmentalism. *Envir. Plan D: Soc. Space.* **1999,** *17*(2), 133–154.

McClatchey, W. Medical Bioprospecting, and Ethnobotanical Research. *Ethnobot. Res. Appl.* **2005,** *3,* 189–190.

Nakatsu, T.; Lupo Jr, A. T.; Chinn Jr, J. W.; Kang, R. K. Biological activity of essential oils and their constituents. *Stud. Nat. Prod. Chem.* **2000,** *21,* 571–631.

Nautiyal, B. P.; Prakash, V.; Bahuguna, R.; Maithani, U.; Bisht, H.; Nautiyal, M. C. Population study for monitoring the status of rarity of three Aconite species in Garhwal Himalaya. *Trop. Ecol.* **2002,** *43*(2), 297–303.

Nayar, M. P.; Sastry, A. R. K. *Red Data Book of Indian Plants.* **1987**.

Newman, D. J.; Cragg, G. M.; Snader, K. M. Natural products as sources of new drugs over the period 1981– 2002. *J. Nat. Prod.* **2003,** *66*(7), 1022–1037.

Nishteswar, K. Depleting medicinal plant resources: A threat for survival of Ayurveda. *Ayu.* **2014,** *35*(4), 349.

Öztürk, A.; Özbek, H. The Anti-Inflammatory Activity of Eugenia Caryophllata Essential Oil: An animal model of anti-inflammatory activity. *Eur. J. Gen. Med.* **2015,** *2*(4), 159–163.

Palevitch, D. Agronomy applied to medicinal plant conservation. *Conserv. Med. Plants.* **1991,** 168–178.

Pelagic, V. Pelagic folk teacher. *Beograd: Freedom.* **1970,** 500–502.

Petrovska, B. B. Historical review of medicinal plants' usage. *Pharmacog. Rev.* **2012,** *6*(11), 1.

Phondani, P. C.; Maikhuri, R. K.; Saxena, K. G. The efficacy of herbal system of medicine in the context of allopathic system in Indian Central Himalaya. *J.Herb. Med.* 2014,4(3), 147–158.

Phumthum, M.; Srithi, K.; Inta, A.; Junsongduang, A.; Tangjitman, K.; Pongamornkul, W.; Balslev, H. Ethnomedicinal plant diversity in Thailand. *J. Ethnopharmacol.* **2018,** *214,* 90–98.

Prasad, R.; Bhattacharya, P. Sustainable harvesting of medicinal plant resources. *Contemporary studies in natural resource management in India.* **2003,** 168–198.

Pudziuvelyte, L.; Stankevicius, M.; Maruska, A.; Petrikaite, V.; Ragazinskiene, O.; Draksiene, G.; Bernatoniene, J. Chemical composition and anticancer activity of Elsholtziaciliata essential oils and extracts prepared by different methods. *Ind. Crops Prod.* **2017**, *107*, 90–96.

Rajeswara, R.; Syamasundar, K. V.; Rajput, D. K.; Nagaraju, G.; Adinarayana, G. Biodiversity, conservation and cultivation of medicinal plants. *World.* **2012,** 422000(77000), 18–2.

Rates, S. M. K. Plants as source of drugs. *Toxicon.* **2001,** *39*(5), 603–613.

Rattray, G. N. The Enola bean patent controversy: biopiracy, novelty, and fish-and-chips. *Duke Law Technol. Rev.* **2002,** *1*(1), 1–8.

Reichert, J. M.A guide to drug discovery: Trends in development and approval times for new therapeutics in the United States. *Nat. Rev. Drug Disc.* **2003,** *2*(9), 695.

Reichling, J.; Schnitzler, P.; Suschke, U.; Saller, R. Essential oils of aromatic plants with antibacterial, antifungal, antiviral, and cytotoxic properties–an overview. Complement. *Med. Res.* **2009,** *16*(2), 79–90.

Ribeiro-Santos, R.; Andrade, M.; Sanches-Silva, A. Application of encapsulated essential oils as antimicrobial agents in food packaging. *Curr. Opin. Food Sci.* **2017,** *14,* 78–84.

Robbins, C. S. Comparative analysis of management regimes and medicinal plant trade monitoring mechanisms for American ginseng and goldenseal. *Conserv. Biol.* **2000,** *14*(5), 1422–1434.

Saad, B.; Azaizeh, H.; Abu-Hijleh, G.; Said, O. Safety of traditional Arab herbal medicine. *Evid. Based Compl. Alt. Med.* **2006,** *3*(4), 433–439.

Sahai, S.; Pavithran, P.; Barpujari, I. Biopiracy-Imitations, Not Innovations. *Khanpur, New Delhi, India: Gene Campaign.* 2007.

Sahoo, A. K.; Dandapat, J.; Dash, U. C.; Kanhar, S. Features and outcomes of drugs for combination therapy as multi-targets strategy to combat Alzheimer's disease. *J. Ethnopharmacol.* 2017.

Sahoo, N.; Manchikanti, P.; Dey, S. Herbal drugs: standards and regulation. *Fitoterapia.* **2010,** *81*(6), 462–471.

Schippmann, U.; Leaman, D. J.; Cunningham, A. B. Impact of cultivation and gathering of medicinal plants on biodiversity: global trends and issues. Biodiversity and the ecosystem approach in agriculture, forestry, and fisheries. Food and Agriculture, Italy, **2002**.

Shakya, A. K. Medicinal plants: future source of new drugs. *Int. J. Herb. Med.* **2016,** *4*(4), 59–64.

Shu, Y. Z. Recent natural products based drug development: a pharmaceutical industry perspective. *J. Nat. Prod.* **1998,** *61*(8), 1053–1071.

Simoes, C. M. O.; Falkenberg, M.; Mentz, L. A.; Schenkel, E. P.; Amoros, M.; Girre, L. Antiviral activity of south Brazilian medicinal plant extracts. *Phytomedicine.* **1999,** *6*(3), 205–214.

Singh, I. Reserpine in hypertension. *Brit. Med. J.* **1955,** *1*(4917), 813.

Siqueira, C. F.; Cabral, D. L.; PeixotoSobrinho, T.; Amorim, E. L. C.; Melo, J. G.; Araujo, T. A. S.; Albuquerque, U. P. Levels of tannins and flavonoids in medicinal plants: evaluating bioprospecting strategies. *Evid. Based Compl. Alt Med.* **2012,** 1–7.

Sun, Z.; Wang, H.; Wang, J.; Zhou, L; Yang, P. Chemical composition and anti-inflammatory, cytotoxic and antioxidant activities of essential oil from leaves of *Menthapiperita* grown in China. *Plos One.* **2014,** *9*(12), e114767.

Szafer, W.The Vegetation of Poland: International Series of Monographs in Pure and Applied Biology, Botany (Vol. 9). Elsevier, Oxford, **2013**.

Tahir, H. U.; Sarfraz, R. A.; Ashraf, A.; Adil, S. Chemical Composition and Antidiabetic Activity of Essential Oils Obtained from Two Spices (*Syzygiumaromaticum* and *Cuminum cyminum*). *Int. J. Food Prop.* **2016,** *19*(10), 2156–2164.

Thorwald J. Power and knowledge of ancient physicians. Zagreb: August Cesarec; **1991,** 10–255.

Toplak Galle, K. Domestic medicinal plants. *Zagreb: Mozaic book.* **2005,** 60–1.

Toscano-Garibay, J. D.; Arriaga-Alba, M.; Sánchez-Navarrete, J.; Mendoza-García, M.; Flores-Estrada, J. J.; Moreno-Eutimio, M. A.; Espinosa-Aguirre, J. J.; González-Ávila, M; Ruiz-Pérez, N. J. Antimutagenic and antioxidant activity of the essential oils of Citrus sinensis and Citrus latifolia. *Sci. Rep.* **2017,** *7*(1), p. 11479.

Traffic International, Medicinal and aromatic plants trade programme (accessed 29.06.15.) http://www.traffic.org/medicinal-plants/Geneva. **2015,** 2014–2023.

Tripathi, P.; Tripathi, R.; Patel, R. K.; Pancholi, S. S. Investigation of antimutagenic potential of Foeniculumvulgare essential oil on cyclophosphamide-induced genotoxicity and oxidative stress in mice. *Drug Chem. Toxicol.* **2013,** *36*(1), 35–41.

Tumen, I.; Akkol, E. K.; Süntar, I; Keleş, H. Wound repair and anti-inflammatory potential of essential oils from cones of Pinaceae: Preclinical experimental research in animal models. *J. Ethnopharmacol.* 2011,137(3), 1215–1220.

Tyler, V. E. Phytomedicines: back to the future. *J. Nat. Prod.* **1999,** *62*(11), 1589–1592.

UNESCO. Traditional knowledge in Nature and Resource. **1994,** *39*(1) UNESCO, Paris.

Uniyal, R. C.; Uniyal, M. R.; Jain, P. *Cultivation of medicinal plants in India: A Reference Book.* TRAFFIC-India and WWF India, 2000.

Vail III, W. B.; Vail, M. L. *U. S. Patent No. 7, 150,888.* Washington, DC: U. S. Patent and Trademark Office. 2006.

Van Noord, J. A.; Bantje, T. A.; Eland, M. E.; Korducki, L.; Cornelissen, P. J. G. A randomized controlled comparison of tiotropium and ipratropium in the treatment of chronic obstructive pulmonary disease. *Thorax.* **2000,** *55*(4), 289–294.

Vlachojannis, J.; Magora, F.; Chrubasik, S. Willow species, and aspirin: different mechanism of actions. *Phytother. Res.* **2011,** *25*(7), 1102–1104.

Walter, K. S.; Gillett, H. J. *IUCN Red List of Threatened Plants.* IUCN, **1998**.

Wani, M. C.; Taylor, H. L.; Wall, M. E.; Coggon, P.; McPhail, A. T. Plant antitumor agents. VI. Isolation and structure of taxol, a novel antileukemic and antitumor agent from *Taxusbrevifolia. J. Am. Chem. Soc.* **1971,** *93*(9), 2325–2327.

World Health Organization (WHO). *Traditional Medicine Strategy*, **2013**.

Yang, C.; Chen, H.; Chen, H.; Zhong, B.; Luo, X.; Chun, J. Antioxidant and anticancer activities of essential oil from Gannan navel orange peel. *Molecules.* **2017,** *22*(8), 1391.

Yoo, K.Y.; Park, S.Y. Terpenoids as potential anti-Alzheimer's disease therapeutics. Molecules, 2012, 17(3), 3524–3538.

CHAPTER 2

Diversity Analysis of Indian Mangrove Organisms to Explore Their Potential in Novel and Value-Added Biomolecules

ANGANA SARKAR*, SUSHANT PRAJAPATI, AMULYA SAI BAKSHI, ASMA KHATOON, RAGHAVARAPU SWATHI, SIDDHARTH KUMAR, ARPITA BEHERA, and RAHUL PRADHAN

Department of Biotechnology and Medical Engineering, National Institute of Technology, Rourkela, Odisha, 769008, India, Tel.: 06612462295;

**E-mail: sarkar.angana@gmail.com; sarkara@nitrkl.ac.in*

ABSTRACT

India treasures diverse flora and fauna. The five Indian mangroves are the home for different species of animals, plants, and microorganisms. Indian mangroves cover an area of 4,628 sq.km along the coastline. This covers 3% of the world's mangroves forest consisting about 1,600 plants and about 3,700 faunal species. Sundarbans is the largest mangroves in the world present in Bay of Bengal consisting of 1,336 number of species of organisms. Bhitarkanika mangrove of Odisha state covers 0.14% of India's geographical area is the home of over a thousand different species. Similarly, Godavari Krishna, Pichavaram, and Baratang also treasure a remarkable diversity of plants and animals. These all mangrove areas have different special features which makes them unique in terms of biological environment. A large class of bioactive compounds is extracted from the organisms of these mangroves including *Actinomycin, Proactinomycin, Streptothricin, Apratoxins*, and much more which have crucial applications. This immense variety of organisms not only contributes to maintaining the ecosystem but also the potential source of different biomolecules of pharmaceutical, therapeutic, enzymatic and medicinal uses. This chapter describes the diversity of living organisms in Indian mangroves and their perspective for exploration of novel biomolecules and value-added products.

2.1 INTRODUCTION

Mangroves are the areas that are situated at the interface between marine and terrestrial environment. These mangroves are highly productive to the surrounding microbiota and the organisms due to high nutrient contents. They are a rich source of versatile organisms and considered to have immense therapeutic and other uses. The Indian coastline length is about 7,516.6 km including the island territories (Anonymous, 1984), spreading the mangroves over an area of about 6,749 sq. km, making it the fourth largest mangroves consisting country in the world (Naskar & Mandal 1999). These mangroves are distinguished among three zones; East coast habitats with a coastline of about 2,700 km along Bay of Bengal, West coast habitats which are facing towards the Arabian sea and the coastline extending up to 3,000 km. All the Island Territories are grouped into another zone which consists of the coastline of about 1,816.6 km. The long coastlines of these mangroves are responsible for the biodiversity. Mangrove once covered three-quarter of the world's tropical coastline, with Southeast Asia hosting the greatest diversity within a given mangrove forest different species occupy a different niche (American Museum of Natural History) (Science Bulletins). There are five different mangroves spread throughout India namely: Sundarban, Bhitarkanika, Baratang, Pichavaram, and Krishna-Godavari. These five regions are the home of a variety of plant and animal species.

Sundarbans is listed as the largest mangrove in the UNESCO World Heritage site. It has a National park, Tiger conservation zone, and a Biosphere reserve park (Aziz and Paul, 2015). Next comes Bhitarkanika Mangroves which is the second largest forest located in Odisha (Chauhan et al., 2008). The delta of Brahmani and Baitarani Rivers and an important Ramsar Wetland in India encloses Bhitarkanika Mangroves (Chauhan et al., 2008). Likewise, the delta formed by Godavari and Krishna rivers includes Godavari Krishna mangroves located in Andhra Pradesh. Mangroves are under protection for climate, animals and bird sanctuary. Pichavaram mangroves are located at Pichavaram in Tamil Nadu. Pichavaram is the most famous tourist spot for its scenic beauty and also houses many species of Aquatic birds (Kathiresan, 2000). The mangroves of Baratang Island are beautiful swamp located between Middle and South Andamans in Portblair at Great Andaman and the Nicobar Islands. There are about 80 different species of mangrove trees (Raghunathan et al., 2010). The mangroves have low oxygen content, and the water flow in those waterbodies is also very less which accumulates fine sediments. Mangrove forest only grows at tropical and subtropical latitudes near the equator where India falls among few of them and expected to contain various genres of organisms (National Ocean Service, Homepage). With this context, this review will focus on exploring and understanding the presence of a diversity of organisms and their potential to produce therapeutics, enzymes, pharmaceuticals, and value-added products in Indian Mangrove.

2.2 CHARACTERISTICS OF THE MANGROVES

Mangroves forests have their ground covered by dense tangles of the roots which are not found in any other forests. This appearance of the mangroves attracts fish and other aquatic organisms as a source for food and shelter (National Ocean Service).

2.2.1 MANGROVE AREAS IN INDIA

The total area of mangrove in India is 4,662.56 sq. km which is 0.14% of the country total geographical area (Tanu Agritechportal Forestry) and 3% of world's mangroves forest (Based on recent data available from the state of forest report 2011 of the forest survey of India, Dehra Dun). The salt-loving flora *growing in brackish to saline tidal waters* distinguish between mangroves. The estuaries contain these wetlands, where an impenetrable woody region is formed when salt water is melted by fresh water (American Museum of Natural History). The five major mangroves of India are (Figure 2.1):

1. Sundarban mangroves
2. Bhitarkanika mangroves
3. Baratang mangroves
4. Pichavaram mangroves
5. Godavari–Krishna mangroves

FIGURE 2.1 Geographical location of Indian mangroves.

2.3 BRIEF INSIGHT TO THE INDIAN MANGROVES

2.3.1 SUNDARBAN MANGROVES

Sundarban mangrove is present in West Bengal state and also occupies a part of Bangladesh. This is considered the largest mangroves in the world (Aziz and Paul, 2015). These cover a land area of 9630 sq.km in India. It lies between 21°30′N to 22°15′N latitudes and 88°10′E and 89°10′E longitudes (Aziz and Paul, 2015). It is a deltaic forest formed 7000 years ago by the deposition of sediments from the Himalayas foothills through the Ganges river system. Sundarban mangrove forest (SF) of West Bengal has the highest variety of organisms which includes 69 species, 49 genera, 35 families. *Scyphiphora hydrophyllacea* and *Atalentia corea* have been first discovered in these mangroves (Mandal et al., 1995). The Sundarbans alone acquires 35 species out of the 50 true mangrove plant species listed throughout the planet, with an abundance of Sundry (*Heritiera fomes*), and Gewa (*Excoecaria agallocha)* (Neogi et al., 2017). Several species are restricted such as *Sonneratia apetala, H. fomes, Aegialitis rotundifolia,* and *S. griffithii*. Phytoplanktons were observed to be diatoms dominated (Bacillariophyceae) followed by Chlorophyceae and Pyrrophyceae (Dinoflagellates). Total 46 taxa classed to 6 groups including the dominants mentioned above, and also Euglenophyceae, Chrysophyceae, and Cyanophyceae have been noted (Neogi et al., 2017).

The forests lie in the delta of Bay of Bengal with waters coming from Ganges, Hooghly, Padma, Brahmaputra and Megha rivers (Aziz and Paul, 2015). The island has fresh water swamps seasonally, therefore, the salinity of the soil and the adjacent water bodies are governed by the number of fresh waters. It also has a tidal waterways network, small islands of salt-tolerant mangrove forests and mudflats.

2.3.2 BHITARKANIKA MANGROVE

Bhitarkanika mangrove of Odisha state covers an area of 650 sq.km, which is 0.14% of India's geographical area. This is located at 20°40′–20°48′ N latitude and 86°48′–87°50′ E longitude at the delta of Brahmani and Baitarani rivers in Kendrapara District, Odisha. Bhitarkanika mangrove forest (BF) has a bio taxa of 57 species, 37 genera, 29 families of organisms approximately (Naskar and Mandal, 1999). Water in Bhitarkanika mangroves has a high percentage of salinity. Saline plays a major role in water chemistry. However, the quality of water is reduced and is polluted by the agricultural and industrial wastes. The heavy metal percentage in these mangroves is high as compared to all other mangroves.

2.3.3 GODAVARI-KRISHNA MANGROVES

The Godavari-Krishna mangroves are present in Coromandel Coast over an area of 7,000 sq. km in Orissa, Tamil Nadu and Andhra Pradesh where the largest mangrove area is covered in Andhra Pradesh, where 36 species, 26 genera, 21 families are habitats. The water parameters favored the growth of the aquatic organisms. It contains the waters from both Krishna and Godavari rivers which are the freshwater river. However, there is some salinity creeping into the water of Krishna which duly affects the salt content in the water. Fungus plays the predominant role in the mangroves.

2.3.4 PICHAVARAM MANGROVES

Pichavaram mangroves of Tamil Nadu state are present about 250 km south of Chennai city and 10 km south of Parangpettai. It falls between latitude 11°29′ N and longitude 79°46′ E. It is between the Vellar and the Coleroon Estuarine complexes with a total area of 1100 ha having 51 islets, ranging 10 sq.m in size separated by intricate waterways connecting the vellar estuary and the Coleroon Estuary in the north and south parts respectively. Pichavaram mangrove is habituated by 35 species, 26 genera and 20 families including a new species *RhizophoraannamalayanaKathir*, the hybrid of *R. apiculata* and *R. mucronata* (Kathiresan, 1995). This proves the existence of important shells and finfishes. These mangroves are also famous for large bird varieties like egrets, storks, herons, storks, spoonbills, snipes, cormorants, and pelicans. A total of about 177 species of birds containing in 15 orders and 41 families have been recorded at the mangroves. The fertility and productivity of the waters of Bay of Bengal are increased when the mangrove water of Pichavaram of high nutritional value is mixed with them by tidal ebb or flow (Kathiresan, 2000). The microbial population of the mangroves is around seven times higher than that of Bay of Bengal. And also, among the microbes, the population ratio of bacteria to fungus is 1:7000 (Kathiresan, 2000). Radiation in these mangroves is higher compared to others. Due to poor functioning of osmoregulatory facility and shallowness of the waterways which would restrict the movements of the adult marine fishes due to the presence of prop roots causing them not survive in these mangroves (Krishnamurthy and Jeyaseelan, 1981). The mangrove consists of neritic water from the adjacent Bay of Bengal through Chinnavaikkal (Marine zone), brackish water from vellar and Coleroon Estuaries and fresh water flowing from an irrigation channel as well as from the main channel of the Coleroon river. This mangrove is predominated by a diverse group of diatoms.

2.3.5 BARATANG ISLAND MANGROVES

Baratang mangroves are present on an island of Andaman and Nicobar island in North and middle Andaman administrative districts. The island lies 150 km (93 miles) north of Port Blair. It is situated between the 12°18′ N latitude and 92°80′ E longitude and covers an area of 242.6 sq.km. Andaman & Nicobar Islands (A&N) are the house of 61 species distributed in 39 genera and 30 families, including two new species, *R. lamarkii* and *R. stylosa* (Singh et al., 1987). 50% of the total mangrove species is found in Baratang mangroves. This is a place where tsunami, earthquakes are too common. Tsunami causes migration of various other species to Andaman and Nicobar islands (Ragavan et al., 2016). The leaves are shed throughout the year which is considered as a special feature of the mangroves (Figure 2.2). The litter is not consumed by the herbivorous animals directly but is being degraded by the microbes first and consumed later. This also imparts nutritional values to the mangroves. The mangroves of Baratang consist of water from various water bodies like Indian Ocean and Bay of Bengal. The waters are of high salinity as the waters are from oceans. These mangroves are dense and are free from human intervention with creeks and mud formation in the river mouths.

2.3.6 PHYSICOCHEMICAL PROPERTIES OF THE INDIA MANGROVE

Each mangrove has different physicochemical properties, which makes them a potential place for the survival of specific microorganisms. Sundarban is the largest mangrove has the highest rainfall throughout the year ranging from 1640–2000 mm (Table 2.1). The temperature here goes to 35°C which is the highest among all mangroves. Humidity varies from 70–80% and has a basic pH level of the soil and water. Some specific type of ions is commonly found here in a significant concentration, namely potash and calcium oxide. Salinity ranging 8.66–26 psu differs according to the area providing an optimum condition for certain rare organism to grow and proliferate. Bhitarkanika mangrove is a noticeable mangrove because of its salinity and dissolved oxygen concentration. The salinity level ranging 3.25–8.58 psu (Table 2.1) is quite higher here as compared to other mangrove making it a suitable region for the growth of halophiles (Naskar et al., 1999). It has different metals, ions and compounds namely hydrogen carbonate, nitrate, nickel, zinc, lead, cobalt, etc., present in soil and water of this region. It has a mild climate

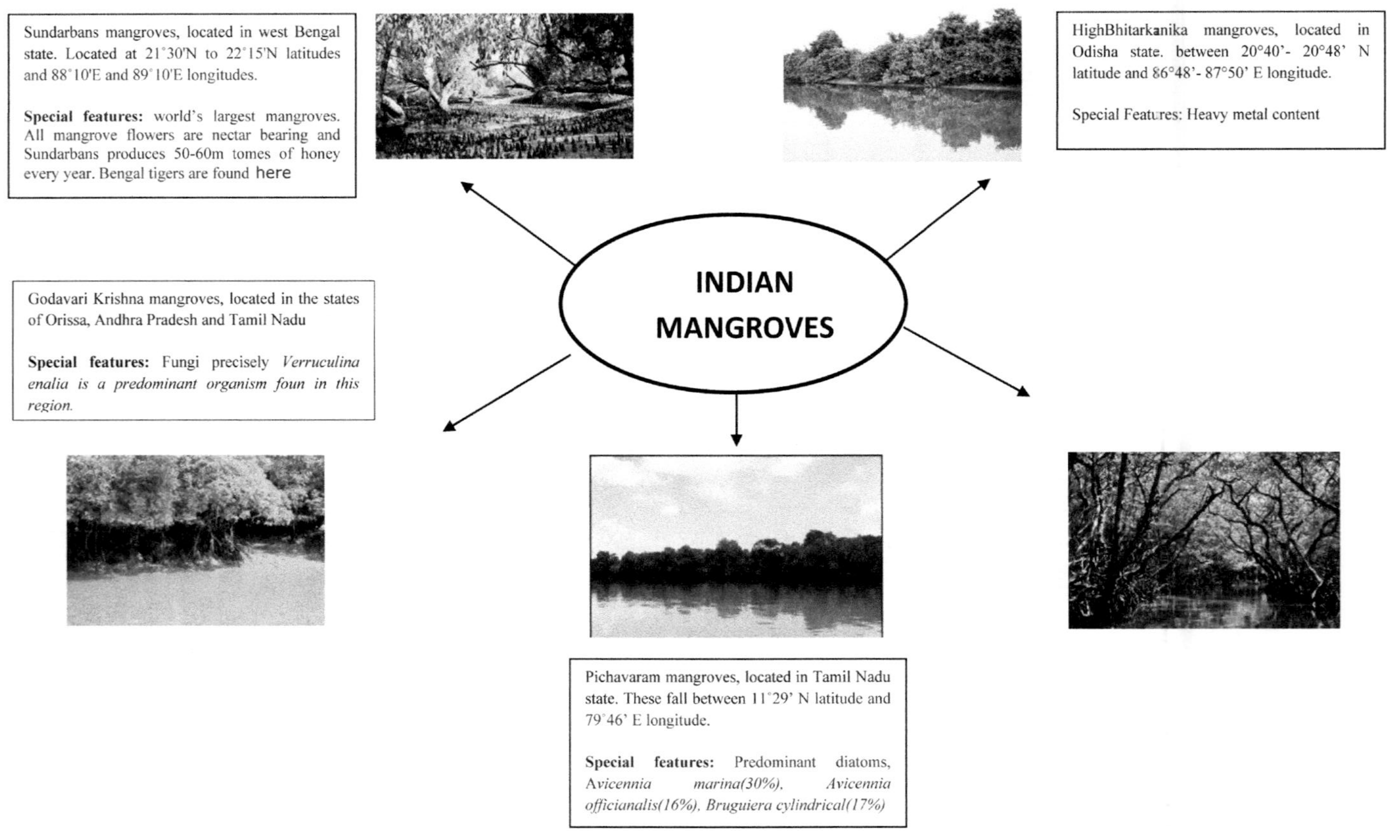

FIGURE 2.2 Indian Mangroves and their key features.

with the temperature of summer 20–30°C and winter 15–20°C and rainfall around 1670 mm every year. It has almost a neutral pH level and consistent humidity of 75–80% (Table 2.1). Godavari Krishna has a basic pH level and a moderate salinity. Rainfall is similar to Bhitarkanika mangrove of about 1160 mm and has a moderate temperature ranging from 26–33.8°C. It has a significant concentration of ammonia (0.05–1.5 ppm) and nitrite (01–1 ppm) which is rare in other mangroves. On the other hand, Pichavaram has a high salinity and an acidic pH level making it a favorable region for the proliferation of halophiles and acidophilus. Temperature is mild throughout the year ranging from 30–34.8°C (Table 2.1). It also has certain rare ions and compound in considerable concentration like nitrites and reactive silicates. Batarang is the mangrove of lowest humidity (12–45%) as compared to the other four mangrove regions. The pH level is 6–8.2 making it a suitable place for organisms which do not grow in high acidic or basic concentration. It has a mild temperature the year around. These special features make this mangrove organism-specific and home to much rare flora and fauna (Table 2.1).

2.4 MANGROVE ORGANISMS AND THEIR POTENTIAL APPLICATIONS

The mangroves in India are the rich source of different kinds of phytoplankton, zooplankton, and microorganisms (Table 2.2). Sundarban mangrove holds 50 phytoplankton species, *Oocystis pusilla* being the dominant one—50 species of Diatoms, 30 species of green algae, about 18 species of blue-green algae and 24 microbes species have been recorded so far. Besides *O. pusilla*, many other species have also been recorded here such as *Staurastrum orbiculare*, *Uronema* sp., *Astasia cylindrical*, *Euglenas pathirhyncha*, *Biddulphia dubia*, *Chaetoceros pendulus*, *C. socialis*, *Tintinnid protozoa*, *Calanoid copepoda*, etc., (Barik and Chowdhury, 2014). Pichavaram is home of a variety of species including 94 species of phytoplankton among which diatoms are the predominant groups. Apart from phytoplankton diversity, it also consists of 52 bacterial species, 23 fungal species, and 95 zooplankton species. The most commonly found among them are *Copenoids*, *Metapenaeus*, *Lutjanus* spp., *Penaeus* spp., *Scatophagus argus*, *Siganus javus*, *Etroplus suratensis*, *Brachionus* spp., *Ciliates*, and many others. Bhitarkanika mangrove is known for its rare diversity of organisms. Till date, 51 phytoplankton species, 29 bacterial species, 101 zooplankton species, 27 fungal species, and 11 algal species have been recorded. Some of the common species here are *B. subtilis*, *Rhizophora*, *Rhizophora*, *A. Officinalis*, *S.apetala*, *Sonneratia apetala*, *Duttaphrynus*

TABLE 2.1 Physicochemical Properties of Indian Mangroves

Mangrove Names	Ion, compound, element, heavy metal concentration	Temperature (°C)	Humidity (%)	Rainfall (mm)	PH	Dissolved Oxygen (mg/l)	Salinity (psu)
Sundarban mangrove	Nitrogen: 0.05–0.9 Phosphate: 0.06–0.1 Potash: 0.3–1 Carbon: 0.5–1 CaO: 0.6–3	Winter 12–24 Summer 25–35	70–80%	1640–2000	7.0–8.9	3.90–7.03	8.66–26
Bhitarkanika mangrove	(mg/L) Sodium: 575 Potassium: 133 Calcium: 110 Sulfate: 370 Phosphate: 0.2 Nitrate: 1 Chlorine: 5900 Hydrogen carbonate: 69 Silicon dioxide: 27 Ammonium: 0.16 Iron: 7.19 Manganese: 0.15 Nickel: 0.04 Zinc: 0.14 Chromium: 0.35	Summer: 20–30 Winter: 15–20	75–80%	1670	7.5	3.25–8.58	9–33.8

TABLE 2.1 *(Continued)*

Mangrove Names	Ion, compound, element, heavy metal concentration	Temperature (°C)	Humidity (%)	Rainfall (mm)	PH	Dissolved Oxygen (mg/l)	Salinity (psu)
	Cobalt: 0.14 Cadmium: 0.14 Lead: 0.41						
Godavari-Krishna mangrove	Ammonia: 1.2–0.05 ppm Nitrite 1–0.01ppm Calcium: 250–80% Magnesium: 450–180%	26–33.8	N/A	1160	7.15–8.5	4–7.8	0–24
Pichavaram mangroves	Nitrate: 7–36.23 µM Nitrite: 0.31–5.46 µM Phosphate: 12.26–56.63 µM Reactive silicates: 22.1–89.8 µM	30–34.8	N/A	N/A	7.2–8.6	3.2–6.5	9.6–35.4
Baratang Island mangroves	Organic carbon: 0.4–0.88% Phosphorus: 2.2–4.4 ppm Potassium: 235–500 ppm	Summer: 30.2°C Winter: 23.0°C	12–45%	N/A	6–8.2	N/A	N/A

melanostictus, D. stomaticus, Euphlyctis cyanophlyctis, E. hexadactylus, B. subtilis, A. aerogenes, E.coli, etc. Baratang, in the same way, has a large variation flora and faunal diversity. It holds 449 species of zooplanktons, 20 fungal species, 103 bacterial species, and 120 algal species. Some of them are *Rhizophora spp., Xylocarpus granatum, Xylocorpus molluccensis, Excoercaria agallocha, Pteropus* spp., *Cynopterus sphinx Nectarina jugularis klossi, Streptoverticillium* spp. *Streptosporangia* spp., *Nocardia* spp., *Micromonospora* spp. and many others (Misra, 1986). Godavari Krishna is also rich in terms of flora and fauna. More than 60 species of Phytoplanktons dominated by diatoms, 55 species of fungi are reported. Some of them are listed below: *Avicennia marina, Bruguiera* spp, *Suaeda* spp, *Rhizophora* spp., *Bruguieraspp, Molluscs, Saltwater crocodiles, Pelecanusphilippenes, Microalgae, Penicillian, Microalgae, Streptoverticillium* spp., *Hemidiscus* spp. and many other species are also dominant here (Sarma et al., 2001).

The mangrove organisms are helpful in producing different kinds of pharmaceuticals, therapeutic agents and many products that can be used in our daily lives. The secondary metabolites of the organisms produce enzymes that are used in several industries and laboratories. Sundarbans, mangroves having marine plants that pave the way to produce therapeutic agents like cellular regeneration, maintaining glucose levels and few enzymes such as digestive enzymes, protease, amylase, etc. (Aziz and Paul, 2015) (Table 2.2). The Bhitarkanika mangrove of Odisha is known for the high salinity in water. There are many organisms that grow in high salt conditions and serve mankind with their various uses like bacteriotherapy, healing of open surgical wounds in the pharma sector; they serve as antibacterial, antioxidant agents (Nabeel et al., 2010; Takemaru et al., 2000). The important species that are found in these mangroves are *Rhizophora* and *Bruguera.* However, the predominant pharmaceutical application is shown by the organisms of Baratang mangroves like curing sore mouth, treating dysentery, diarrhea, abdominal troubles, pneumonia, and asthma. They also provide poison antidote. The enzymatic application of the mangrove organisms is very poor. Pichavaram mangroves have various kinds of uses. The noticeable uses of the organisms are their use as food for marine organisms (Støttrup et al., 1997; Shyam et al., 2013), as biocatalyst and biosensors. It also plays an important role in cell signaling and is also capable of drug delivery (Table 2.2). The enzymatic uses of these mangroves are high compared to the others. Godavari-Krishna delta mangrove organisms are quite useful in the area of value-added products. For example, the compounds can be used as an insect repellent, and it has properties like antibacterial and antiviral activities (Dahdouh et al., 2006, Chaudhuri et al., 2015). The pharmaceutical

TABLE 2.2 Diversity of Indian Mangrove Organisms

Mangroves	Phytoplankton	Zooplankton	Microorganisms	Pharmaceutical and therapeutic product uses	Enzymatic product uses	Value-added products	Reference
Sunderbans	*Coscinodiscus Ditylum Ceratium Biddulphia, Chaetoceros, Thalassiothrix, Rhizosolenia, Thalassionema*	Tintinnid protozoa Calanoid copepoda	Blue-green algae	Depression symptoms Cellular regeneration Immune system boost Supports cardiovascular health Maintain a healthy glucose level	Digestive enzymes protease, amylase, lipase, lactase, bromelain, and papain	Omega 3 polyunsaturated fatty acids (PUFA), Carotenoids	Aziz and Paul, 2015
Bhitarkanika	*B. subtilis, Rhizophora A. officinalis, S.apetala, Sonneratia apetala,* E. agallocha, *Ceriops, A.alba, Bruguiera*	*Duttaphrynusmelano stictus D. stomaticus Euphlyctis cyanophlyctis E. hexadactylus Fejervarya cancrivora F. orissaensis* F. syhadrensis Hoplobatrachus crassus H. tigerinus Sphaerotheca rolandae Polypedates maculatus Hylarana tytleri	*B. subtilis A. aerogenes E. coli*	Bacterioprophylaxis of gastrointestinal disorders, antitumor, oral bacteriotherapy, antimicrobial, wound-healing agents, herbal medicine, antioxidant activity, gut colonization prevention.	Antibacterial and antioxidant activities. Bioflocculant, defecating the trona suspension, salt stress response based on molecular approach	Silver nanoparticles (AgNPs) synthesis, production of phytochemicals	Nabeel et al., 2010; Takemaru et al., 2000
Baratang	*Rhizophora spp., Xylocarpus granatum, Xylocorpus molluccensis, Excoercaria agallocha, Clerodendrum spp, Lumnitzera racemose,*	*Pteropus spp., Cynopterus sphinx, Nectarinitl jugularis klossi, Accipiter virgatus gularis, Ardeola striatus spodiogas/er, Crocodilus porosus, Varanus salvator andamanensis, Goniocephalus subcristatus, Rana*	*Streptoverticillium* spp	Cures sore mouth, treats dysentery, diarrhea, abdominal troubles, treats swellings of the breast and elephantiasis, itchy skin, treats pneumonia or asthma, acts as poison antidote, cures constipation, treatment of coughs, venereal infections, skin diseases,	N/A	Preservatives, tannins, dyes, pencils, illuminant, hair oil, used for piles, poles, house posts, ties, paving blocks, bridges, ship planks, decks, handles, and cabinetry.	Kar et al., 2014; Saranya et al., 2015; Simlai et al., 2016

TABLE 2.2 *(Continued)*

Mangroves	Phytoplankton	Zooplankton	Microorganisms	Pharmaceutical and therapeutic product uses	Enzymatic product uses	Value-added products	Reference
	Acanthus ilicifolius, Heriteria littoralis, Cynoemetra ramiflora, Avicienna marina, Ceriops Decandra	*limnocharis spp., Rhizophora spp., Philopona* spp., *Hospitalotermes blaire, Oeeophylla spp.*	*Streptosporangia* spp. *Nocardia* spp. *Micromonospora* spp. *Actinoplanes* spp. *Actinomadura* spp	rheumatism, tropical burns, vermifuge, febrifuge, malaria, used as anti-diabetic, anti-hypertensive and sedative, treatment of diabetes, leprosy, neuralgia, paralysis, ringworm, rheumatism, snakebite, stomachache, leucorrhea, leukemia, cytotoxic/anticancer activity, anti-bacterial, anti-inflammatory, anti-cancer, antiprotozoal, antimicrobial, osteoblastic, hypercholesterolemia			
Pichavaram	Copenoids, *Metapenaeus, Lutjanus spp.*	*Lutjanus spp.* Copenoids *Metapenaeus Penaeus spp* Scatophagus argus Siganus javus Etroplus suratensis Terapon jarbua Mugli cephalus	*Brachionus* spp *Ciliates*	Antibacterial, drug delivery, biosensor, and biocatalyst, cell signaling	Used to do a clinical assay or allergen specific-IgE, supports the growth of bacteria as a nitrogen source, Used for bioassay of the toxicity of insecticides, Bioindicator.	Food products	Støttrup et al., 1997; Shyam et al., 2013; Srikandace et al.; 2017; Qi et al., 2017; Venkatesan et al., 2007
Godavari-Krishna	*Avicennia marina*	Molluscas	Microalgae	Treatment of smallpox lesions, HIV-1 and HSV, treats sore eyes, burns,	N/A	Insect repellent, dye, strong antiproliferative,	Dahdouh et al., 2006; Chaudhuri

TABLE 2.2 *(Continued)*

Mangroves	Phytoplankton	Zooplankton	Microorganisms	Pharmaceutical and therapeutic product uses	Enzymatic product uses	Value-added products	Reference
	Bruguiera spp *Suaeda spp* Rhizophora spp. Bruguiera spp	Saltwater crocodiles *Pelecanus philippenes*	Penicillian Microalgae Streptoverticillium spp. Hemidiscus spp	bleeding, lower blood pressure, chronic pain, treats epilepsy, scabies, treat skin diseases, antifungal, antibacterial, antiparasitic, treat bacterial infections, antibiotics effective against *H. influenzae*, *E. coli*, treat dental infections, skin and soft tissue infection, snake bite and to remove placenta after childbirth		moderate cytotoxic activities, antibacterial, antidote, anodyne, sodium carbonate, wine, tannins, vegetable, timber is used for building poles, firewood, charcoal, perfume handbags, high end garments, fashionable items, tools, musical instruments, decorations, hair clips, belt buckles, buttons. Food additives, biofuels, herbicides, cosmetics, natural beverages, syrup, pickles, carbohydrates, proteins, Beta carotene, lipids, carotenoids or vitamins.	et al., 2015; Sridhar et al., Karuna et al., 2007; Sahu et al., 2015

applications are the treatment of smallpox lesions and help in lowering blood pressure (Table 2.2). A wide range of metabolites such as proteins, lipids, carbohydrates, carotenoids or vitamins, Beta-carotene, astaxanthin, PUFA such as DFA and EPA and a polysaccharide such as Beta glucan dominate (Table 2.3). Different allelopathic capability of the wetland invasive plants *Aegiceras corniculatum*, *Spartina alterniflora Bruguiera gymnorrhiza*, and *Kandelia candel* was noted (Duan et al., 2015). Many show a high efficiency (majorly phytoplanktons) in nitrogen fixation released in the environment through pollutants (Ramesh et al., 2017).

The land-sea interface location of mangroves shield the coastal areas against natural hazards like cyclones and tsunamis they recycle nutrients and retain terrestrial sediments, therefore, providing support to clear offshore waters, which in turn favors the phytoplankton's photosynthetic activity as well as growth and robustness of seagrass beds, reef fish and coral reefs groups, they function as an essential nursery, refuge and habitat, supplying food for countless living organisms including humans (Ghosh et al., 2015).

Mangrove organisms such as sponge have many useful compounds such as cytarabine (Ara C), Vidarabine (Ara-A), Eribulin Mesylate (E7389) which are all FDA approved and targets dangerous diseases such as cancer (Mayer et al., 2016; Newman and Cragg, 2016) (Table 2.3). Other mangrove organisms such as Tunicate, Mollusca, Nudibranch also targets the disease, cancer. The omega 3 fatty acid ethyl esters from fish can be used for the treatment of Hypertriglyceridemia. Some important compounds such as Brentuximab Vedotin (SGN-35) Elisidesin, Glembatumumab Vedotin (CDX 011), SGN-75, ASG-5ME can be obtained from Mollusca majorly found in Baratang and Krishna Godavari mangroves. The bacterium *Bryozoa* gives Bryostatin, which can treat cancer and Alzheimer's disease. The soft Coral produces pseudopterosins whose molecular target is Eicosanoid Metabolism and helps in wound healing. The compound Trabectedin (ET 743), which belongs to the alkaloid class targets the minor, grooves of DNA and helps in the treatment of cancer. The compound, Eribulin Mesylate (E7389) obtained from the sponge targets the microtubules for the treatment of cancer. Zincoside, obtained from cone nail is used as an analgesic drug. The compound DMXBA (GTS-21) obtained from worm targets the alpha-7 Nicotinic Acetylcholine Receptor and helps in the treatment of cognition schizophrenia (Table 2.3). Mangrove pharmaceuticals can be used as cytostatic drugs, antiviral drugs, analgetic drugs, antihyperlipidemic drugs and can also be used as diagnostic and experimental tools. The cosmetics industry is also stepping towards the sea, so as to find new ingredients (Fenical et al., 2009; Martins et al., 2014).

TABLE 2.3 Biomolecules Derived from Indian Mangroves

Clinical status	Compound name	Marine organism	Chemical class	Molecular target	Diseases area	Reference
FDA approved	Cytarabine (Ara C)	Sponge	Nucleoside	DNA polymerase	Cancer	Martins et al., 2014
	Vidarabine (Ara A)	Sponge	Nucleoside	Viral DNA polymerase	Antiviral	
	ziconotide	Cone nail	Peptide	N type calcium channel	Analgesic	
	Erybulin Mesylate (E7389)	Sponge	Macrolide	Microtubules	Cancer	
	Omega 3 fatty acid ethyl esters	Fish	Omega 3 fatty acids	Triglyceride	Hypertriglyceridemia	
	Trabectedin (ET 743)	Tunicate	Alkaloid	Minor groove of DNA	Cancer	
Phase I	Marizomib (Sali-nosporamide A)	Bacterium	Beta-Lactone-Gamma Lactam	20S Proteasome	Cancer	Martins et al., 2014
	PM01183	Tunicate	Alkaloid	Minor Groove of DNA	Cancer	
	SGN-75	Mollusca	Antibody Drug Conjugate (MM Auristatin F)	CD70 and Microtubules	Cancer	
	ASG-5ME	Mollusca	Antibody Drug Conjugate (MM Auristatin E)	ASG-5 and Microtubules	Cancer	
	Hemiasterlin (E7974)	Sponge	Tripeptide	Microtubules	Cancer	
	Bryostatin 1	Bryozoa	Polyketide	Protein Kinase C	Cancer, Alzheimer's	
	Pseudopterosins	Soft coral	Diterpene glycoside	Eicosanoid Metabolism	Wound healing	
Phase II	DMXBA (GTS-21)	Worm	Alkaloid	Alpha-7 Nicotinic Acetylcholine Receptor	Cognition, Schizophrenia	Martins et al., 2014

TABLE 2.3 *(Continued)*

Clinical status	Compound name	Marine organism	Chemical class	Molecular target	Diseases area	Reference
	Plinabulin (NPI 2358)	Fungus	Diketopiperazine	Microtubules and JNK Stress Protein	Cancer	
	Elisidepsin	Mollusca	Depsipeptide	Plasma Membrane Fluidity	Cancer	
	PM00104	Nudibranch	Alkaloid	DNA-Binding	Cancer	
	Glembatumumab Vedotin (CDX-011)	Mollusca	Antibody Drug Conjugate (MM Auristatin E)	Glycoprotein NMB and Microtubules	Cancer	
Phase III	Brentuximab Vedotin (SGN-35)	Mollusca	Antibody-Drug Conjugate (MM Auristatin E)	CD30 and Microtubules	Cancer	Martins et al., 2014
	Plitidepsin	Tunicate	Depsipeptide	Rac1 and JNK Activation	Cancer	

2.5 CHALLENGES FACED DURING NATURAL PRODUCTS DEVELOPMENT

Mangrove region is full of life, having metabolically and structurally diverse flora and fauna. Accessing the mangrove system is very difficult and robotic, and engineering technology is required to check the value of biodiversity. Very little is known about the mangrove organisms; therefore, further exploration occurs by a random selection procedure. The following three sampling techniques are mostly used:

i) Scrutinizing of the unexploited taxa to increase the probability of finding new molecules.
ii) Examining new taxonomical groups where there is a vast chemical diversity.
iii) Combining both of the above-mentioned strategies. In both the techniques, knowledge of bio- and chemo-diversity is a prerequisite (Martins, 2014).

2.6 MARKET AND ACCESS

There is a huge scarcity of access of drugs derived from mangrove microorganism. This points out to the fact that its approbation is expected to take up a lot of time (8–15 years approximately) and might also involve a lot of capital (Approximately US$900 million on an average) from discovery to the market (Deszca et al., 1999; Snelgrove et al., 2016). Failures have always been an obstacle, and several promising marine compounds have failed during the development processes due to different reasons like toxicity, insignificant availability, etc. Interdisciplinary clubbing of chemists, biologists, pharmacists, biotechnologists, doctors and also the universities, hospitals, and companies dictates the success of pharmaceutical products available from marine sources. Table 2.3 gives an idea about the different products derived from mangrove organisms.

Cost of the mangrove-based pharmaceuticals can be varying and certainly high due to the supply problem. It causes hindrance to the development of Marine Natural Products (MNPs). Natural product screening is essentially one of the most useful avenues for bioactive discovery. Previously, studies on MNP have been concentrated critically on macro-organisms such as sponges, corals, etc. Majority of the new chemical entities do not surpass the pharmaceutical pre-clinical trials and only a very few ends up being marketed as pharmaceutical products. Apart from these, there are many difficulties of MNPs such as sustainable source, scale up related issues. The potential for innovation from vastly unexploited mangrove habitats needs to be further looked into it, and if the market entry success rates are improved by 10% of the present situation, then there could be hundreds of hits reaching the market (Martins et al., 2014).

2.7 CONCLUSION AND OUTLOOK

The proof of the massive potential of MNPs is shown by the growth of mangrove organism derived drugs. Many useful drugs, particularly in the field of cancer chemotherapy can be developed from the mangrove which offers a rich source of biodiversity. Currently, seven FDA or EMA approved drugs from marine sources, most of which are antitumor drugs, are potentially increasing in the market. About 20 are presently in clinical trials Phase I, I/II, II or III and a very promising candidate is salinosporamide A (marizomib) which is an active β-lactone produced by actinomycete Salinispora tropica. It is an inhibitor of the 20S proteasome and is currently in clinical development for the treatment of hematological malignancies (mainly of multiple myeloma). Although cancer is the main indication, an extension of the range of indications can also be expected, for example, a candidate against Alzheimer disease which targets nicotinic acetylcholine receptors is in Phase II (DMXBA). The global mangrove pharmaceutical pipeline includes eight Food and Drug Administration (FDA) or European Medicines Agency (EMEA) approved drugs and several compounds in different phases of the clinical pipeline. The market is continuously blooming for mangrove nutraceuticals and cosmetics.

KEYWORDS

- **bioactive compounds**
- **biomolecules**
- **enzymes**
- **mangrove**
- **organisms**
- **pharmaceuticals**
- **therapeutics**
- **value-added products**

REFERENCES

Abdel-Wahab, M. A., (2005). Diversity of marine fungi from Egyptian Red Sea mangroves. *Bot. Mar., 48,* 248–355.

Alias, S. A., Kuthubutheen, A. J., & Jones, E. B. G., (1995). Frequency of occurrence of fungi on wood in Malaysian mangroves. *Hydrobiologia, 295,* 97–106.

Alongi, D., M., (1988). Bacterial productivity and microbial biomass in tropical mangrove sediments. *Microb. Ecol., 15,* 59–79.

Aziz, A., & Paul, A. R., (2015). Bangladesh Sundarbans: Present status of the environment and biota. *Diversity, 7,* 242–269.

Banerjee, A., (1998). *Environment, Population, and Human Settlements of Sundarban Delta.* Concept Publishing Company.

Belley, R., & Snelgrove, P. V., (2016). Relative contributions of biodiversity and environment to benthic ecosystem functioning. *Front. Marine Sci., 3,* 242–268.

Borodulina, U. C., (1935). The interrelation between soil *Actinomycetes* and *B. mycoides. Microbiology, 4,* 561–586.

Chaudhuri, P., Ghosh, S., Bakshi, M., Bhattacharyya, S., & Nath, B., (2015). A review of threats and vulnerabilities to mangrove habitats: with special emphasis on east coast of India. *J. Earth. Sci. Clim. Change, 6,* 1–9.

Chauhan, R., & Ramanathan, A. L., (2008). Evaluation of water quality of Bhitarkanika mangrove system, Orissa, east coast of India. *IJMS, 37,* 153–158.

Dahdouh-Guebas, F., Collin, S., Seen, D. L., Rönnbäck, P., Depommier, D., Ravishankar, T., & Koedam, N., (2006). Analyzing the ethnobotanical and fishery-related importance of mangroves of the East-Godavari Delta (Andhra Pradesh, India) for conservation and management purposes. *J. Ethnobiol. Ethnomed., 2,* 24–34.

Deszca, G., Munro, H., & Noori, H., (1999). Developing breakthrough products: Challenges and options for market assessment. *J.O.M., 17,* 613–630.

Duan, L., Liang, S., Li, F., & Zhou, Q., (2015). Comparison of the leaf allelopathic potential of the invasive wetland plant Spartina alterniflora and three native mangrove plants. *Journal of Guangxi Normal University-Natural Science Edition, 33*(2), 109–114.

Fenical, W., Jensen, P. R., Palladino, M. A., Lam, K. S., Lloyd, G. K., & Potts, B. C., (2009). Discovery and development of the anticancer agent salinosporamide A (NPI–0052). *Bioorganic & Medicinal Chemistry, 17*(6), 2175–2180.

Gerwick, W. H., & Moore, B. S., (2012). Lessons from the past and charting the future of marine natural products drug discovery and chemical biology. *Chem. Biol., 19,* 85–98.

Ghosh, A., Schmidt, S., Fickert, T., & Nüsser, M., (2015). The Indian Sundarban mangrove forests: History, utilization, conservation strategies, and local perception. *Diversity, 7*(2), 149–169.

Kar, P., Goyal, A., Das, A., & Sen, A., (2014). Antioxidant and pharmaceutical potential of Clerodendrum L.: An overview. *Int. J. Green. Pharm., 8,* 210–221.

Karuna, C. H. L. D., Bapuji, M., Rath, C. C., & Murthy, Y. L. N., (2009). Isolation of mangrove fungi from the Godavari and Krishna Deltas of Andhra Pradesh, India. *J. Ecobiolo., 24,* 91–96.

Kathiresan, K., (1995). Rhizophora Annamalai: A new species of mangroves. *Environ. & Ecol., 13,* 240–240.

Kathiresan, K., (2000). A review of studies on Pichavaram mangrove, southeast India. *Hydrobiologia, 430,* 185–205.

KM, V. K., & Kumara, V., (2016). Diversity of true mangroves and their associates in the Kundapura region, Udupi district, Karnataka, Southwest coast of India. *Current Botany, 3,* 70–78.

Krassilnikov, M., & Koreniako, A. I., (1939). The bactericidal substance of the actinomycetes. *Microbiologia, 8,* 673–685.

Krishnamurthy, K., & Jeyaseelan, M. P., (1981). The early life history of fishes from Pichavaram mangrove ecosystem of India. In: *The Early Life History of Fish: Recent Studies* (Vol. 1, pp. 2–4). The second ICES symposium. Conseil International Pour L'Exploration de la Mer Palegade.

Kriss, A., (1940). On the lysozyme of actinomycetes. *Microbiologia, 9,* 32–39.

Madhyastha, H. K., Radha, K. S., Sugiki, M., Omura, S., & Maruyama, M., (2006). Purification of c-phycocyanin from *Spirulina fusiformis* and its effect on the induction of urokinase-type plasminogen activator from calf pulmonary endothelial cells. *Phytomedicine, 13,* 564–569.

Martins, A., Vieira, H., Gaspar, H., & Santos, S., (2014). Marketed marine natural products in the pharmaceutical and cosmeceutical industries: Tips for success. *Marine Drugs, 12,* 1066–1101.

Misra, J. K., (1986). Fungi from mangrove muds of Andaman Nicobar Islands, *IJMS, 15.*

Mitra, A., Chowdhury, R., Sengupta, K., & Banerjee, K., (2010). Impact of salinity on mangroves. *J. Coast. Env., 1,* 71–82.

Mullai, P., Rene, E. R., & Sridevi, K., (2013). Biohydrogen production and kinetic modeling using sediment microorganisms of Pichavaram mangroves, India. *Bio. Med. Res. Int., 1,* 1–9.

Nabeel, M. A., Kathiresan, K., & Manivannan, S., (2010). Antidiabetic activity of the mangrove species *Ceriopsde candra* in alloxan-induced diabetic rats. *Journal of Diabetes, 2,* 97–103.

Naskar, K., & Mandal, R., (1999). *Ecology and Biodiversity of Indian Mangroves*. Daya Books.

Neogi, S. B., Dey, M., Kabir, S. L., Masum, S. J. H., Kopprio, G., Yamasaki, S., & Lara, R., (2017). Sundarban mangroves: Diversity, ecosystem services, and climate change impacts. *Asian Journal of Medical and Biological Research, 2*(4), 488–507.

Newman, D. J., & Cragg, G. M., (2016). Drugs and drug candidates from marine sources: An assessment of the current "state of play." *Planta Medica, 82*, 775–789.

Parikh, P., Mani, U., & Iyer U., (2001). Role of Spirulina in the control of glycemia and lipidemia in type 2 diabetes mellitus. *J. Med. Food., 4,* 193–199.

Qi, Y., Wang, J., Wang, X., Cheng, J. J., & Wen, Z., (2017). Selective adsorption of Pb (II) from aqueous solution using porous biosilica extracted from marine diatom biomass: Properties and mechanism. *App. Surface Sci., 396*, 965–977.

Ragavan, P., Saxena, A., Mohan, P. M., Ravichandran, K., Saravanan, S., & Vijayaraghavan, A., (2016). A review of the mangrove floristics of India. *Taiwania, 61*(3), 224–242.

Raghunathan, C., & Sivaperuman, C., (2010). *Recent Trends in Biodiversity of Andaman and Nicobar Islands*. Zoological Survey of India.

Rajkumar, M., Perumal, P., Prabu, V. A., Perumal, N. V., & Rajasekar, K. T., (2009). *Phytoplankton Diversity in Pichavaram Mangrove Waters from the South-East Coast of India.*

Ramesh, R., Selvam, A. P., Robin, R. S., Ganguly, D., Singh, G., & Purvaja, R., (2017). Nitrogen assessment in Indian coastal systems. The Indian nitrogen assessment: Sources of reactive nitrogen, environmental, and climate effects. *Management Options, and Policies, 23,* 365–367.

Rao, N. K., (2016). Status of mangrove wetlands of Nellore and Prakasam districts, Andhra Pradesh, India. *Indian Forester, 142,* 471–480.

Sahu, S. C., Suresh, H. S., Murthy, I. K., & Ravindranath, N. H., (2015). Mangrove area assessment in India: Implications of loss of mangroves. *J. Earth Sci., & Climate Change, 6,* 1.

Saranya, A., Ramanathan, T., Kesavanarayanan, K. S., & Adam, A., (2015). Traditional medicinal uses, chemical constituents and biological activities of a mangrove plant, *Acanthus ilicifolius* Linn. A brief review. *American-Eurasian J. Agric. Environ. Sci., 15,* 243–250.

Sarma, V. V., Hyde, K. D., & Vittal, B. P. R., (2001). Frequency of occurrence of mangrove fungi from the east coast of India. *Hydrobiologia, 455,* 41–53.

Selvam, V., Gnanappazham, L., Navamuniyammal, M., Ravichandran, K. K., & Karunagaran, V. M., (2002). *Atlas of Mangrove Wetlands of India: Part 1.* Tamil Nadu.

Shyam, S. S., Ignatius, B., Suresh, V. K., Pushkaran, K. N., Salini, K. P., & Abhilash, P. R., (2013). Economic analysis on the hatchery technology and grow out of pearl spot (*Etroplus suratensis*). *Journal of Fisheries, Economics, and Development, 14,* 1–20.

Simlai, A., Mukherjee, K., Mandal, A., Bhattacharya, K., Samanta, A., & Roy, A., (2016). Partial purification and characterization of an antimicrobial activity from the wood extract of mangrove plant *Ceriopsde candra. EXCLI Journal, 15*(1), 103.

Singh, V. P., Garge, A., Pathak, S. M., & Mall, L. P., (1987). Pattern and process in mangrove forests of the Andaman Islands. *Vegetation, 71,* 185–188.

Srikandace, Y., Priatni, S., Pudjiraharti, S., Kosasih, W., & Indrani, L., (2017). Kerong fish (Teraponjarbua) peptone production using papa in enzyme as a nitrogen source in bacterial media. In: *IOP Conference Series: Earth and Environ. Sci., 60,* 012005.

Støttrup, J. G., & Norsker, N. H., (1997). Production and use of copepods in marine fish larviculture. *Aquaculture, 155,* 231–247.

Takemura, T., Hanagata, N., Sugihara, K., Baba, S., Karube, I., & Dubinsky, Z., (2000). Physiological and biochemical responses to salt stress in the mangrove, *Bruguiera gymnorrhiza. Aquatic Botany, 68,* 15–28.

Torres-Duran, P. V., Ferreira-Hermosillo, A., & Juarez-Oropeza, M. A., (2007). Antihyperlipemic and antihypertensive effects of *Spirulina maxima* in an open sample of Mexican population: A preliminary report. *Lipids Health Dis., 6,* 33.

Venkatesan, R., Karthikayen, R., Periyanayagi, R., Sasikala, V., & Balasubramanian, T., (2007). Antibacterial activity of the marine diatom, *Rhizosolenia alata* (Brightwell, 1858) against human pathogens. *Res. J. Microbiol., 2,* 98–100.

CHAPTER 3

Ecological and Biomass Assessment of Vegetation Cover of a University Campus

KAKOLI BANERJEE*, GOBINDA BAL, GOPAL RAJ KHEMENDU, NIHAR RANJAN SAHOO, GOPA MISHRA, CHITRANGADA DEBSARMA, and RAKESH PAUL

Department of Biodiversity and Conservation of Natural Resources, Central University of Orissa, Landiguda, Koraput–764021, India,
**E-mail: banerjee.kakoli@yahoo.com*

ABSTRACT

Any ecosystem whether aquatic or terrestrial is exposed to various anthropogenic and natural threats which directly or indirectly affects the ecological habitat of the different species whether plants or animals making them rare or endangered. Therefore, researchers have tried for their different conservation habitats, protecting them from the major threats like overgrazing, deforestation, bush fires, shifting cultivation and road construction which are thought to be the major causes of biodiversity. Natural forest conservation plays a vital role in the provision of environmental and social services such as productivity, nutrient biogeochemical cycling, carbon dioxide mitigation problems, litter biomass, etc.

India is bestowed with vast forest resources, which play a pivotal role in social, cultural, historical, economic and industrial development of the country and in maintaining its ecological balance. Forests act as an important resource base for the sustenance of its population and a storehouse of biodiversity. Other land use practices, such as agriculture and animal husbandry are benefitted by forests. The worldwide destruction of the natural environment in the recent past has led to a tremendous loss of biological diversity. This alarming situation has necessitated the in-depth research on the current status of different life forms. Two important attributes of biodiversity: species richness and endemism need particular attention in this regard. As projections indicate that species in ecosystems will be at maximum risk from human activities during the next few decades, critical taxonomic evaluation of life forms should be the highest priority. With reference to Article 7 of the Convention of Biological Diversity (CBD), there is an urgent need to prepare an inventory of plant and animal biodiversity through surveys or on ground practicals which will serve as an archive for future biodiversity studies. In this background, an attempt was made to study the ecological and biomass assessment of the present status of the vegetation cover during 2016–17 and also to indicate the change in vegetation cover over the last decade of Central University of Orissa (CUO), Main Campus, Koraput, Odisha which is located in between the two major industries of HAL (Hindustan Aeronautics Limited) and NALCO (National Aluminum Company Limited) of Govt. of India and is hence an important biodiversity study site.

3.1 INTRODUCTION

Biodiversity is actually the study of life forms starting from microorganisms to macro-organisms both plant and animal along with the ecological processes that influence the assemblages of such organisms in a particular habitat. The term sustainable development usually speaks of optimum use of the biological resources so that the diversity is maintained for future generations. However, the multiple threats that are operating on the biodiversity of an area (specially the anthropogenic and climatic factors) are the major causes of the decline in biodiversity. The long-term conservation of plant bioresources requires a good understanding of the ecological and natural processes operating in the area, and the challenge becomes more difficult when long-term environmental factors come into play. Thus in a conspicuous forest ecosystem diversity study, the understory cover particularly the herbs and shrubs and also the tree saplings play a major role in the regeneration of the forest and their inter- and intra-specific competition. This understanding also supports the conservation of forest ecosystem whether at the local or regional scale. Additional research in *ex-situ* conservation techniques like cryopreservation and slow growth storage is considered as determining factors for the understanding of the biological mechanisms of species for their conservation. In rural India, around 200 million people partially or wholly depend upon the forest products for their livelihood (Khare et al., 2000). Forests not only contribute timber but they also constitute a major share in NTFP products (non-timber forest products). In order to strengthen the role of the timber industry in biodiversity conservation, there is an utmost need of providing alternative livelihood for the local communities who live in the fringe areas and are partially or solely dependent on the forest products. The present era of green wood technology would call for innovative ideas of preparing timber wood from biomass wastes.

Ecosystems often become degraded due to pollution or overexploitation leading to elimination or severely damaged condition of forests which has a direct influence on species interaction, habitat and indirectly on human beings (Pearson and Rosenberg, 1976; Vitousek and Melillo, 1979).

The plant communities act as the determinants of ecosystem properties such as productivity, carbon sequestration, water relations, nutrient cycling and storage, litter quality and resistance and resilience to perturbations (Huston, 1997; Aarssen, 1997). The ecosystem function is directly related to total plant biomass. Palmer (1995) stated that species diversity is a function of genetic diversity and community diversity which are the two components of biodiversity.

One of the key goals of ecology is to explain the distribution and abundance of the species (Harte et al., 1999a; Kunin et al., 2000). Diversity studies of a plant community can be assessed by quadrant sampling method and relative abundance data (Fischer et al., 1943; Preston, 1948) or by a variety of non-parametric measures (Simpson, 1949; Shannon and Weiner, 1963). Due to the complex nature and lack of theoretical justification for statistical sampling theory, the non-parametric measures have also gained a great deal of popularity in the recent past (Krebs, 1989).

Biomass is a major source of energy for nearly 50% of the world's population (Karekezi and Kithyoma, 2006). In case of forest ecosystem studies, herbs, shrubs, and litter are mainly ignored for biomass estimation, although forest floor (having important herb and shrub species along with their biomass) consequently plays an important role in carbon sequestration studies. For management of herbivores, the biomass of herbs and shrubs are of utmost importance. However, due to the lack of accurate methodology and difficulty in estimation, they are omitted for biomass calculation studies (Karki, 2002; Khanal, 2001). There always exists an intra-specific competition between trees, herbs and shrubs for nutrients, light and moisture (Knoop and Walker, 1985; Anderson and Sinclair, 1993) and herbs and shrubs being understory have maximum competition under the canopy (Grunow et al., 1980; Sandford et al., 1982; Puri et al., 1992; Pandey et al., 2000).

Grasses also form an important component of understory in the forest ecosystem. Grasses inhabit the earth in greater abundance than any comparable group of plants. They are cosmopolitan in distribution and are recorded from the warm, humid, tropical climates, polar-regions, and deserts where the annual precipitation is 5 inches or less. Gramineae (Poaceae), the family of grasses is one of the largest of flowering plants comprising about 10,000 species and 651 genera (Clayton and Renvoize, 1986). The grasses not only serve as food for human consumption, but also hold greater values, like pastures provide grazing land for livestock, good sand binders, and some yield essential oils, still some have medicinal value. The woody grasses in maximum cases form the basis for paper industry and construction activity. By any reckoning, the grasses are a successful family, in which three themes constantly recur: their adaptability to changeable environments; their ability to co-exist with grazing herbivores and with man; and their position of distinctive life forms, in which fidelity to a single architectural scheme is counter-balanced by the endless ingenuity of its variations (Clayton and Renvoize, 1986). Over-increasing human and livestock populations have caused a serious stress on the grassland resources. Investigation of species composition and sociological interaction of species in communities are an integral part of vegetation ecology (Mueller-Dombois and Ellenberg, 1974).

The quantity of biomass in a forest determines the potential amount of carbon (Brown et al., 1999). Carbon sequestration is the phenomenon of storage of carbon in the plant biomass which is actually the outcome of photosynthesis in vegetation but also by the soil in organic matter (Vashum and Jayakumar, 2012). Carbon storage potential varies with respect to ecosystems like forests, grasslands and agricultural systems (DOE, 1999). According to Lal, (1999) the rate of carbon sequestration is also a function of species composition, region, climate, topography, and management practices.

Koraput, being the initial peak of the Eastern Ghats, also represents quite a variability in altitudinal forest types. Banerjee et al., (2016) conducted an Environmental Impact Assessment (EIA) study on the plant diversity (herbs, shrubs, climbers and trees) in two sampling stations of Koraput district comprising the old and new campus of Central University of Orissa during 2013–2014 through random quadrant method employing different indices like Shanon-Weiner species diversity index, Richness index, Index of Dominance, Evenness Index and Threat Index. The frequency, density, and abundance were also calculated simultaneously. The study revealed a total of 61 plant species comprising almost 31 different families recorded from both the studied locations, of which 15 were trees, 9 were shrubs, 34 were herbs and rest were climbers. Asteraceae and Poaceae were the dominant families followed by Fabaceae. Temperature and rainfall data was also plotted over a decade to organize the results with respect to climate change. Among all the species, *Mimosa pudica* L. and *Cynodon dactylon* (L.) Pers. were found to be dominant (80%) at both the studied locations which show their adaptability to the changing environmental conditions like temperature, rainfall or soil characteristics. In the present study, an attempt has been made to assess the diversity of trees, herbs, shrubs, grasses and estimate their biomass along with litter, which will help to pinpoint the health of the forest presently existing in the Central University of Orissa main campus, where a lot of construction activities are going on. Apart from diversity studies, the ambient environment with respect to soil and existing water body was also assessed in order to monitor the effect of the soil and water parameters in sustaining the biomass of the existing forest ecosystem.

3.1.1 IMPORTANCE OF TROPICAL FORESTS

Tropical forests include the most floristically diverse habitats on the planet. A defining component of tropical forests is the richness of the plant community (Gentry, 1990; Leopold and Salazar, 2008). Plant diversity is variously distributed according to the parallels of latitude across the globe with tropical America approximately 93,500 plant species followed by Asia and Africa with 62,000 and 20,000 species, respectively (Primack and Corlett, 2005). These differences are often attributed to changes in past climatic fluctuations and contrasting developments over a period of time (Kissling et al., 2012). Moreover, there is a marked a change in diversity and distribution of local, regional, and inter-continental forests (Corlett and Primack, 2006). However, researchers and previous studies have very less explained regarding the quantitative assessment of understory cover, e.g., herbs and shrubs which also contribute significantly to tropical forest biomass (Poulsen and Pendry, 1995; Annaselvam and Parthasarathy, 1999; Upadhaya et al., 2015). But, in addition to the trees, the understory of tropical forests has a distinct array of species which include the shrubs and herbs, different from the overstory and is an integral part of the tropical forest community (Bhat and Murali, 2001).

Biodiversity is usually affected by habitat loss and fragmentation on biodiversity (Fahrig, 2003); however, few studies have focused on the relevance of these changes in terms of ecological group and life forms, and their effect on conservation measures of tropical rain forest at the landscape scale.

3.1.2 BIODIVERSITY IN INDIAN AND LOCAL CONTEXT

The Biological Diversity Act (WRI, 1994) states the variability in life forms and the ecological complexes of which they are part and includes diversity within species or between species and of ecosystems. The National Forest Policy (NFP), 1988 realized the need for assessing the forests cover maintaining a minimum of 33% of the country's geographical area. Three mega centers of endemic plants in India are (i) Eastern Himalaya harboring 9,000 species of plants with 3,500 endemic species; (ii) the Western Ghats possessing 5,800 plant species with about 2,000 endemics; and (iii) Western Himalayas having 4,500 species with 1,195 endemic species of plants. The Andaman and Nicobar Islands harbor about 83% endemic species. The vegetation and forest types have been analyzed by Champion and Seth (1968), National Remote Sensing Agency (NRSA, 1979), Forest Survey of India (SFR, 2003).

India ranks tenth in the world and fourth in Asia in terms of plant diversity, with over 45,500 plant species, India represents nearly 11% of the world's known floral diversity. As elsewhere in the world, many organisms, especially in lower groups such as bacteria, fungi, algae, lichens, and bryophytes, are yet to be described, and remote geographical areas are to be comprehensively explored (CBD, 2009).

Forests are the most important resource among all the natural resources (Karia et al., 2001). As per FAO (2010), India's forest cover area has been estimated to be about 68 million hectares or 24% of the country's area. The forest cover in Odisha state based on interpretation of satellite data of Oct-Dec 2006 is 48,855 km^2, which is 31.38% of the state's geographical area (Indian State of Forest Report, 2009). The forests of Koraput district can be broadly classified into two type groups, e.g., Group 3: Tropical Moist Deciduous Forests and Group 5: Tropical Dry Deciduous Forests out of 16 types according to Champion and Seth's classification of forest types of India (1968) (FSI, 1989). But the forests of this district have been extensively damaged by the practice of shifting cultivation, and this practice has reduced the rich forests of this district to open forests and scrub jungles over large areas along with forest fires and damage were done by domestic animals, near human habitations (FSI, 1989).

The benefits of multi-functionality in grassland agriculture focuses towards the current and future management goals to provide large number of ecosystem services, like yield, decomposition, nutrient leaching, pollination, soil conservation, resistance to weed invasion, carbon sequestration for mitigation of greenhouse gases and land conservation (Sanderson et al., 2004; Sanderson et al., 2007; Lemaire et al., 2005).

3.2 METHODOLOGY

3.2.1 DESCRIPTION OF THE STUDY SITE

Koraput is an integral part of Eastern Ghats and constitutes the central part of it which is one of the four Biodiversity hotspots in India. The Eastern Ghats located along peninsular India extends over 1750 km between 77°22′ to 85°20′ E longitude and 09°95′ to 20°74′ N latitude. It consists of a series of discontinuous low hill ranges running in a northeast-southwest direction parallel to the coast of the Bay of Bengal between river Mahanadi of Odisha in north and Vagai river of Tamil Nadu in the south. The Eastern Ghats passes through states like Orissa, Andhra Pradesh, and Tamil Nadu. These Ghats are ripped through by rivers like Godavari, Mahanadi, Krishna and Cauvery, the four major rivers of South India. The Eastern Ghats are older than the Western Ghats. The elevation of Eastern Ghats is lower than the Western Ghats. It covers a total area of around 75,000 sq. km. One of the biggest characteristics of Eastern Ghats is that it is extremely fertile. In fact, the Ghats is said to be the watershed of many rivers as the Ghats get a higher average waterfall. Due to higher rainfall, the fertile land results in better crops. Often referred to as "Estuaries of India," Eastern Ghats gift its inhabitant the popular profession of fisheries, as its coastal area is full of fishing opportunity. The diversified ecological niches and environmental situation provides habitat for rich fauna. The Eastern Ghats is home to the largest number of Asiatic elephants in the world. Other large animals include Nilgiri Tahr, Leopards, Gaurs, Sambar, and Tigers abound the landscape. Apart from this, these Ghats are known for the wide variety of bird species. There are many wildlife sanctuaries and national parks of Odisha which comes under the Eastern Ghats such as the Simlipal National Park, Baissipalli Wildlife Sanctuary and Satkoshia Gorge Sanctuary. The Eastern Ghats also holds the rich floral system along with a large number of medicinal plants. Last but not the least climate change is emerging as a new threat to the whole ecosystem.

The selected biodiversity study area is situated in the hill ranges of Eastern Ghats peninsular regions. The northern portion of Eastern Ghats is located in Odisha state. The area lies between 19°62′ and 20°74′ N latitude and covers most parts of the districts of Khandamal, Kalahandi, Ganjam, Nayagarh, Boudh and few hilly areas of Rayagada, Khurda, Bolangir, Cuttack, and Angul. The geographical location with the climatic condition made Koraput district a heaven for biological diversity. The climate in Koraput is tropical in nature. In winter, there is less rainfall than in summer, and this climate is considered to be "Aw" according to the Köppen-Geiger climate classification (Kottek et al., 2006). The average annual temperature is 23.1°C and the rainfall averages 1604 mm in Koraput. The driest month is February. There is 3 mm of precipitation in February. With an average of 437 mm, the most precipitation falls in the month of July (www.climate-data.org). The average altitude of the hilly terrain ranges from 900 to 1400 m AMSL (GWIB, 2013). It is regarded as one of the epicenters for the origin of rice. The district is predominantly a tribal dominated one. More than 64% of the total population are tribals (Pattanaik et al., 2006).

Central University of Orissa main campus is located at Sunabeda, adjacent to the Naval Armament Depot (NAD) which joins the campus from N.H. 26 at the point of Sunabeda Junction-1 (about 6 km) and is situated very close to Hindustan Aeronautics Limited (HAL), between the coordinates 82° 47′ 55" E and 18° 44′ 09" N. The campus is located in village Chikapar and Chakarliput coming under Sunabeda NAC (Urban area). Damanjodi railway station is at distance of 15 km, and Koraput railway station is at a distance of 25 km. The land size is 450.09 acres spreading over village Chikapar (310.96 acres) and the adjoining Chakarliput (139.13 acres). There are a lot of construction works going on in the campus since its establishment in 2009. A study was undertaken in the campus for studying the biomass of the vegetation cover along with the atmospheric, water and soil parameters in forest areas of the campus (Figure 3.1). The study was carried out from April to September 2017 for a period of 6 months.

3.2.2 ANALYSIS OF WATER SAMPLES

The ambient water samples were collected from the existing water bodies in the study sites in clean tarson bottles, for *in-situ* analysis of water temperature, pH, transparency and dissolved oxygen. The surface water temperature was noted by collecting samples using sampler bottles. The sampler bottles were opened at the particular water level, and samples were collected and closed immediately. The values were measured using a digital thermometer (MEXTECH multi stem handheld portable LCD digital thermometer with sensor probe –50°C to 300°C or –58 °F to 572 °F). The values for pH were measured using a pH meter (Oakton eco-testr). The transparency and dissolved oxygen of the water bodies were also measured using Multi-parameter water analysis kit (Model No. 1026G).

3.2.3 ANALYSIS OF SOIL PARAMETERS

Soil samples were collected randomly from all the sampling locations for *in-situ* analysis of soil temperature, soil pH, soil moisture, and soil organic carbon. Soil temperature was measured by using a digital thermometer, soil pH and soil moisture were measured using soil pH meter (Model No. Touch10B006). Soil samples were collected and brought to the laboratory for organic carbon analysis by wet digestion method of Walkley and Black (1934).

3.2.4 ANALYSIS OF ATMOSPHERIC PARAMETERS

The atmospheric parameters were recorded using the CO_2 meter (Model No. SUPCO's IAQ 55) for the analysis of atmospheric temperature, humidity, and CO_2. The readings were taken during the daytime from 9 am to 12 noon.

3.2.5 VEGETATION STRUCTURE AND COMPOSITION OF THE CAMPUS

The quadrant method was employed in the open and scrub forest patch of the University. By this method, the structure and organization of plant communities can be studied. The random stratified sampling method was chosen for sampling. For sampling purpose 0.1ha plot was taken for trees, 5 m x 5 m and 1 m x 1 m quadrants were taken for shrubs, herbs and litters respectively inside each of the bigger quadrants. A total of 10 quadrants were taken for trees, and 20 were taken for the shrubs and herbs during the study, in each site. The relative abundance of trees, shrubs and herbs species were determined as per Curtis and McIntosh (1950). The plants were identified as per the standard protocol. The unidentified plants were collected in polythene bags and taken into the laboratory of IMMT (Institute of Mineral and Material Technology), Bhubaneswar Herbarium as well as to the Herbarium of M. S. Swaminathan Research Foundation (MSSRF) for their identification.

Relative abundances of the species were calculated as per the formula,

$$\text{Relative Abundance} = \frac{\text{Total number of individuals of a species in all quadrants}}{\text{Total number of quadrants in which the species occurred}} \times 100$$

3.2.6 ESTIMATION OF ABOVE GROUND BIOMASS OF TREES

Laser range finder was used for measurement of tree heights and diameters at breast height (DBH) at 1.37 m from tree base, and all the information of that site were also noted. For easy sampling and to save time a bamboo stick of 1.37 m long was prepared. For the unknown plant species, local names were collected from the local inhabitants. The sampling points were fixed by marking the trees in each and every sampling plot (0.1 ha) one by one (Figure 3.1).

In the process of stem biomass estimation, estimation of the volume of each tree was done. Since the non-destructive approach was adopted, the wood volume was estimated using diameter at breast height (DBH) and tree height. The product of volume and species-specific density gave the biomass of stem.

For Leaf and Branch biomass estimation, the total number of branches and leaves were counted irrespective of size (based on basal diameter into three groups, viz. <6 cm, 6–10 cm and >10 cm) on each of the sample trees. The leaves on the branches were removed by hand and oven-dried at 70°C overnight in hot air oven in order to remove moisture content (if any) present in the branches. To determine the dry weight of the branch, two branches from each size group were recorded separately using the equation of Chidumaya (1990). The average biomass of the leaves per branch of each tree was multiplied with the number of branches in that tree to obtain the leaf biomass. Finally, the total leaf biomass of the selected species (for each region) was recorded as per the standard procedure.

3.2.7 ESTIMATION OF ABOVEGROUND BIOMASS OF SHRUBS

Shrubs are the plants that attain a maximum height of about 5 m at maturity. Two sub-plots of 5 m x 5 m were taken within the 0.1 ha plot on the opposite corners or randomly. The circumference of the plant or selected tiller (if the shrub species forms a rosette) at 30 cm from base needs to be measured. Following steps were used to optimize the sampling and measurement:

i. The number of bushes was counted in a 5m x 5m plot for different species.
ii. Three to four bushes of each species were selected for further investigation representing the smallest to the largest bush.
iii. The number of tillers in representative small, medium and large bushes/plant was counted.

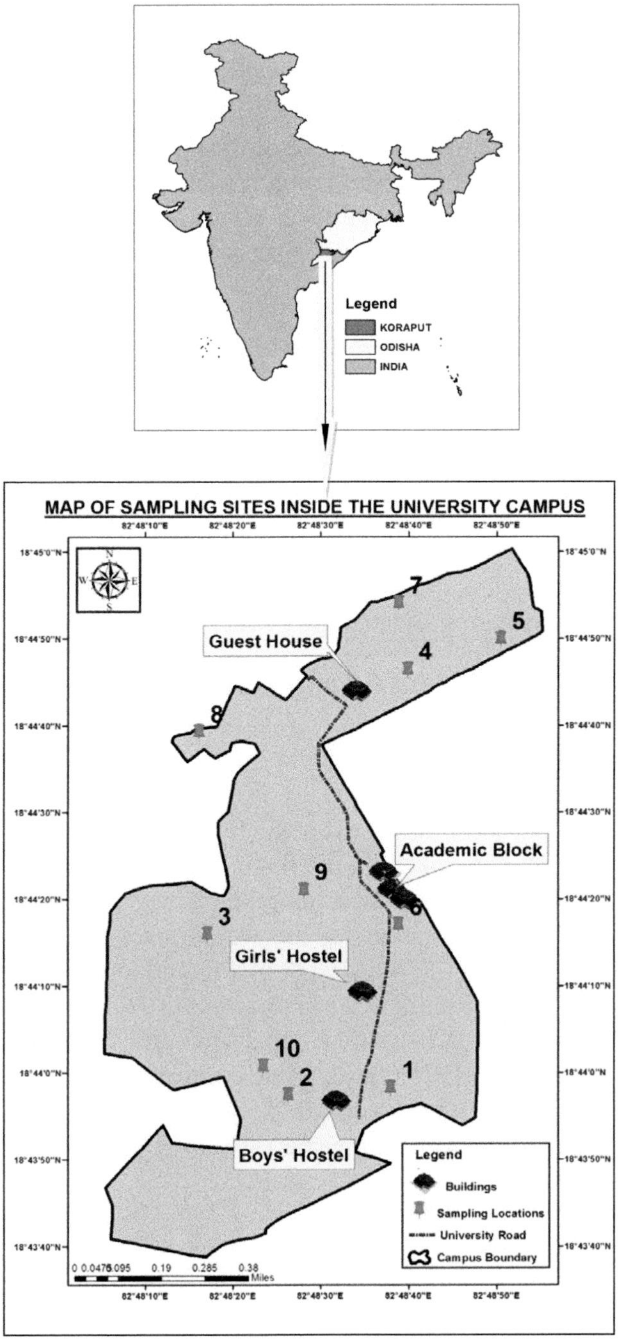

FIGURE 3.1 (See color insert.) Map of the sampling locations.

iv. In each bush, the number of thickest, medium and thinnest tillers/plant were counted.
v. Height/length and diameter of one plant/tiller of thick, medium and thin circumference were measured using a caliper.
vi. The same entire tiller was harvested for fresh weight, and in possible cases, the whole plant was brought to the laboratory for dry weight measurement. In case of thicker stems three pieces of 15 cm long base, middle and top were brought for dry weight measurement.

3.2.8 ESTIMATION OF ABOVE GROUND BIOMASS OF HERBS (INCLUDING GRASSES)

Information on biomass in grassland or herbaceous layer is lacking except in some cases where attempts have been made to do carrying capacity analyses by the researchers. Such studies are very limited. The biomass present in the layer is significant, particularly in view of the food availability to wild animals. We saw this as a great opportunity to gather information on the biomass in the herbaceous layer (Figure 3.11).

i. Five plots of 1 m x 1 m were taken (four in the corner and one in the center).
ii. All the species and individuals were enumerated as per the format in five plots.
iii. All the plots were harvested using a cutter, and fresh weights of all the samples were taken on the field.
iv. Out of five plots, harvested material of one representative plot was brought to the laboratory for dry weight measurement or harvested material from all the plots were mixed and known amount of fresh materials were brought to the laboratory for dry weight measurement.
v. The electric balance of 50 kg capacity with 1 to 5 g least count was also taken to the field for accuracy in measurement.
vi. Fresh materials were stored in polythene bags and brought to the laboratory. Details about the site-ID, plot-ID, fresh weight, etc., were noted down.

3.2.9 ESTIMATION OF BIOMASS OF LITTER

Litter and humus contain a large quantity of organic carbon. The thickness of litter and humus varies greatly in dry deciduous to moist deciduous forests and evergreen forests. The litter consists of dead branches, leaves and non-decomposable matter, which were collected carefully by hands in five plots of 1 m x 1 m.

i. Fresh weight of all the components (twigs, branches, fruits, and leaves) were taken as per the format in all the plots.
ii. Either whole material from one plot or part of material with known fresh weight from a mix of all the plots was brought in polythene bags with proper notations.
iii. The material was brought in polythene bags for dry weight measurement.

3.2.10 VEGETATIONAL CHANGE DETECTION

ArcGIS 10.1 and ERDAS IMAGINE 2015 softwares were used to perform this analysis. Satellite imageries were geo-corrected, and image processing was carried out. Supervised Maximum Likelihood image classification method was followed for the image classification. Satellite imageries procured from National Remote Sensing Center (NRSC) Hyderabad, and United States Geological Survey (USGS) of 30 m x 30 m during the month January–December were used for the analysis. The kmz file prepared from Google Earth was used for the preparation of shape file of the University Campus boundary which was further used in clipping the satellite imageries. NDVI was calculated for the strength and health of the vegetation for the years 2009 and 2017. And finally, a change detection map (2009–2017) was also produced using remap function in ArcGIS 10.1 software to detect the temporal increase and decrease in the vegetation in the past 8 years.

3.3 RESULTS AND DISCUSSION

3.3.1 BIODIVERSITY STUDY

The studies of biodiversity have now assumed greater importance is owing to the need of the time as ecologists have tried to document global biodiversity in the face of unprecedented perturbations, habitat loss, and extinction rates. Diversity studies are usually understood by indices. The floristic studies are considered as the backbone for assessment of phytodiversity, conservation management and sustainable utilization (Jayanthi and Rajendran, 2013). For understanding the ecosystem function and its conservation, preparation of the flora of smaller regions like districts, protected areas, unexplored areas, etc. is essential which also acts as the pre-requisite for the revision of the flora of the vast country like India. Hence, floristic studies form a vital component of any natural resource management and planning activities at the local, regional and global levels.

The checklist for the relative abundance of trees, shrubs, and herbs are listed in Tables 3.1–3.3 and Figures 3.9–3.11, respectively. The data suggests that there are 18, 16 and 35 number of species among trees, shrubs, and herbs respectively. Considering all the 10 sampling plots of the campus, the highest abundance in trees were recorded in the species *Diospyros melanoxylon* Roxb. (23.17%) > *Simarouba glauca* DC. (10.98%)> *Buchanania lanzan* (8.54%)> *Aegle marmelos* (L.) Corrêa (8.54%) and lowest in the species *Caesalpinia pulcherrima* (L.) Sw. (1.22%) and *Magnolia champaca* (L.) Baill. (1.22%), respectively. Similarly, the trend of relative abundance in shrubs showed the highest values in the species *Lantana camara* L. (31.01%) > *Clerodendrum bungii* Steud. (16.28%)> *Anogeissus latifolia* (Roxb. ex DC.) Wall. and lowest in cases of *Chromolaena odorata* (4.91%) and *Wrightia arborea* (Dennst.) Mabb. (4.65%), respectively. The trend of relative abundance in herbs varied as per the order: *Cynodon dactylon* (L.) Pers. (13.36%) followed by *Mimosa pudica* (5.79%), *Sesleria autumnalis* (Scop.) F.W. Schultz (4.56%), *Borreria articularis* (L.f.) F.N. Williams (4.37%) and found lowest in case of *Atylosia cajanifolia* Haines (0.80%) and *Bidens pilosa* L. (0.80%) respectively. Several studies have been conducted to analyze the floristic composition of the wall habitats in India and abroad (Brandes, 1995; Krigas et al., 1999; Altay et al., 2010; Bilge, 2001; Ocak and Ture, 2001; Ture and Bocuk, 2001; Turgut, 1996; Parthipan et al., 2016).

The great wealth of our country is the tremendous biological diversity in tropical regions because of the excellent prevailing climatic conditions (Parthipan et al., 2016). Recently interests have increased on the study of biodiversity as the forests are almost affected by the anthropogenic activities (Merigot et al., 2007). The factor on which the floristic composition is depended on is, however, the composition on regional scales and further reflects both anthropogenic and natural disturbances (Ward, 1998; Ayyappan and Parthasarathy, 1999). Urbanization is one of the major reasons for the destruction of the natural vegetation, including deforestations and degradation of other natural areas (Kumar et al., 2010; Von der Lippe and Kowarik, 2007; 2008). Urbanized areas can also harbor a high number of threatened species (Sodhi et al., 2010). Koraput, is also a growing urban town in the southernmost district of Odisha with two industrial areas HAL and NALCO

TABLE 3.1 Checklist of Tree Species Composition

Sl. No	Name of the Species	Family	Number of species	Relative Abundance (%)
1	*Aegle marmelos* (L.) Corrêa	Rutaceae	7	8.54
2	*Bombax ceiba* L.	Bombacaceae	2	2.44
3	*Buchanania lanzan* Spreng.	Anacardiaceae	7	8.54
4	*Caesalpinia pulcherrima*(L.) Sw.	Fabaceae	1	1.22
5	*Diospyros melanoxylon*Roxb.	Ebnaceae	19	23.17
6	*Ficus religiosa* L.	Moraceae	2	2.44
7	*Holarrhena antidysenterica* (Roth) Wall.	Apocynaceae	4	4.88
8	*Magnolia champaca* (L.) Baill.	Magnoliaceae	1	1.22
9	*Melia azedarach* Linn.	Meliaceae	6	7.32
10	*Phyllanthus emblica* L.	Euphorbiaceae	3	3.66
11	*Pongamia glabra* Vent.	Papilionaceae	4	4.88
12	*Psidium guajava* L.	Myrtaceae	2	2.44
13	*Santalum album* L.	Santalaceae	2	2.44
14	*Schleichera oleosa* (Lour.) Merr.	Sapindaceae	2	2.44
15	*Simarouba glauca* DC.	Simaroubaceae	9	10.98
16	*Syzizium cuminii* (L.) Skeels	Myrtaceae	2	2.44
17	*Ziziphus jujuba* Mill.	Rhamnaceae	3	3.66
18	*Ziziphus oenoplia* (L.) Mill	Hamnaceae	2	2.44
			78	**100%**

TABLE 3.2 Checklist for Shrub Species Composition

Sl. No	Name of the Species	Family	Number of species	Relative Abundance (%)
1	*Anisomeles indica* (L) R. Br.	Lamiceae	20	5.17
2	*Anogeissus latifolia* (Roxb. ex DC.) Wall.	Combretaceae	26	6.72
3	*Calea zacatechichi* Schltdl.	Compositae	12	3.10
4	*Chromolaena odorata* (L.) R.M.King and H.Rob.	Asteraceae	19	4.91
5	*Cipadessa baccifera*	Meliaceae	4	1.03
6	*Clerodendron bungii* Steud.	Lamiaceae	63	16.28
7	*Clerodendrum infortunatum L.*	Lamiaceae	6	1.55
8	*Crotalaria mucronata* Desv.	Fabaceae	6	1.55
9	*Diospyros peregrina* (Gaertn.) Gürke	Ebenaceae	21	5.43
10	*Matteuccia struthiopteris* (L.) Tod.	Onocleaceae	7	1.81
11	*Indigofera pulchella* Roxb.	Fabaceae	16	4.13
12	*Ipomoea abyssinica* (Choisy) Schweinf.	Convolvulaceae	9	2.33
13	*Lantana camara* L.	Verbenaceae	120	31.01
14	*Murraya koenigii* (L.) Spreng.	Rutaceae	16	4.13
15	*Nerium indicum* Mill.	Apocynaceae	24	6.20
16	*Wrightia arborea* (Dennst.) Mabb.	Apocynaceae	18	4.65
			387	**100**

TABLE 3.3 Checklist for Herb Species Composition

Sl. No	Name of the species	Family	Number of species	Relative Abundance (%)
1	*Cynodon dactylon* (L.) Pers.	Poaceae	217	13.36
2	*Mimosa pudica* L.	Fabaceae	94	5.79
3	*Sesleria autumnalis* (Scop.) F.W. Schultz	Poaceae	74	4.56
4	*Borreria articularis* (L.f.) F.N. Williams	Rubiaceae	71	4.37
5	*Sida cordifolia* L.	Malvaceae	70	4.31
6	*Ageratum conyzoides* (L.) L.	Asteraceae	66	4.06
7	*Centipeda minima* (L.) A. Braun and Asch.	Asteraceae	65	4.00
8	*Commelina benghalensis* L.	Commelinaceae	65	4.00
9	*Tridax procumbance* L.	Asteraceae	64	3.94
10	*Sonchus asper* (L.) Hill	Asteraceae	63	3.88
11	*Sorghastrum nutans* (L.) Nash	Poaceae	63	3.88
12	*Senna tora* (L.) Roxb.	Fabaceae	56	3.45
13	*Evolvulus alsinoides* (L.) L	Convolvullaceae	49	3.02
14	*Stachytarpheta indica* (L.) Vahl	Verbenaceae	49	3.02
15	*Spermacoce hispida* L.	Rubiaceae	44	2.71
16	*Argemone mexicana* L.	Papaveraceae	43	2.65
17	*Achyranthes aspera* L.	Amaranthaceae	41	2.52
18	*Alocasia macrorrhizos* (L.) G.Don	Araceae	41	2.52
19	*Spermacoce articularis* L.f.	Rubiaceae	41	2.52
20	*Arundinella setosa* Trin.	Poaceae	37	2.28
21	*Alternanthera sessilis* (L.) R.Br. ex DC.	Amaranthaceae	35	2.16

TABLE 3.3 *(Continued)*

Sl. No	Name of the species	Family	Number of species	Relative Abundance (%)
22	*Blumea lacera* (Burm.f.) DC.	Asteraceae	29	1.79
23	*Cyperus rotundus* L.	Cyperaceae	25	1.54
24	*Ajuga reptans* L.	Lamiaceae	23	1.42
25	*Oxalis corniculata* L.	Oxalidaceae	23	1.42
26	*Scoparia dulcis* L.	Plantaginaceae	23	1.42
27	*Datura stramonium* L.	Solanaceae	21	1.29
28	*Tephrosia purpurea* (L.) Pers.	Papilonaceae	21	1.29
29	*Euphorbia hirta* L.	Euphorbiaceae	20	1.23
30	*Echinochloa crus-galli* (L.) P. Beauv.	Poaceae	18	1.11
31	*Centella asiatica* (L.) Urb.	Mackinlayaceae	17	1.05
32	*Sida acuta* Burm.f.	Malvaceae	16	0.99
33	*Leucas aspera* (Willd.) Link	Lamiaceae	14	0.86
34	*Atylosia cajanifolia* Haines	Fabaceae	13	0.80
35	*Bidens pilosa* L.	Asteraceae	13	0.80
			1624	100%

around, still harbors many patches of tropical dry and tropical moist deciduous forests. Hence it is necessary to document the floristic diversity in order to know its wealth and also to identify those plant species that are in urgent need of conservation.

3.3.2 BIOMASS STUDY

Forest cover comprises of all woody and perennial tree species including herbs and shrubs which are actually the reflectance of healthy tree vegetation over a land cover.

3.3.2.1 BIOMASS OF TREES

Biomass is a major source of energy for nearly 50% of world's population (Karekezi and Kithyoma, 2006) and wood biomass is a major renewable energy source in developing world, representing a significant proportion of rural energy supply (Hashiramoto, 2007). Forest biomass is an important source of food, fodder, and fuel, and its exploitation leads to forest degradation (Rawat and Nautiyal, 1988). As per the records of Forest Survey of India (FSI, 2003), the area under forest was 102.68 m ha in 1880, which has been reduced to 67.83 m ha in 2003. Since 2003, carbon stocks in Indian forests are continuously decreasing. The data reveals that in India, forests are under excessive anthropogenic pressures (Rai and Chakrabarti, 2001). It is an established fact that the tree biomass is a function of tree density, height and basal area at any given location. These parameters contribute to the aboveground biomass which differs with site, habitat, forest succession stage, the composition of the forest, species variability and varying tree density, etc. (Brunig, 1983; Joshi and Ghose, 2014; Whitmore, 1984). Variation in biomass at various sites can be attributed to some internal and external factors, such as the type of forest, site-to-site variations, disturbances, total annual rainfall and geographical location of the forests (Terakunpisut et al., 2007).

Depending upon the DBH (Diameter at Breast Height) mean values and the height, the aboveground biomass in the tropical moist and dry deciduous forests varies. These parameters contribute to the aboveground biomass which differs with site, habitat, forest successional stage, the composition of the forest, species variability and varying tree density, etc. (Brunig, 1983; Joshi and Ghose, 2014). Out of the 18 species of the trees taken in the 10 quadrants in each site, the trend of mean above ground biomass of the trees varied as per the order *Caesalpinia pulcherrima* (L.) Sw. (30991.32±2.5 t/ha) > *Ficus religiosa* L. (3919.66±3.27 t/ha) > *Bombax ceiba* L. (2045.12±1.38 t/ha) > *Syzizium cuminii* (L.) Skeels (706.38±5.75 t/ha) > *Schleichrea oleosa* (542.55±2.62 t/ha) > *Pongamia glabra* Vent. (256.95 ±2.20 t/ha) > *Phyllanthus emblica* L. (238.19±1.36 t/ha) among the most dominant species (Figure 3.2). It has also been noticed that some of the trees such as *Mangifera indica* present in the campus during the previous study by Banerjee et al., (2016) have been cut due to the construction activities that are going on in the University Campus.

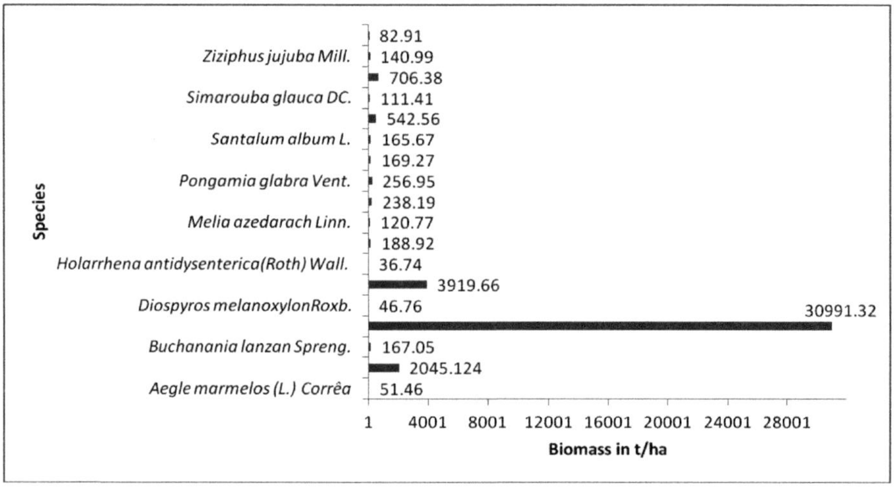

FIGURE 3.2 Biomass of trees in t/ha.

The ecosystem services of trees in terms of carbon sequestration and carbon dioxide recycling is poorly documented (Mitra et al., 2011) and although considerable storage and sequestration of carbon dioxide occur by trees during their normal process of growth. The storage of carbon varies spatially and temporally based on the abundance of the species, growth rate of the species and site conditions (Mitra et al., 2015). Net annual carbon sequestration is positive for growing forest with considerable Above Ground Biomass (AGB) of the species. The net sequestration, however, becomes negative during periods of forest decline and or loss when carbon emissions from dead trees occur through decomposition or forest fire that exceeds the carbon uptake by the live trees. However, an overall estimate fixes that the relative change of climatic and edaphic factor contributes significantly to carbon sequestration by tree species.

3.3.2.2 BIOMASS OF SHRUBS

The height of understory varied among sampling sites, depending on the structure and composition of live and dead vegetation type as well. The biomass of understory vegetation usually changes with the season, which is at its peak in the rainy season. The trend of mean above ground biomass of the shrubs varied as per the order *Lantana camara* L. (294±2.5 t/ha) > *Clerodendron bungii* Steud. (85.05±3.22 t/ha) > *Nerium indicum* Mill. (36±1.38 t/ha) > *Diospyros peregrina* (Gaertn.) Gürke (31.5±1.75 t/ha) > *Anogeissus latifolia* (Roxb. ex DC.) Wall. (31.25±1.01 t/ha) > *Wrightia arborea* (Dennst.) Mabb. (26.1 ±1.21 t/ha)> *Murraya koenigii* (L.) Spreng. (25.6±1.16 t/ha) among the most dominant species (Figure 3.3). The shrubs are mostly dominated by the exotic species suppressing the native species and gradually converting the hilly lands into xerophytic scrublands.

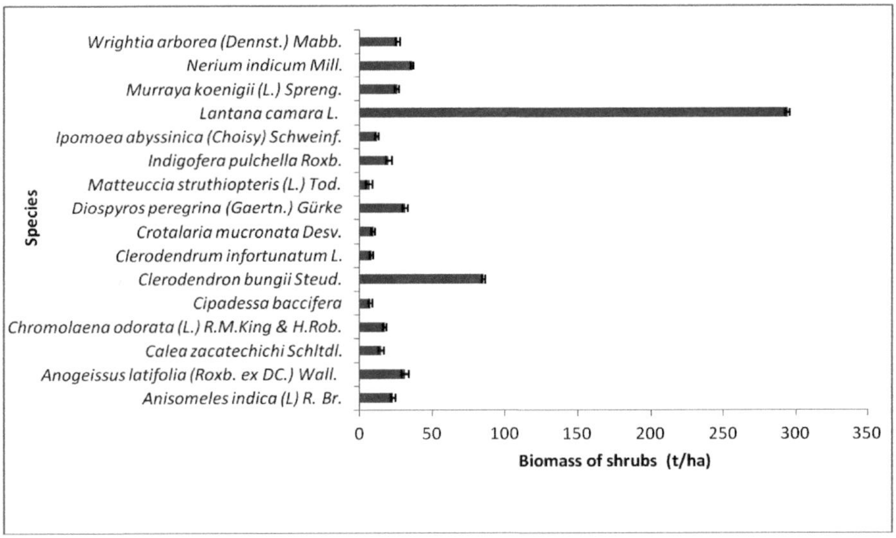

FIGURE 3.3 Biomass for shrubs in t/ha.

3.3.2.3 BIOMASS OF HERBS

The trend of mean above ground biomass of the herbs which were recorded in all the 20 quadrants studied showed a variation as per the order quadrant 1 (0.123 kg) > quadrant 3 (0.120 kg) > quadrant 9 (0.107) > quadrant 10 (0.106 kg) > quadrant 5 (0.103 kg) > quadrant 2 (0.102) > quadrant 4 (0.101) > quadrant

6 (0.095 kg) > quadrant 13 (0.091 kg) > quadrant 11 (0.090 kg) > quadrant 7 (0.090 kg) > quadrant 14 (0.086 kg) > quadrant 12 (0.082 kg) > quadrant 8 (0.082 kg) > quadrant 18 (0.081 kg) > quadrant 19 (0.076 kg) > quadrant 16 (0.076 kg) > quadrant 20 (0.071 kg) > quadrant 17 (0.070 kg) > quadrant 15 (0.064 kg), respectively. The total herb biomass of the 20 quadrants was found to be 1.818 kg and the total average biomass of the herbs in the campus was calculated as 172.884 t/ha (Figure 3.4).

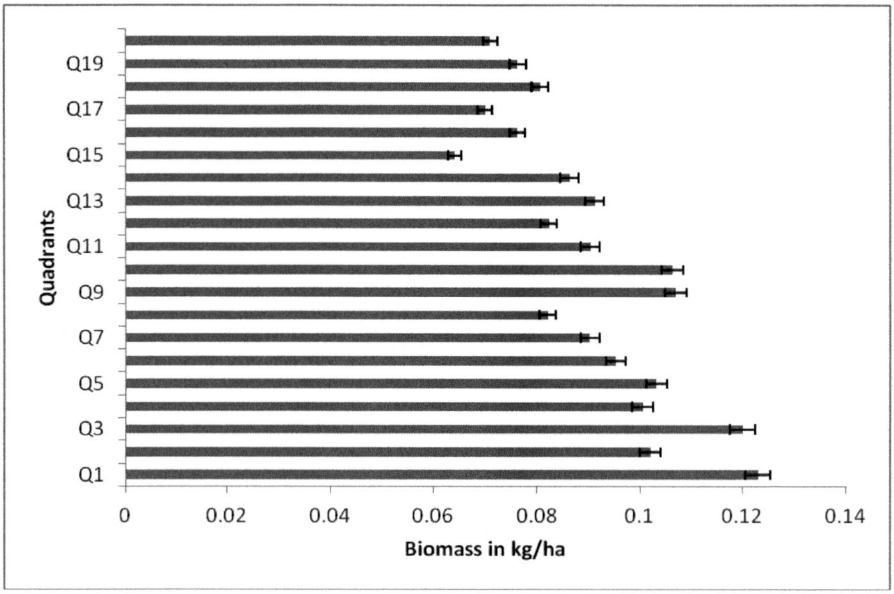

FIGURE 3.4 Quadrant wise biomass for herbs.

3.3.2.4 *BIOMASS OF THE LITTER*

The litter biomass of the 10 quadrants taken for study varied as per the order quadrant 5 (0.106 kg) > quadrant 4 (0.090 kg) >= quadrant 2 (0.090 kg) >= quadrant 7 (0.090 kg) > quadrant 9 (0.080 kg) > quadrant 3 (0.079 kg) > quadrant 8 (0.077 kg) > quadrant 10 (0.071 kg) > quadrant 1 (0.070 kg) > quadrant 6 (0.050 kg) (Figure 3.5). The litter biomass was found highest in the quadrant having the highest number of trees and gradually decreased in the quadrants having lesser biomass in descending order due to more amounts of leaf falls by more number of trees.

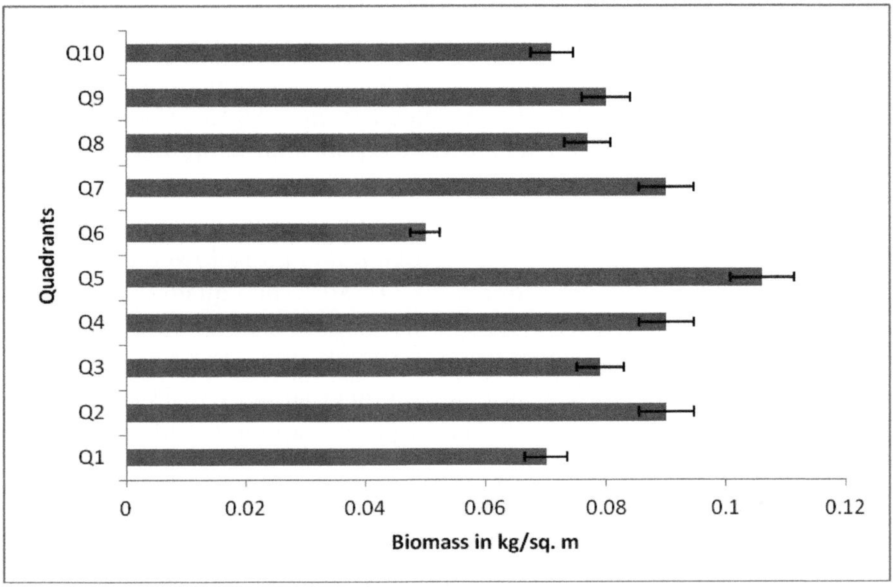

FIGURE 3.5 Quadrant wise biomass for litters.

3.3.2.5 ANALYSIS OF THE WATER, SOIL, AND ATMOSPHERIC PARAMETERS

There were only three water bodies found on the campus having very less and stagnant water. The temperature varied as per the order 33.1°C in quadrant 2 followed by 30.8°C and 25.8°C in quadrants 3 and 1, respectively. The pH also didn't show much variation among the three water bodies. Water transparency and the dissolved oxygen were highest in the second water body (21 cm and 12 mg/l) which had less dissolved materials as compared to the other two water bodies which showed 18 cm and 15 cm of transparency and 8 and 7 parts of dissolved oxygen respectively having more turbid water (Figure 3.6). The water transparency and dissolved oxygen were found to be optimum for the growth of aquatic flora and fauna.

From the graphs, it is quite clear that the temperatures of soil water and atmosphere didn't vary much within the campus as it is only 470 acres in area. The highest soil temperature was found to be 33.8°C in the quadrants 3 and 4 followed by quadrant 1 (33.1°C), quadrant 9 and 2 (32.1°C) and quadrant 8 (30.1°C) respectively (Figure 3.7). The soil pH was found the highest in quadrant 2 (7.0) and lowest in the quadrant

7 (5.8). The pH of the quadrant 2 may be higher because it is near to the hostel area of the University and the water discharge from the hostel may have made the soil alkaline. The soil moisture content was found to be very low in the study area as it ranged from 10% to 15% in all the 9 quadrants except quadrant 7 which showed 80% moisture because of the presence of a water body. The soil organic carbon was found highest in the quadrant 5 (1.50%) which might probably be due to the presence of maximum number of trees in the particular quadrant followed by quadrant 3 (1.02%), quadrant 6 (1.0%), quadrant 2 (0.80%) and quadrant 7 (0.78%) respectively (Figure 3.7).

In the case of the atmospheric parameters, the humidity was found to be very low ranging from as low as 15.4% in quadrant 6 to 36.6% in quadrant 1 of the studied area. Similarly, the atmospheric carbon dioxide varied from 391 ppm in quadrant 6 to the highest value of 413 ppm in quadrant 4. The temperature showed less variation among all the 10 quadrants with the highest value of 39.8°C in the quadrant 9 and lowest in the quadrant 2 with 30.8°C temperatures (Figures 3.8–3.11).

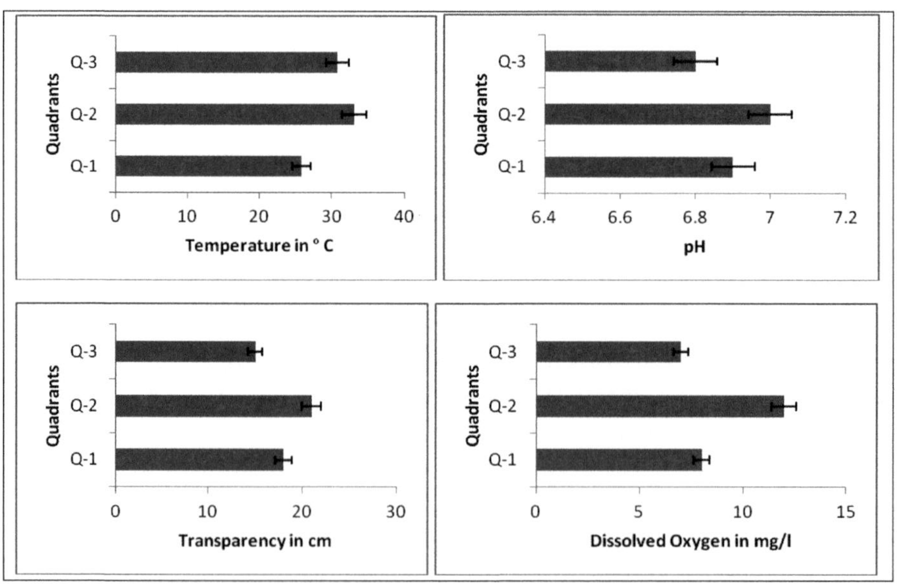

FIGURE 3.6 Variation in water parameters in the study area.

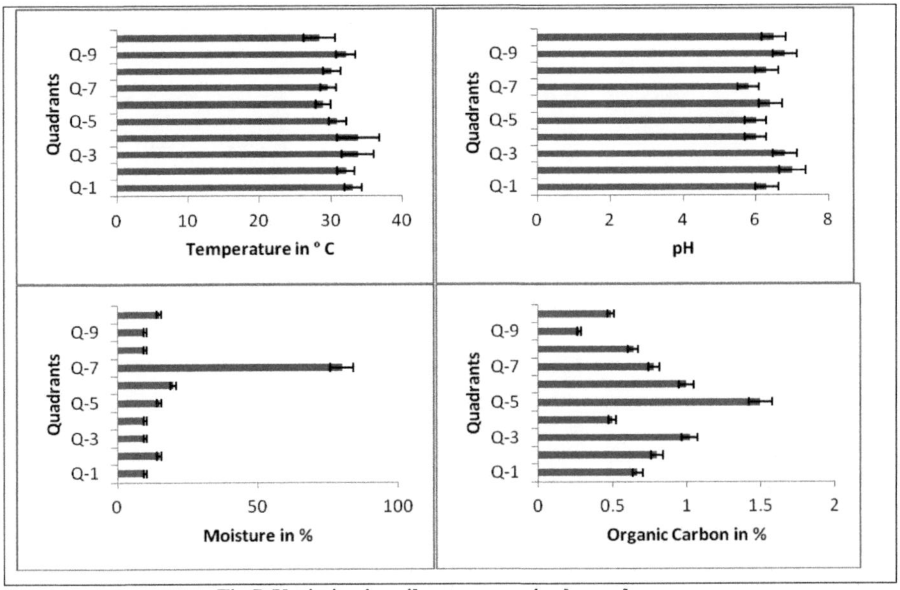

FIGURE 3.7 Variation in soil parameters in the study area.

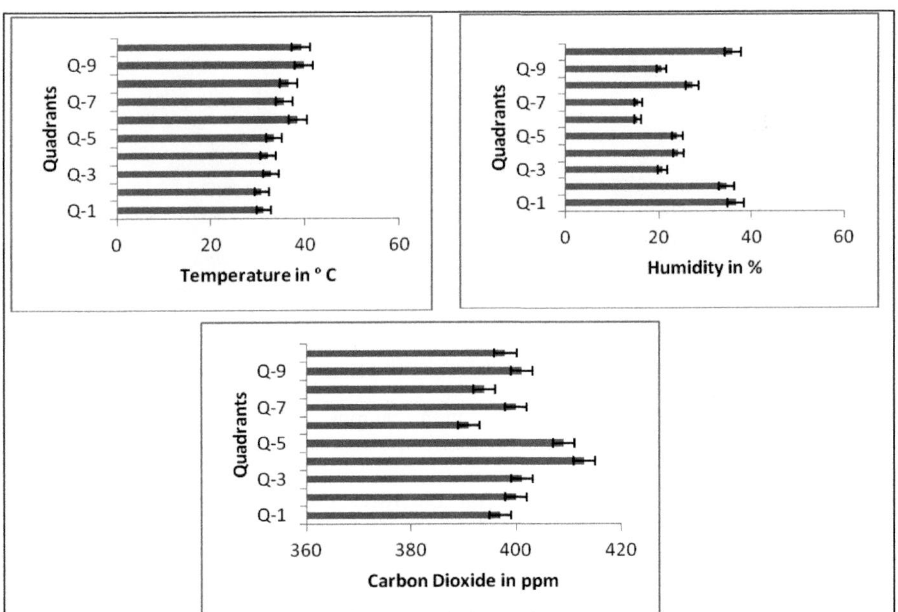

FIGURE 3.8 Variation in atmospheric parameters in the study area.

Diospyros melanoxylon Roxb.

Bombax ceiba L.

Syzizium cuminii (L.) Skeels

Aegle marmelos (L.) Corrêa

Psidium guajava L.

Schleichera oleosa (Lour.) Merr.

Santalum album L.

Phyllanthus emblica L.

Buchanania lanzan Spreng.

FIGURE 3.9 Dominant tree species found in the campus.

Various floras from institutional campuses were already reported by various workers (Giles-Lal and Livingstone, 1978; Gopi, 2008; Natarajan and Gopi, 2010; Parthasarathy et al., 2010; Udayakumar, et al., 2011; Rekha and Paneerselvam, 2014; Rajendran, et al., 2014; Irwin et al., 2015). The main reason behind the degradation of forest cover may be due to some anthropogenic activities going on in the campus such as construction of new buildings or structures. So this is the right time for the floristic study in the campus for the documentation of phytodiversity, its conservation, management and sustainable utilization (Jayanthi and Rajendran, 2013). The campus flora of an institution is a unique opportunity for an outdoor botanical and ecological learning for the students and researchers.

Clerodendron bungii Steud.

Lantana camara L.

Anogeissus latifolia (Roxb. ex DC.) Wall.

Nerium indicum Mill.

Anisomeles indica (L) R. Br.

Chromolaena odorata (L.) R.M.King & H.Rob.

Diospyros peregrina (Gaertn.) Gürke

Wrightia arborea (Dennst.) Mabb.

Murraya koenigii (L.) Spreng.

FIGURE 3.10 (See color insert.) Dominant shrub species found in the campus.

Cynodon dactylon (L.) Pers.

Mimosa pudica L.

Sesleria autumnalis (Scop.) F.W.Schultz

Borreria articularis (L.f.) F.N.Williams

Sida cordifolia L.

Ageratum conyzoides (L.) L.

Centipeda minima (L.) A.Braun & Asch.

Commelina benghalensis L.

Tridax procumbance L.

FIGURE 3.11 Dominant herb and grass species found in the campus.

The overall correlation coefficient of above ground biomass of trees, shrubs, herbs, and litter with water parameters have shown the significant positive relationship in case of water pH, transparency and dissolved oxygen. However, with water temperature, the relationship is significant in the case of trees and shrubs. With soil temperature and biomass of trees, shrubs, herbs, and litter, the biomass has shown a significant negative relationship, which shows the adverse impact of soil temperature on the biomass of trees, shrubs and litters respectively. However, in the case of herbs, the relationship is reversed, which shows that the biomass of herbs is not affected with the change in temperature. Soil pH and moisture have shown a positive relationship with the biomass of trees and shrubs only showing the positive relationship between the two. However, in context to herbs and litter biomass, the relationship between soil pH and moisture is not so significant. Soil organic carbon, which is the result of decomposed biomass of trees, shrubs, herbs, and litters, has shown the insignificant relationship in case of trees and shrubs whereas a vice-versa relationship between herbs and litter. This clearly speaks that with an increase of biomass of herbs (being soft plants), is faster than in the case of trees and shrubs. Similarly, the significant inverse relationship between litter biomass and soil organic carbon shows that with the decrease in litter biomass the soil organic carbon increases which are absolutely correct in the present geographical scenario. Atmospheric temperature plays a vital role in the biomass of woody plants (particularly trees and shrubs), hence with the increase in temperature, the biomass of trees and shrubs have decreased and vice-versa. The same relationship has been shown by litter biomass and atmospheric temperature, which proves that with an increase in atmospheric temperature, the biomass of litter decreases as the litter biomass is directly related with the number of leaf falls of trees and shrubs respectively. Atmospheric CO_2 concentration has a significant negative effect on the increase of biomass in cases of trees, shrubs, and herbs respectively which proves that with the increase in biomass of the floral types, more amount of CO_2 is absorbed from the atmosphere and hence the relationship is significantly negative. The reverse relationship has been shown in the biomass of trees, shrubs, and litters with atmospheric humidity, where it has shown sign of a positive relationship with atmospheric humidity. This clearly speaks of the increase in atmospheric humidity the biomass of trees, shrubs, and litter increase although it is not so prominent in the case of herbal species (Table 3.4).

The results suggest that this environment of the campus is quite a good habitat for the growth and regeneration of the forest.

TABLE 3.4 Interrelationship Between Ambient Parameters and Biomass of Trees, Shrubs, Herbs, and Litter

Parameters	Combination	*r*-value	*P*-value
Trees	Biomass × Water Temperature	0.775	<0.01
	Biomass × Water pH	0.84	<0.01
	Biomass × Water Transparency	0.84	<0.01
	Biomass × Water DO	0.971	<0.01
	Biomass × Soil Temperature	–0.999	<0.01
	Biomass × Soil pH	0.691	<0.01
	Biomass × Soil Moisture	0.999	<0.01
	Biomass × Soil Organic Carbon	0.137	IS*
	Biomass × Atm Temperature	–0.941	<0.01
	Biomass × Atm CO_2	–0.599	<0.01
	Biomass × Atm Humidity	0.957	<0.01
Shrubs	Biomass × Water Temperature	0.657	<0.01
	Biomass × Water pH	0.919	<0.01
	Biomass × Water Transparency	0.919	<0.01
	Biomass × Water DO	0.998	<0.01
	Biomass × Soil Temperature	–0.993	<0.01
	Biomass × Soil pH	0.56	<0.01
	Biomass × Soil Moisture	0.993	<0.01
	Biomass × Soil Organic Carbon	–0.031	IS*
	Biomass × Atm Temperature	–0.984	<0.01
	Biomass × Atm CO_2	–0.456	<0.01
	Biomass × Atm Humidity	0.992	<0.01
Herbs	Biomass × Water Temperature	0.276	IS*
	Biomass × Water pH	–0.829	<0.01
	Biomass × Water Transparency	–0.829	<0.01
	Biomass × Water DO	–0.601	<0.01
	Biomass × Soil Temperature	0.439	IS*
	Biomass × Soil pH	0.392	IS*
	Biomass × Soil Moisture	–0.439	IS*
	Biomass × Soil Organic Carbon	0.856	<0.01
	Biomass × Atm Temperature	0.682	<0.01
	Biomass × Atm CO_2	–0.500	<0.01
	Biomass × Atm Humidity	–0.643	<0.01

TABLE 3.4 *(Continued)*

Parameters	Combination	*r*-value	*P*-value
Litter	Biomass × Water Temperature	–0.209	IS*
	Biomass × Water pH	0.866	<0.01
	Biomass × Water Transparency	0.866	<0.01
	Biomass × Water DO	0.655	<0.01
	Biomass × Soil Temperature	–0.500	<0.05
	Biomass × Soil pH	–0.327	IS*
	Biomass × Soil Moisture	0.5	<0.05
	Biomass × Soil Organic Carbon	–0.818	<0.01
	Biomass × Atm Temperature	–0.731	<0.01
	Biomass × Atm CO_2	0.439	<0.05
	Biomass × Atm Humidity	0.695	<0.01

*IS: means insignificant.

3.3.2.6 VEGETATIONAL CHANGE ANALYSIS

The satellite imageries were procured from NRSC (National Remote Sensing Center), Hyderabad and also from the USGS (United States Geological Survey) website online to detect the change that has occurred in the vegetation pattern during the year 2009 to 2017 in the campus using the NDVI (Normalized Differential Vegetation Analysis) values of the satellite imageries. The NDVI value ranged from 0.06 to 0.38 in 2009 and from 0.05 to 0.45 in 2017, respectively (Figure 3.12) showing not so healthy vegetation. The higher limit value of NDVI in 2017 may probably be of mature trees that are present during 2017 (Figure 3.12), and hence their mature leaves show better NDVI. However, on the contrary there is a decrease in the lower limit value of the NDVI during 2017, indicating the decrease in the quality of the undergrowth and conversion of the open vegetation into a scrubland during the past 8 years in the campus which is also undergoing a lot many construction activities simultaneously with local invasion, deforestation and also forest fires. The change detection map clearly shows that there is decrease in the vegetation in almost entire campus in the past 8 years excepting only 56 pixels accounting for an area of 5.04 hectare out of the 190.202 hectare of total land (Figure 3.12) which has been possible due to the plantation programmes that has been taken up by the Forest Department and the University Administration.

FIGURE 3.12 (See color insert.) Vegetational change detection of CUO campus using NDVI.

3.4 CONCLUSION

Enumeration of biomass helps in assessing the amount of carbon stored in the plant and will help us to assess the amount of carbon that is assimilated in the selected forest patch. Most of the fringe populations residing within the periphery of the forest are tribals, and they largely depend on the forests for their livelihood. Hence, a large area of the vegetation remains under threat of disturbances. Under the present situation, there is an urgent need to reduce the magnitude of disturbances and their impact on the balance of carbon stock. As per the Millennium Ecosystem Assessment Programme, biomass and carbon stock of vegetation are to be estimated in data deficient areas. Biomass assessment is an important facet for national development planning as well as for scientific studies of ecosystem productivity and carbon budgets (Devagiri et al., 2013). Therefore, the present study focused on biomass estimation in trees in the tropical deciduous forests of the Central University of Orissa, Koraput.

In conclusion, the natural beauty of Central University of Orissa campus is with its indigenous plant varieties (ornamental or cultivated or afforested), aesthetic value, ecological uniqueness, and resource importance. Thus, taking a walk around the campus would enrich the botanical knowledge, ecological consciousness and conservation values, not only of the academia but also the common mass.

ACKNOWLEDGMENT

The authors are thankful to the administrative authority of the Central University of Orissa, Koraput for granting permission and providing infrastructural facility during the tenure of the work. The authors are also grateful to the Soil Testing Laboratory, Semiliguda, Koraput for analysis of soil and water parameters of the study area.

KEYWORDS

- **biological diversity**
- **biomass assessment**
- **change in vegetation cover**
- **CUO campus**

REFERENCES

Altay, A. A.; Ozyigit. I. I.; Yarci, C. Urban flora and ecological characteristics of the Kartal District (Istanbul): A contribution to urban ecology in Turkey. *Sci. Res. Essay.* **2010,** *5*(2), 183–200.

Anderson, L. S.; Sinclair, F. L. Ecological interactions in agroforestry systems. *Agroforest. Abst.* **1993,** *6,* 57–91.

Annaselvam, J.; Parthasarathy, N. Inventories of understorey plants in a tropical evergreen forest in the Anamalais, Western Ghats, India. *Ecotropica.* **1999,** *5,* 197–211.

Ayyappan, N.; Parthasarathy, N. Biodiversity inventory of trees in a large-scale permanent plot of tropical evergreen forest at Varagalaiar, Anamalais, Western Ghats, India. *Biodiv. Conserv.* **1999,** *8*(11), 1533–1554.

Banerjee, K.; Paul, R.; Khemendu, R. G. *Vegetational Diversity Analysis in Koraput District of Odisha, India. In: Conserving Biological Diversity: A Multiscaled Approach.* Mir, A. H.; Bhat, N. A. Eds.: Research India Publications, New Delhi, India. **2016,** 93–112.

Bhat, D. M.; Murali, K. S. Phenology of understory species of tropical moist forest of Western Ghats region of Uttara Kannada district in South India. *Current. Sci.* **2001,** *81,* 799–805.

Bilge, Z. Flora of Middle East Technical University Campus Flora (Ankara). M.Sc. Thesis Middle East Technical University, Ankara. **2001**.

Bolin, B. The carbon cycle. *Sci. Am.* **1970,** *223,* 125–132.

Brandes, D. Flora of old town centres in Europe. In: Urban Ecology as the Basis of Urban Planning Skupp, H.; Numata, M.; Huber, A. Eds.: SPB Academics Publishing, Amsterdam. **1995,** 49–58.

Brown, S. L.; Schroeder, P.; Kern, J. S. Spatial distribution of biomass in forests of the eastern USA. *Forest. Ecol. Manag.* **1999,** *123,* 81–90.

Brunig, E. F. *Structure and growth.* In: *Ecosystems of the World, Tropical Rain Forest Ecosystems: Structure and Function.* Golley, F. B. Ed.: Elsevier Scientific Publication, New York. **1983,** 49–75.

Chaiyo, U.; Garivait, S.; Wanthongchai, K. Carbon Storage in Above-Ground Biomass of Tropical Deciduous Forest in Ratchaburi Province, Thailand. *Int. J. Env. Ecol. Eng.* **2011,** *5*(10), 585–590.

Champion, H. G.; Seth, S. K. A Revised Survey of Forest Types of India, Government of India Press, New Delhi. 1968.

Chidumaya, E. N. Above ground woody biomass structure and productivity in a Zambezian woodland. *Forest. Ecol. Manag.* **1990,** *36,* 33–46.

Clayton, W. D.; Renvoize, S. A. Genera Gramineum: Grasses of the World. Kew Bulletin Additional Series XIII, Royal Botanic Gardens, Kew. 1986.

Curtis, J. T.; McIntosh, R. P. The interrelations of certain analytic and synthetic phytosociological characters. *Ecology.* **1950,** *31,* 434–455.

Department of Energy (DOE). Carbon Sequestration Research Development. Office of Fossil Energy. 1999.

Devagiri, G. M.; Money, S.; Singh, S., Dadhwal, V. K.; Patil, P.; Khaple, A.; Devkumar, A. S.; Hubballi, S. Assessment of above ground biomass and carbon pool in different vegetation types of southwestern part of Karnataka, India using spectral modeling. *Trop. Ecol.* **2013,** *54,* 149–165.

Fahrig, L. Effects of habitat fragmentation on biodiversity. *Annu. Rev. Ecol. Evol. Syst.* **2003,** *34,* 487–515.

Fischer R. A.; Corbet A. S.; Williams C. B. The Relation between the Number of Species and the Number of Individuals in a Random Sample of an Animal Population. *J. Anim. Ecol.* **1943,** *12*(1), 42–58.

Food and Agriculture Organization of the United Nations, Forestry Department. Global Forest Resources Assessment Country Report India, **2010,** 6–11.

Forest Resources of Koraput District of Orissa State, Forest Survey of India, Central Zone, Nagpur. Government of India, New Delhi. 1989.

Gentry, A. H. Four neotropical rainforests. Yale University Press, New Haven, Connecticut, USA. 1990

Giles-Lal, D.; Livingstone, C. Campus flora of Madras Christian College. Madras: The Balussery Press, **1978,** p 78.

Gopi, M. Untapped floral carpet of Guru Nanak College (GNC), Tamil Nadu, India. *J. Theo. Exp.Biol.*, **2008,** *5*(1 and 2), 27–32.

Ground Water Information Booklet (GWIB). Govt. of India, Ministry of Water Resources Central Ground Water Board. South Eastern Region, Bhubaneswar, March **2013**.

Grunow, J. O.; Groeneveld, H. T.; DuToit, S. H. C. Above ground dry matter dynamics of the grass layer of a South African tree Savanna. *J. Ecol.* **1980,** *68,* 877–889.

Harte, J.; Kinzig, A.; Green, J. Self-similarity in the distribution and abundance of species. *Science.* **1999a**, *284*, 334–336.

Huston, M. A. Hidden treatments in ecological experiments: evaluating the ecosystem function of biodiversity. *Oecologia.* **1997,** *110,* 449–460.

Indian State of Forest Report. Forest Survey of India. Ministry of Environment and Forests, Govt. of India. Kaulagarh Road, PO–IPE, Dehradun, **2009,** 11–26.

Irwin, S, J.; Thomas, S; Pand, R; Narasimhan, D. Angiosperm diversity of the Theosophical Society campus, Chennai, Tamil Nadu, India. *Check. List.* **2015,** *11*(2), 1–36.

Jayanthi, P.; Rajendran, A. Life Forms of Madukkarai Hills of Southern Western Ghats, Tamil Nadu, India. *Life. Sci. Leafl.* **2013,** *9,* 57–61.

Joshi, H. G.; Ghose, M. Community structure, species diversity, and aboveground biomass of Sundarbans Mangrove swamps. *Trop. Ecol.* **2014,** *55,* 283–303.

Karekezi, S.; Kithyoma, W. Bioenergy and agriculture: Promises and challenges. Bioenergy and the poor. 2020 Vision for Food, Agriculture, and the Environment. International Food Policy Research Institute, Washington DC, USA. 2006.

Karia, J. P.; Porwal, M. C.; Roy, P. S.; Sandhya, G. Forest change detection in Kalarani round, Vadodara, Gujarat: a Remote Sensing and GIS approach, *J. Indian. Soc. Remote.* **2001,** *29*(3), 129–135.

Karki, D. Economic Assessment of Community Forestry in Inner Terai of Nepal: A Case Study from Chitwan District, Asian Institute of Technology, Thailand. 2002.

Khanal, K. P. Economic Evaluation of Community Forestry in Nepal and its Equity Distribution Effect, The Royal Veterinary and Agricultural University, Denmark. 2001.

Khare, A., Sarin, M., Saxena, N. C., Palit, S., Bathla, S., Vania, F. and Satyanarayana, M. *Joint Forest Management: Policy, Practice and Prospects;* 2000. In *India Country Study;* Mayers, J. and Morrison, E., Eds.; *WWF India and IIED, UK;* New Delhi.

Kissling, W. D.; Dormann, C. F.; Groeneveld, J.; Hickler, T.; Kühn, I.; McInerny, G. J.; Montoya, J. M.; Römermann, C.; Schiffers, K.; Schurr, F. M.; Singer, A.; Svenning, J. C.; Zimmermann, N. E.; O'Hara, R. B. Towards novel approaches to modeling biotic interactions in multispecies assemblages at large spatial extents. *J. Biogeogr.* **2012,** *39,* 2163–2178.

Knoop, W. T.; Walker, B. H. Interactions of woody and herbaceous vegetation in a South African Savanna. *J. Ecol.* **1985,** *73,* 235–253.

Kottak, M.; Grieser, J.; Beck, C.; Rudolf, B. and Rubel, F. World map of the Koppen-Gieger climate classification updated. *Meteorologische Zeitschrift*. **2006,** 15 (3), 259–263.

Krebs, C. J. Ecological Methodology, 1st ed.; Addison-Wesley: Boston, 1989.

Krigas, N.; Lagiou, E.; Hanlidou, E.; Kokkini, S. The vascular flora of the Byzantine walls of Thessaloniki (*N Greece*). *Willdenowia*. **1999,** *29,* 77–94.

Kumar, M.; Mukherjee. N.; Sharma, G. P.; Raghubanshi, A. S. Land use patterns and urbanization in the holy city of Varanasi, India: a scenario. *Environ. Monit. Assess.* **2010,** *167*(1), 417–422.

Kunin, W. E.; Hartley, S.; Lennon, J. J. Scaling down: On the challenge of estimating abundance from occurrence patterns. *Am. Nat.,* **2000,** *156,* 560–566.

Lal, R. Global carbon pools and fluxes and the impact of agricultural intensification and judicious land use. Prevention of Land Degradation, Enhancement of Carbon Sequestration and Conservation of Biodiversity through Land Use Change and Sustainable Land Management with a Focus on Latin America and the Caribbean. World Soil Resources Report 86. FAO, Rome. 1999.

Lemaire, G.; Wilkins, R.; Hodgson, J. Challenge for grassland science: Managing research priorities. *Agric. Ecosyst. Environ.* **2005,** *108,* 99–108.

Leopold, A. C.; Salazar, J. Understory species richness during restoration of wet tropical forest in Costa Rica. *Ecol. Rest.* **2008,** *26,* 22–26.

May, R. M.. *Patterns of species abundance and diversity.* In: *Ecology and Evolution of Communities* Cody, M. L.; Diamond, J. M. Eds.: Harvard University Press, Cambridge, M. A. **1975,** 81–120.

Merigot, B.; Bertrand, J. A.; Mazouni, N.; Mante, C.; Durbec, J. P.; Gaertner, J. C. A multi-component analysis of species diversity of groundfish assemblages on the continental shelf of the Gulf of Lions (north-western Mediterranean Sea). *Estuar. Coast. Shelf Sci.,* **2007,** *73*(1–2), 123–136.

Mitra, A.; Bagchi, J.; Thakur, S.; Parkhi, U. S.; Debnath, S.; Pramanick, P.; Zaman, S. Carbon sequestration in Bhubaneswar City of Odisha, India. *Int. J. Innov. Res. Sci., Eng. Tech.* **2015,** *4*(8), 6942–6947.

Mitra, A.; Sengupta, K.; Banerjee, K. Standing biomass and carbon storage of above-ground structures in dominant mangrove trees in the Sundarbans. *Forest. Ecol. Manag.* **2011,** *261*(7), 1325–1335.

Mueller-Dombois, D.; Ellenberg, E. Aims and Methods of Vegetation Ecology. John Willey and Sons, New York, **1974**.

Natarajan, S; Gopi, M..Herbal wealth of Guru Nanak College, Chennai, India. *J. Theo. Exp. Biol.* **2010,** *7*(1 and 2), 17–27.

National Remote Sensing Agency. Satellite Remote Sensing Survey of Natural Resources of Haryana. Project Report, National Remote Sensing Agency, Secunderabad, India. 1979.

Ocak, A.; Ture, C. The Flora of the Meselik Campus of the Osmangazi University (Eskişehir-Turkey). *Ot. Sist. Bot. Derg.* **2001,** *8*(2), 19–46.

Palmer M. W. How should one count species? A review of the methods for counting and estimating the number of plant species within an area. *Nat. Area. J.* **1995,** *15,* 124–135.

Pandey, C. B.; Pandya, K. S.; Pandey, D.; Sharma, R. B. Growth and productivity of rice (Oryza sativa) as affected by Acacia nilotica in a traditional agroforestry system. *Trop. Ecol.* **2000,** *40*(1), 109–117.

Parthasarathy, N.; Ragasan. A. L.; Muthumperumal, C.; Anbarashan, M.; Flora of Pondicherry University Campus. Puducherry: Pondicherry University Publication, **2010,** p.398.

Parthipan, B.; Rajeeswari, M.; Jeeva, S. Floristic Diversity of South Travancore Hindu College (S. T. Hindu College) Campus, Kanyakumari District (Tamilnadu) India. *Biosci. Discov.* **2016,** *7*(1), 41–56.

Pattanaik, C.; Reddy, C. S.; Murthy, M. S. R.; Reddy, P. M. Ethnomedicinal observations among the Tribal People of Koraput District, Orissa, India. *Res. J. Bot.* **2006,** *1,* 125–128.

Pearson, T. H.; Rosenberg, R. A. comparative study of the effects on the marine environment of wastes from cellulose industries in Scotland and Sweden. *Ambio.* **1976,** *5,* 77–79.

Poulsen, A. D.; Pendry, C. A. Inventories of ground herbs at three altitudes on Bukit Belalong, Brunei, Borneo. *Biodiv. Conserv.* **1995,** *4,* 745–757.

Preston, F. W. The Commonness and Rarity of Species. *Ecology.* **1948,** *29*(3), 254–283.

Primack, R. B.; Corlett, R. T. Tropical Rain Forests: An Ecological and Biogeographical Comparison. Blackwell Science, Oxford. 2005.

Puri, S.; Singh, S.; Khara, A. Effect of wind break on the yield of cotton crop in semiarid region of Haryana. *Agroforest. Syst.* **1992,** *18,* 183–195.

Rajendran, R.; Aravindhan, V.; Sarvalingam, A. Biodiversity of the Bharathiyar University Campus, India. A floristic Approach. *Int. J. Biodiver. and Conser.* **2014,** *6*(4), 308–319.

Rekha, D.: Panneerselvam, A.; Thajuddin, N. Studies on medicinal plants of A. V. V. M. Sri Pushpam College Campus Thanjavur district of Tamil Nadu, Southern India. *World J. Pharma. Res.* **2014,** *3*(5), 785–820.

Sanderson, M. A.; Goslee, S. C.; Soder, K. J.; Skinner, R. H.; Tracy, B. F.; Deak, A. Plant species diversity, ecosystem function, and pasture management-A perspective. *Can. J. Plant Sci.* **2007,** *87,* 479–487.

Sanderson, M. J.; Thorne, J. L.; Wikström, N.; Bremer, K. Molecular evidence on plant divergence times. *Am. J. Bot.*, **2004,** *91,* 1656–1665.

Sandford, W. W.; Usman, S.; Obot, S. E.; Isichei, A. O.; Wari, M. Relationship of woody plants to herbaceous production in Nigerian savanna. *Trop. Agric.* **1982,** *59,* 315–318.

Secretariat of the Convention on Biological Diversity. Connecting Biodiversity and Climate Change Mitigation and Adaptation: Report of the Second Ad Hoc Technical Expert Group on Biodiversity and Climate Change. Montreal, Technical Series No. 41, **2009,** p 126.

Shannon, C. E.; Wiener, W. The Mathematical theory of communication. University of Juionis Press. Urbana, 1963; p 117.

Simpson, E. H. Measurement of Diversity. *Nature,* **1949,** *163,* 688.

Sodhi, N. S.; Posa, M. R. C.; Lee, T. M.; Bickford, D.; Koh, L. P.; Brook, B. W. The state and conservation of southeast Asian biodiversity. *Biodiv. Conserv.* **2010,** *19*(2), 317–328.

State Forest Report. Forest Survey of India. Government of India Ministry of Environment and Forests New Delhi. **2003.**

Terakunpisut, J.; Gajaseni, N.; Ruankawe, N. Carbon sequestration potential in aboveground biomass of Thong Phaphun National Forest, Thailand. *Appl. Ecol. Env. Res.* **2007,** *5,* 93–102.

The Statistic of Forestland in Thailand of Department of Forestry, Ministry of Natural Resources and Environment, Thailand.

Ture, C.; Bocuk, H. The Flora of The Anadolu University Campus (Eskisehir-Turkey). *Anadolu Univ. J. Sci. Technol.* **2001,** *2*(1), 83–95.

Turgut, T. Flora of Abant Izzet Baysal Campus. M.Sc. Thesis, Abant Izzet Baysal University, Bolu. 1996.

Udayakumar, M; Ayyanar, M.; Sekar, T. Angiosperms, Pachaiyappa's College, Chennai, Tamil Nadu, India. *Check. List.* **2011,** *7*(1), 37–48.

Upadhaya, K.; Thapa, N.; Barik, S. K. Tree diversity and biomass of tropical forests under two management regimes in Garo hills of north-eastern India. *Trop. Ecol.* **2015,** *56*(2), 257–268.

Vashum, K. T.; Jayakumar, S. Methods to estimate above ground biomass and carbon stock in natural forest: a review. *J. Ecosyst. Ecogr.* **2012,** *2,* 1–7.

Vitousek, P. M.; Melillo, J. M. Nitrate losses from disturbed forests; patterns and mechanisms. *J. For. Sci.* **1979,** *25,* 605–619.

Von der Lippe, M.; Kowarik, I. Do cities export biodiversity? Traffic as dispersal vector across urban-rural gradients. *Divers. Distributions.* **2008,** *14*(1), 18–25.

Von der Lippe, M.; Kowarik, I. Long-distance dispersal by vehicles as driver in plant invasions. *Conserv. Biol.* **2007,** *21*(4), 986–996.

Walkley, A.; Black, I. A. An examination of Degtjareff method for determining soil organic matter and a proposed modification of the chromic acid titration method. *Soil. Sci.,* **1934,** *37,* 29–37.

Ward, J. V. Riverine landscape: biodiversity pattern, disturbance regimes, and aquatic conservation. *Biol. Conserv.* **1998,** *83*(3), 269–278.

Whitmore, T. C. Tropical Rain Forests of the Far East. Oxford University Press, London. 1984.

Willig, M. R., Kaufman, D. M.; Stevens, R. D. Latitudinal gradients of biodiversity: pattern process, scale and synthesis. *The Annual Review of Ecology, Evolution, and Systematics,* **2003**, *34*, 273–309.

WRI (World Resources Institute). World Resources 1994–95. Oxford University Press, New York, **1994,** p. 400.

CHAPTER 4

Edible *Solanum* Species Used by the Ethnic and Local People of North-Eastern India

MOUMITA SAHA* and B. K. DATTA

Plant Taxonomy and Biodiversity Laboratory Department of Botany, Tripura University Suryamaninagar–799022, Tripura, India,
**E-mail: sahamou1987@gmail.com*

ABSTRACT

In the present study, the ethnobotanical observation of wild, as well as cultivated *Solanum* of Tripura state has been recorded. The state bearing a rich diversity of both wild and planted species of *Solanum* which are consuming more or less by the different community of the state. The present paper highlights on eight species of *Solanum* of which three of them are cultivated. The species with their botanical names, local name(s), distribution, availability, flowering, and fruiting period and their mode of utilization by local people has been enumerated in alphabetical order.

4.1 INTRODUCTION

Solanum Linn. is a large and diversified genus belonging to the economically important cosmopolitan family Solanaceae with over 2700 species (Olmstead and Bohs, 2007) in the world and distributed in the tropics and subtropics of both the old and new world (Figure 4.1). This cosmopolitan family is commonly scattered throughout tropical and temperate regions with Latin America and Australia as main dispersion centers (Barroso et al., 1991) and their centers of diversity occurring in Central and South America and Australia (Edmonds, 1978; D'Arcy, 1991). The family Solanaceae contains a number of the important agricultural plant as well as many toxic plants. *Solanum* is the largest genus in the family *Solanaceae* which has approximately 1500 species and is ideal for combined wired taxonomy in puzzling tropical plant groups (Knapp et al., 2004). Like other families of the angiosperms, the family *Solanaceae* is well known for its economic value. The genus *Solanum* is known worldwide for its traditional and modern uses (see Table 4.1). They have diverse bioactive compounds (Table 4.2; Figures 4.2–4.4). A total of 15 *Solanum* species were recorded from Tripura of which three are cultivated. The inhabitants use the majority of the species in one system or the other. Among them, fruits of eight *Solanum* species mentioned in the current paper are edible which are commonly used by the people of Tripura.

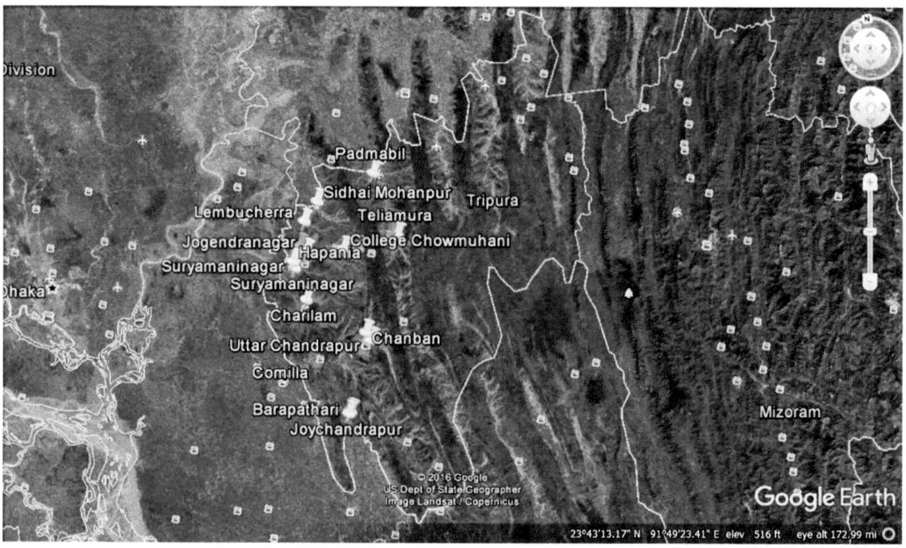

FIGURE 4.1 Location of population survey of *Solanum* sp.

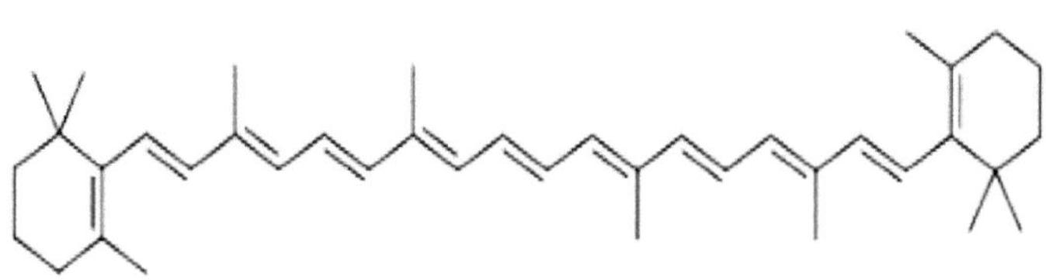

β-Carotene

FIGURE 4.2 β-Carotene (www.wildflowerfinder.org.uk).

TABLE 4.1 The Medicinal Uses of Enumerated Species and Their Associate Species of Tripura, India

Botanical Name	Local name	Parts used	Uses	Source(s)
S. aethiopicum L.	Khamka	Fruits	Fruits contained higher levels of beneficial nutrients which are nutritionally and therapeutically valuable and also have the potential of providing precursors for the synthesis of useful drugs	Chinedu et al., (2011)
S. anguivi Lamk.	Phutki begun	Fruit	It is a nourishing vegetable commonly used in soup and medically to control high blood pressure and diabetes	Schipper, (2000)
S. melongena L.	Begun/Phantak		Various parts of the plant are useful in the treatment of inflammatory conditions, cardiac debility, neuralgias, bronchitis, and asthma	Mutalik et al., (2003)
S. ovigerum Dunal	Anda begun	Fruit, Leaves	A fruit vegetable; young leaves are used by the tribal people in treating diarrhea	Saha and Datta, (2017)
S. surattense Burm *f.*	Kantakari	Whole plant	The plant is widely studied for the various pharmacological activities like antiasthmatic, hepatoprotective, cardiovascular, hypoglycemic and mosquito repellent properties	Reddy and Reddy, (2014)
S. torvum Swartz	Khamka Sikum, Brihati begun	Whole plant	The plant is sedative and diuretic, and the leaves are used as hemostatic	Jaiswal and Mohan, (2012)
S. tuberosum L.	Alu	Tuber	Mainly used as staple food. A juice made from the tubers, when taken in moderation, can be helpful in the treatment of peptic ulcers, bringing relief from pain and acidity	Yadav et al., (2016); Chevallier, (1996)
S. violaceum Ortega	Khamkha	Root, Fruits, Seeds	Fruits are eaten as a vegetable; also diuretic and expectorant. Root paste is applied for poison. Seeds are used for the preparation of oil used in case of cough and bronchial diseases	Silja et al., (2008)

TABLE 4.2 Bioactive Compounds Available in the Enumerated *Solanum* Species

Plant name	Bioactive compounds	References
Solanum aethiopicum	β-carotene	Mibet et al., (2017)
Solanum anguivi	Anguiviosides A–C	Ramamurthy et al., (2012)
Solanum melongena	Saponin	Vohora et al., (1984)
Solanum torvum	β-carotene	Ramamurthy et al., (2012)
Solanum tuberosum	Thiamine	Chowanski et al., (2016)
Solanum violaceum	Steroidal alkaloids	Chowanski et al., (2016)
Solanum surattense	Steroidal alkaloid glycosides	Lu et al., (2011)

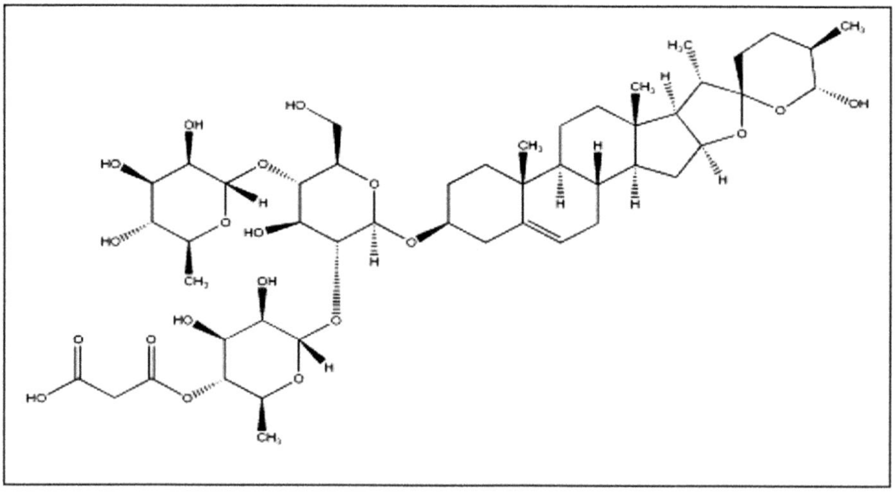

FIGURE 4.3 Saponin (www.genome.jp).

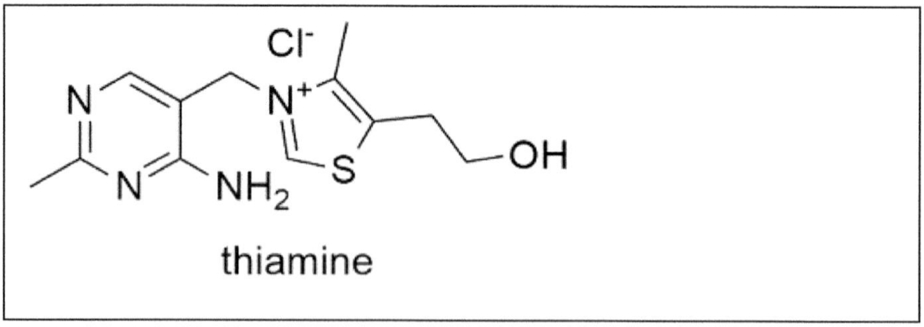

FIGURE 4.4 Thiamine (www.softscools.com).

Tripura, the third smallest state of India is a former princely state of North East India, covering an area of 10,491.69 sq. km. It extends from 22°56′ N to 24°32′ N and 91°09′ E to 92°20′ E and is bordered by Bangladesh to the west, north and south and the Indian states of Assam to the northeast and Mizoram to the east. The state measures about 184 km from north to south and 113 km east to west up to its maximum extent and comprises of eight districts namely Dhalai, North Tripura, South Tripura, West Tripura, Khowai, Unakoti, Sipahijala, and Gomati. The pleasant growth of various types of forests spread all over the state from hilly track to plain has been supported by this suitable tropical climate. About 19 ethnic groups are primarily existing in the forests of Tripura. These groups have individual culture, food-habit, language, and socio-religious behavior (Sajem and Gosai, 2008; De et al., 2010; WHO, 2001). Most of the ethnic peoples live in the remote areas with close harmony to the adjacent forest areas. Being a resident of forested areas these ethnic communities are reliant on biotic resources immediately available to them for subsistence livelihood. Considering this fact, an effort has been prepared to document the edible fruits resources, especially which are consumed raw and locally available in wild condition in different forests of Tripura.

4.2 METHODS OF SURVEY

The present investigation is the outcome of several field trips carried out in and around Tripura during 2013–2015 (Figure 4.1). Information regarding the utilization of edible *Solanum* species in their natural habitat as well as cultured was obtained through interview, field observation, group discussion and market survey from different communities of Tripura. Specimens were collected by consulting with the local informants. Herbarium specimens were dried and preserved according to the methodology proposed by Jain and Rao (1977). Plants were identified with the assistance of standard treaties viz. Hooker (1872–1897), Prain (1903), Kanjilal et al. (1934–1940) and Deb (1981, 1983). The voucher specimens are deposited in the Herbarium of the Department of Botany, Tripura University.

4.3 ENUMERATION

The *Solanum* species are alphabetically arranged below alongside with their botanical name, local name(s), availability, flowering, and fruiting periods and their usage (Plate 4.1).

1. S. aethiopicum L.

An undershrub; branches and leaves are without prickles; flowers white; fruit red.

Distribution: Dhalai and North Tripura.

Availability: Occasional.

Exsiccate: Dhalai, Saha 1582 (TUH), dated 19.5.2014.

Flowering and fruiting period: April–June.

Uses: The immature fruits of *Solanum aethiopicum* are sometimes eaten as raw and mostly used as cooked vegetables.

2. S. anguivi Lamk.

A spiny undershrub or shrub; Leaves and stem stellately tomentose; Prickles short, curved, yellowish; Calyx prickly; Flowers white, lobes reflexed; Fruit globose, smooth, green.

Local name: Phutki begun.

Distribution: South and North Tripura.

Availability: Common.

Exsiccate: Khowai, Saha & Datta 1573 (TUH), dated 23.10.2014.

Flowering and Fruiting period: February–April.

Uses: Fruits are edible.

3. S. melongena L.

A bushy cultivated undershrub; Leaves large, elliptic and lobed; Flowers blue; Fruit violet, large, globose and are of varying sizes and forms.

Local name: Begun/Phantak

Distribution: Throughout the state.

Availability: Common.

Exsiccate: Dhaleswar, Saha 1501 (TUH), dated 05.04.2013.

Flowering and fruiting period: Cultivated all over the year.

Uses: Fruits are used as a vegetable.

4. S. ovigerum Dunal

A medium sized shrub; Branches armed and tomentose and prickly; Leaves lobed, stellate and prickly on the petiole; Flowers purple; Fruit white resembles hen's egg; yellow when ripe.

Local name: Anda begun

Distribution: Cultivated all over the year.

Availability: Common.

Exsiccate: Jogendranagar, Saha 1670 (TUH), dated 12.05.2014.

Flowering and fruiting period: Cultivated all over the year.

Uses: Young leaves are used by the tribal people in treating diarrhea.

5. *S. surattense* Burm *f.*

Herb with a woody base and densely prickly all over, with very sharp and straight yellow prickles; Leaves lobed, stellate pubescent; lobes acute; Flowers blue; Fruit green and yellow when ripe.

Local name: Kantakari.

Distribution: Sadar.

Availability: Rare.

Exsiccate: Sabroom, Saha & Datta 1517 (TUH), dated 02.08.2014.

Flowering and fruiting period: March-June.

Uses: Fruits are eaten as anthelmintic and for indigestion.

6. *S. torvum* Swartz

A tall shrub; Branches sparingly armed and tomentose; Leaves lobed, densely stellate, sparsely prickly on the petiole; Flowers white; Fruit bright green with persistent calyx.

Local name: Khamka Sikum/Brihati begun

Distribution: Throughout the state

Availability: Very Common

Exsiccate: Charilam, Saha 1609 (TUH), dated 19.04.2014.

Flowering and fruiting period: Throughout the year.

Uses: Fruits are cooked to prepare a fried vegetable.

7. *S. tuberosum* L.

A cultivated herb with tuberous underground branches; Leaves large, unequally and irregularly pinnatisect, pubescent; Flowers blue.

Local name: Alu

Distribution: Throughout the state

Availability: Common

Exsiccate: Charilam, Saha 1600 (TUH), dated 29.04.2013.
Flowering and fruiting period: Cultivated extensively in the cold season.
Uses: Used as a vegetable.

8. S. violaceum Ortega

An armed undershrub; Branches minutely stellate-hairy on young parts, glabrate when mature; Leaves deeply lobed and prickly; Flowers purple; Fruit globose, orange.
Local name: Khamkha.
Distribution: Sadar and North Tripura
Availability: Common.
Exsiccate: Padmabil, Saha 1634 (TUH), dated 18.06.2014.
Flowering and fruiting period: Almost throughout the year.
Uses: Ripe berries are edible.

4.4 DISCUSSION

The present communication recorded eight species of wild and cultivated edible *Solanum* species. These fruits are sold in remote village markets depending on availability. The utilization of wild fruits continues to benefit society even today, from processes mostly involving local experimentation through indigenous and local knowledge. The availability of such edible fruits in good amount helped these people to depend on the indigenous plant resources of Tripura as an integral part of their livelihood. This also expresses the rich biodiversity of Tripura forests. Thus sustainable utilization of wild edible fruits may provide an effective incentive to conserve those wild races and forest ecosystem and will enhance species community structure and composition. The necessity to take care of such ecosystems as a unique key to maintain biological diversity as to conserve soil and giving livelihood services.

4.5 RECOMMENDATION FOR THE CONSERVATION OF ENUMERATED *SOLANUM* SPECIES

In the present work, it was found that in the majority of the villages of Tripura the fruits of *S. violaceum*, *S. torvum*, and *S. ovigerum* are consumed as well as marketed by the ethnic groups and therefore considered as an

PLATE 4.1 (See color insert.) Enumerated *Solanum* species of Tripura, India, 1: *S. torvum*, 2: *S. sisymbriifolium*, 3: *S. tuberosum*, 4: *S. erianthum*, 5: *S. violaceum*, 6: *S. viarum*, 7: *S. nigrum*, 8: *S. aethiopicum*, 9: *S. ovigerum*, 10: *S.violaceum*, 11: *S. ovigerum*, 12: *S. melongena*, 13: *S. aethiopicum*, 14: *S. torvum*.

important crop of subsistence especially in tribal areas. As the local inhabitants are dependent on the natural vegetation so overutilization of these wild species may affect the local diversity and lower possibilities of availability of these species in the natural habitat. Since the above-mentioned species are not cultivated and managed (except *S. ovigerum*) like other plants in their home gardens, so the availability of those wild edible *Solanum* species depends on their wild collection only. Thus, these species need to use in a sustainable manner and should be in corporate immediately under managed cultivation or homestead gardens. The sustainable utilization of these edible, as well as medicinal species of *Solanum,* may provide an efficient enticement to conserve those wild races in natural and managed condition.

ACKNOWLEDGMENTS

The first author is thankful to the Department of Science and Technology (DST), New Delhi, Govt. of India for providing INSPIRE (Innovation in Science Pursuit for Inspired Research) Fellowship in order to carry out the work. Authors are also thankful to P. D. Singh, Ambika Prasad Research Foundation, Regional Center, Imphal for the help in manuscript preparation.

KEYWORDS

- **edible**
- ***Solanum***
- **Tripura**

REFERENCES

Barroso, G. M.; Peixoto, A. L.; Costa, C. G.; Ichasso, C. L. F.; Guimaraes, E. F.; Lima, H. C. *Sistematica das Angiospermas do Brasil*. In: *Viçosa, Imprensa Universitaria;* **1991**, Vol. 2.

Chevallier. A. *The Encyclopedia of Medicinal Plants,* Dorling Kindersley, London, **1996**.

Chinedu, S. N., Olasumbo, A. C., Eboji, O. K., Emiloju, O. C., Arinola, O. K.; Dania, D. I. **2011**. Proximate and Phytochemical Analyses of *Solanum aethiopicum* L.; *Solanum macrocarpon* L. Fruits. *Research Journal of Chemical Sciences. 1*(3), 63–71.

Chowanski, S.; Adamski, Z.; Marciniak, P.; Rosinski, G.; Buyukguzel, K.; Falabella, P.; Scrano, L.; Ventrella, E.; Filomena, L.; Bufo, S. A. A review of bioinsecticidal activity of Solanaceae alkaloids. *Toxins*, **2016**, *8*(3), 2–28.

D'Arcy, W. G. The Solanaceae since 1976, with a review of its biogeography. *In*: Hawkes, J. G., Lester, R. N., Nee, M., Estrada, N. (Eds.), Solanaceae III: taxonomy-chemistry-evolution, Royal Botanical Gardens Kew, London, **1991**, pp. 71–137.

De, B.; Debbarma, T.; Sen, S.; Chakraborty, R. Tribal life in the environment and biodiversity of Tripura, India. *Current World Environment*. **2010**, *5*, 59–66.

Deb, D. B. *The Flora of Tripura State*. Vols. I and II. Today and Tomorrow Printers and Publisher. New Delhi, **1981**, **1983**.

Edmonds, J. M. *Solanaceae*. In: Heywood VH (ed.), *Flowering Plants of the World*. Oxford University Press, Oxford, **1978**. pp. 228–229.

Hooker, J. D. *The Flora of British India*. Vols. I–VII. L. Reeve and Co., London, **1872–189.**

Jain, S. K.; Rao, R. R. A Handbook of Field and Herbarium Methods. Today and Tomorrow Printers and Publisher. New Delhi, **1977**.

Jaiswal, B. S.; Mohan, M. Effect of *Solanum torvum* on the contractile response of isolated tissues preparation in fructose fed rat. *International Journal of Pharma and Bio Sciences*. **2012**, *3*(3), 161–169.

Kanjilal, U. N., Kanjilal, P. C. Das, A., De, R. N.; Bor, N. L. *Flora of Assam*. Vols. I–IV. Government Press, Shillong, **1934–1940**.

Knapp, S., Bohs, L., Nee, M.; Spooner, D. M. Solanaceae: a model for linking genomics and biodiversity. *Comparative and Functional Genomics*. **2004**, *5*, 285–291.

Lu, Y.; Luo, J.; Kong, L. Steroidal alkaloid saponins and steroidal saponins from *Solanum surattense*. *Phytochemistry*, **2011,** *72*(7), 668–673.

Mibei, K. E.; Ambuko, J.; Giovannoni, J. J.; Onyango, A. N.; Owino, W. O. Caratenoid profiling of the leaves of selected African eggplant accessions subjected to drought stress. *Food Science & Nutrition*. **2017**, *5*(1), 113–122.

Mutalik, S., Paridhavi, K., Rao, C. M.; Udupa, N. Antipyretic and analgesic effect of leaves of *Solanum melongena* Linn. in rodents. *Indian J Pharmacol*. **2003**, *35*, 312–315.

Olmstead, R. G.; Bohs, L. **2007**. A summary of molecular systematic research in Solanaceae, 1982–2006. *In:* Spooner, D. M., Bohs, L., Giovannoni, J., Olmstead, R. G.; Shibata, D. (Eds.), *Solanaceae VI: Genomics Meets Biodiversity. Proceedings of the Sixth International Solanaceae Conference*. Acta Horticulturae 745. International Society for Horticultural Science, Leuven, pp. 255–268.

Prain, D. *Bengal Plants* (Vols. 1 and 2). Repd. ed., BSI., Calcutta, **1903**.

Ramamurthy C. H.; Kumar, M. S.; Mareeswaran, R.; Thirunavukkarasu, C. Evaluation of antioxidant, radical scavenging activity and polyphenolics profile in Solanum torvum L. fruits. *Journal of Food Sciences*, **2012**, *77*(8), 907–913.

Reddy, N. M.; Reddy R. N. *Solanum xanthocarpum* Chemical Constituents and Medicinal Properties: A Review. *Scholars Academic Journal of Pharmacy*. **2014,** *3*(2), 146–149.

Saha, M.; Datta, B. K. Diversity of *Solanum* L. (Solanaceae) in Tripura (India) including new records. *Pleione* **2017,** *11*(1), 85–96.

Sajem, A.; Gosai, K. Ethnobotanical investigations among the *Lusai* tribes in North Cachar Hills district of Assam, Northeast India. *Indian Journal of Traditional Knowledge*. **2008**, *9*(1), 108–113.

Schipper, R. R. *Natural Resources Institute*, Chatham, UK. **2000**, p. 214.

Silja, V. P., Varma, K. S.; Mohanan, K. V. Ethnomedicinal plant knowledge of the Mullu Kuruma tribe of Wayanad district, Kerala. *Indian Journal of Traditional Knowledge*. **2008**, *7*(4), 604–612.

Vohora, S. B.; Kumar, I.; Khan, M. S. Effect of alkaloids of Solanum melongena on the central nervous system. *Journal of Ethnopharmacology,* **1984**, *11*(3), 331–336.

WHO. General Guidelines for Methodologies on Research and Evaluation of Traditional Medicine. Geneva, Switzerland, **2001**.

Yadav, R., Rathi, M., Pednekar, A.; Rewachandani, Y. A detailed review on solanaceae family. *European J. Pharmac. Med. Res.* **2016,** *3*(1), 369–378.

CHAPTER 5

Medicinal Plant Diversity in Urban Areas

SANJEET KUMAR,[1] P. DEVANDA SINGH,[1] RAJNDRA K. LABALA,[2] L. AMITKUMAR SINGH,[2] and SUNIL S. THORAT[1,2,*]

[1]*Bioresource Database and Bioinformatics Division, Institute of Bioresources and Sustainable Development, Govt. of India, Takyelpat, Imphal–795001, Manipur, India*

[2]*Distributed Information Sub-Centre, Institute of Bioresources and Sustainable Development, Govt. of India, Takyelpat, Imphal–795001, Manipur, India, *E-mail: sunilthorat@gmail.com*

ABSTRACT

Imphal is the capital city of Manipur and known as the city of Kangala. It is situated between 24.8074 °N and 93.9384 °E having an average elevation of 786 m. The city enjoys a humid subtropical climate and gets an average rainfall of 1581 mm. The prevailing climatic conditions of the city provide favorable conditions for the rich diversity of medicinal plants. Despite all, very less information was available on the medicinal plants of the city. Keeping this in mind an extensive survey was made in the year 2017 and enumerated the medicinal plants from the field and world's famous Ima Keithal. Results revealed that 82 common medicinal plants available in the wild having rich diversity and sound medicinal potentials. Distribution of selected 10 medicinal plants was presented in the map of the city for bringing attention towards the conservation of such valuable urban resources. The present study also highlights the economic values of urban medicinal plants and their role in the sustainable development and health care of the populace of the Imphal, India.

5.1 INTRODUCTION

Urbanization is spreading the green cover at an alarming rate specifically in developing countries like India (Imam and Banerjee, 2016). As per census

1901, 11.4% of the total population of India is residing in urban areas which crossed 31% in the 2011 census (Kavita and Gayathri, 2017). The most urbanized states of the country are Maharashtra followed by Tamil Nadu and Kerala as per census 2001 (Kadi and Nelavigi, 2015). These reflect the development and economy of the country but also lead to hazardous issues including water and air pollution, biodiversity loss and hence disbalance the environment (Kelly and Fussell, 2015). At the present scenario, most part of the country is facing the negative impacts of urbanization, including the green patch of the country, the North-Eastern region of India (Yadav and Sharma, 2013). This has lead in developing two biodiversity hotspots in the region namely; Indo-Burma Biodiversity Hotspot and Eastern Himalaya Biodiversity Hotspot (Singh, 2016). North-East India hosts about 13,500 plant species in the Indo-Burma biodiversity hotspot region of India and about 10,000 species in the Eastern Himalayas biodiversity hotspot of which many are endemic in both areas (Mahanta, 2015). The people of North-Eastern region use food and medicines from the plant and animal wealth of the respective state (Mao et al., 2009). The traditional food and medicine are obtained by nearby forest areas and used in various forms. The rural and tribal communities of the region use to sell them for their livelihood. These foods are good for health as per the climatic conditions (Debbarma et al., 2017). The rapid urbanization in these areas is going to vanish the traditional food and medicine which are intimately connected to their sociocultural, economical, spiritual life (Satterthwaite et al., 2010). Among the eight states of the northeast region of the country, Manipur is well known for its ethnic food comprising few medicinal plants in its recipes (Lokho, 2012). The state is bounded by Nagaland, Mizoram, Assam, and Myanmar having 22,327 km^2 with 3 million populations along with Meitei, Loi, Yaithibi, Kuki and Naga communities (Mathur, 2011). The ethnic group Meitei represents 53% of the total population of state, and most are settled in the capital city Imphal (Ningombam and Singh, 2014). The women of the Meitei community are known for their role in the conservation and utilization of bioresources. They run markets in Imphal city with the name of "Ima" where they sell all the indigenous food in the market thereby playing a significant role in the conservation and sustainability of the bioresources of Manipur (Devi et al., 2010). Imphal city or palace of Kangla is situated between 24.8074°N to 93.9384°E at an average elevation of 786 m and enjoys the humid subtropical climate to nurture the rich floral resources (Devi et al., 2010). At the present scenario, the city is growing and facing the negative impact of urbanization (Dociu and Dunarintu, 2012). Keeping in view all the above facts; an attempt has been made to enumerate the most common medicinal plants available in wild and urban areas and document the most common traditional vegetable or plant parts available in the Ima market which has food and medicinal value. The present study highlights the following for urban areas:

a) Importance of urban medicinal plants;
b) Importance of Ima market in the conservation of traditional food and medicines;
c) Role of woman in the conservation of urban bioresources of Manipur;
d) Importance of urban bio wealth for the screening of nutraceuticals to fight against various diseases and disorders of the state.

5.2 METHODOLOGY

The fieldwork was conducted with the Meitei community of Imphal city and Ima market, Imphal during 2017. The methodology framework was followed as per the standard technique of ethnobiological approaches of Christian and Brigitte (2004). The information on plant use as food and traditional medicine were collected through a questionnaire. Plant species were confined using Bentham and Hooker system.

5.3 RESULTS AND DISCUSSION

During the field survey, 82 plant species in 69 genera and 42 families were recorded from Ima market and urban area of Imphal city (Figure 5.1). Result revealed that the rich diversity of medicinal plant species in the urban area of Imphal used against various disease (Figure 5.2). The enumerated plant species belongs to following dominant families such as Lamiaceae (6), Solanaceae (4), Euphorbiaceae (3), Fabaceae (3), Cucurbitaceae (3) and Zingiberaceae (2). It was observed that among the 48 plant species were commonly traded. The most common species are *Saccharum officinarum, Parkia javanica* (Plate 5.2P), *Hedychium flavum, Alocasia cucullata, Neptunia oleracea* (Plate 5.2A), *Capsicum annum* (Plate 5.2B), *Oryza sativa* (Plate 5.2C), *Sechium edule* (Plate 5.2D), *Nelumbo nucifera* (Plate 5.2E), *Pinus roxburgaii* (Plate 5.2F), *Psophocarpus tetragonolobus* (Plate 5.2G), *Allium odorum* (Plate 5.2H), *Phyllanthus embelica* (Plate 5.2J), *Ananas comosus* (Plate 5.2K), *Elaeocarpus floribundus* (Plate 5.2L) *Manihot escalenta* (Plate 5.2M), *Elsholtzia blanda* (Plate 5.2N), *Phlogacanthus thyrsiflorus* (Plate 5.2O), etc. It was noted that among the 34 common medicinal plants found

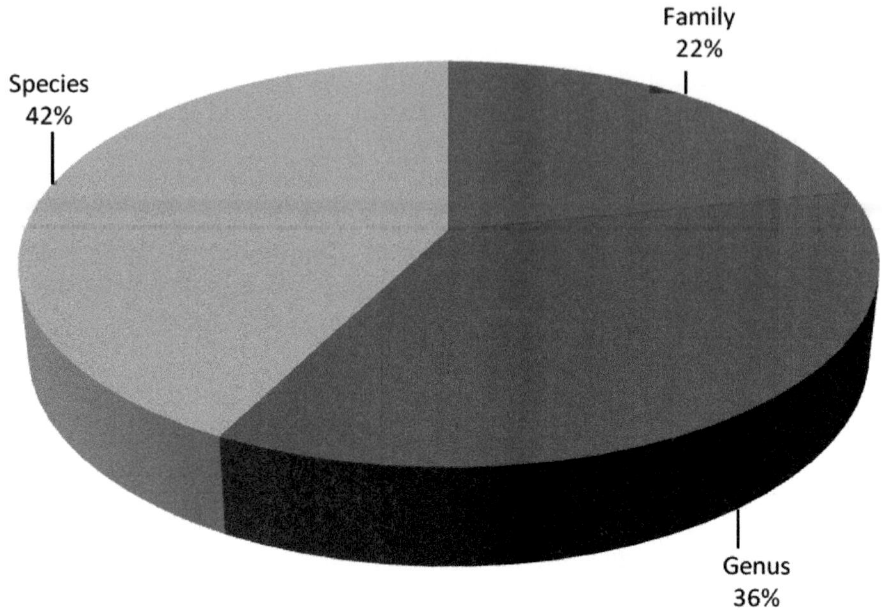

FIGURE 5.1 **(See color insert.)** The diversity of enumerated plant species.

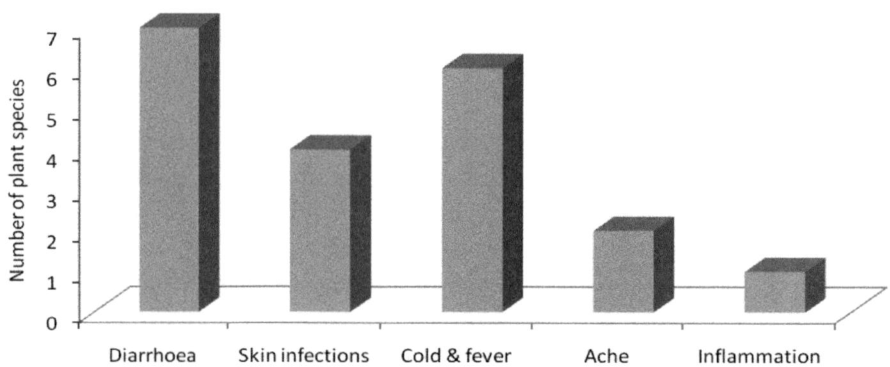

FIGURE 5.2 **(See color insert.)** Plants used against some most common diseases/disorders.

in the urban areas of Imphal (Table 5.1; Plate 5.1), *Clerodendrum serratum, Agrimonia eupotoria* (Plate 5.1A), *Eryngium foetidum, Ocimum basilicum, Oroxylum indicum, Polygonum orientale, Polygonum posumbu, Psophocarpus tetragonolobus, Rhussemi alata, Sesbania grandiflora, Zanthoxylum acanthopodium, Bamboo spp.* (Plate 5.2I), *Fragaria vesca, Sonerila erecta, Nicandra physalodes* (Plate 5.1R), *Utricularia aurea, Scutellaria discolour*

(Plate 5.1B), *Dioscorea alata* (Plate 5.1C), *Spilanthes paniculata* (Plate 5.1D), *Utricularia aurea* (Plate 5.1 E), *Urena lobata* (Plate 5.1F), *Cissampelos pareira* (Plate 5.1G), *Mukia madraspatana* (Plate 5.1I), *Costus speciosus* (Plate 5.1J),) *Fragaria vesca* (Plate 5.1K), *Glochidion manipurensis* (Plate 5.1L), *Merremia hederacea, Dioscorea alata, Plumbago zeylanica* (Plate 5.1M), *Lindernia ruellioides* (Plate 5.1N), *Lobelia nummularia* (Plate 5.1O),) *Ludwigia adscendens* (Plate 5.1P), *Merremia hederacea* (Plate 5.1Q) are encountered frequently. The uses of the common medicinal plant species are given in Tables 5.1 and 5.2. In the present study, we also collected the data for the distribution of 10 common medicinal plants available in the urban area of Imphal. The results showed that most of the selected plants are found in Khongampat area of Imphal (Table 5.3). In the year 2010, Yumnam and Tripathi reported 64 plant species which were used as medicines whereas Devi et al. 2010 reported 71 plant species used for medicinal and food purposes. Recently, in the year 2014, Leisangthem and Sharma reported 50 medicinal plants belongings to 26 families from Imphal East. They reported *Acorus calamus, Cassia alata, Andrographis paniculata, Kaempferia galanga, Zanthoxylum acanthopodium, Eryngium foetidum, Eupatorium birmanicum, Adhatoda vasica*, etc. as common medicinal plants available in this region whereas the present study found species namely *Oroxylum indicum, Polygonum orientale, Ludwigia adscendens, Costus speciosus, Glochidion manipurensis, Spilanthes paniculata, Cissampelos pareira, Urena lobata, Merremia hederacea, Dioscorea alata, Plumbago zeylanica,* etc.

5.4 CONCLUSION

The exclusive study highlights on the social importance of medicinal plant in the urban area of Imphal as a significance of folk medicine in the treatment of many diseases like diarrhea, skin disease, anti-inflammatory, cold and fever and associated symptom. The local inhabitants of Manipur have a wide knowledge about the uses of traditional bioresources either as medicine or food in their day to day life. The indigenous people collect this medicinal plant species from forest and hill and sell them in the market for their livelihood. The Ima market or Ima keithel is a unique marketplace exclusively run by women. The Imphal city is a mixture of different communities, yet the people believe in the ethnobotanical and traditional use of plant species for its varied uses. The authors were of strong view for conducting an appropriate awareness program for the conservation and sustainable use of plant species by making the people aware about the plant values and their significance.

PLATE 5.1 Some common medicinal plants available in the urban areas of Imphal, India: (A) *Agrimonia eupotoria*, (B) *Scutellaria discolour*, (C) *Dioscorea alata*, (D) *Spilanthes paniculata*, (E) *Utricularia aurea*, (F) *Urena lobata*, (G) *Cissampelos pareira*, (H) *Oroxylum indicum*, (I) *Mukia madraspatana*, (J) *Costus speciosus*, (K) *Fragaria vesca*, (L) *Glochidion manipurensis*, (M) *Plumbago zeylanica* (N) *Lindernia ruellioides*, (O) *Lobelia nummularia*, (P) *Ludwigia adscendens*, (Q) *Merremia hederacea*, (R) *Nicandra physalodes*

PLATE 5.2 Common medicinally important plant parts available at Ima, Keithel, Imphal, India: (A) *Neptunia oleracea,* (B) *Capsicum annum,* (C) *Oryza sativa,* (D) *Sechium edule,* (E) *Nelumbo nucifera,* (F) *Pinus roxburgaii,* (G) *Psophocarpus tetragonolobus,* (H) *Allium odorum,* (I) *Bamboo spp.,* (J) *Saccharum officinalis,* (K) *Phylanthus embelica,* (L) *Ananas comosus,* M) *Nelumbo nucifera,* (N) *Elaeocarpus floribundus,* (O) *Manihot escalenta,* (P) *Elsholtzia blanda,* (Q) *Phlogacanthus thyrsiflorus,* (R) *Parkia javanica.*

TABLE 5.1 Common Medicinal Plants Available in the Urban Areas of Imphal, Manipur, India

Name	Family	Local Name	Use(s)
Cinnamomum tamala	Lauraceae	Tejpat	Dizziness
Alocasia indica	Araceae	Pan	Purify blood
Andrographis paniculata	Acanthaceae	Bhubati	Chronic fever
Eryngium foetidum	Apiaceae	Awa-phadigom	Arthritis
Blumea balsamifera	Asteraceae	Langthrei	Epilepsy
Euphorbia hirta	Euphorbiaceae	Pakhangleiton	Diarrhea
Jatropha gossypifolia	Euphorbiaceae	Kege-manbi	Eczema
Ocimum basilicum	Lamiaceae	Naosek-le	Fever
Piper longum	Piperaceae	Tabopi	Laxative
Gynura nepalensis	Asteraceae	Terapaibi	Against stomach ulcer
Cedrela toona	Meliaceae	Tairen	Skin diseases
Hedychium coronarium	Zingiberaceae	Takhellei-angouba	Cough and vomiting
Artocarpus heterophyllus	Moraceae	Theibong	Fever
Plumbago zeylanica	Plumbaginaceae	Telhidak	Piles and bronchitis
Smilax ovalifolia	Liliaceae	Kwa-mana-manbi	Skin diseases
Tinospora cordifolia	Menispernaceae	Ninthou-khong-lee	Diarrhea
Sida rhombifolia	Malvaceae	Broom jhutesida and Uhal	Urinary disorder
Oroxylum indicum	Bignoniaceae	Shamba	Gastric ulcer and tonsil
Drynaria quercifolia	Polypodiaceae	—	Hectic fever
Melothria perpusilla	Cucurbitaceae	Lamthabi	Jaundice and kidney affection
Mimosa pudica	Fabaceae	Kangphal-ikaithab	Piles and jaundice
Meriandra benghalensis	Lamiaceae	Kanghuman	Cough
Solanum xanthocarpum	Solanaceae	Singkhangkha	Mouth ulcer
Solanum nigrum	Solanaceae	—	Stomachache
Stellaria media	Caryophyllaceae	Yerum keirum	Nasal bleeding
Celosia argentea	Amaranthaceae	Hou lei	Swelling
Elaeocarpus floribundus	Elaeocarpaceae	Chorphon	Hypertension
Syzygium cumini	Myrtaceae	Jam	Diabetes and digestion disorder
Zanthoxylum acanthopodium	Rutaceae	Mukthrubi	Purification of blood
Citrus aurantiifolia	Rutaceae	Champra	Bronchitis
Azadirachta indica	Meliaceae	Neem	Toothache
Centella asiatica	Umbelliferae	Paruck	Reduce high blood pressure
Leucas aspera	Lamiaceae	Mayanglambum	Fever-reducing
Cissumpleous periara	Menispermaceae	—	Skin diseases

TABLE 5.2 Common Medicinal Plant or Plant Parts Available at Ima Keithel, Imphal, Manipur, India

Name	Family	Local Name	Medicinal Values
Sechium edule	Cucurbitaceae	Maiscot	Leaves tea used to dissolve kidney stone
Saccharum officinarum	Poaceae	Chu	Stem juice is used to treat sore throats
Phylantus embelica	Phyllanthaceae	Heikru	Source of vitamin C and digestive agents
Ananas comosus	Bromeliaceae	Kihom	Anti-inflammatory
Oryza sativa	Poaceae	Chahou	Stopping diarrhea
Nelumbo nucifera	Neloumbonaceae	Thamchet	Helps stop diarrhea
Nelumbo nucifera	Neloumbonaceae	Thambou	Astringent
Parkia javanica	Fabaceae	Yongchak	Bleeding piles
Pinus roxburgaii	Pinaceae	Ou chand	Antiseptic
Hedychium flavum	Zingiberaceae	Loklei	Aromatic
Elsholtzia blanda	Lamiaceae	Lomba	Aromatic
Phlogacanthus thyrsiflorus	Acanthaceae	Nongmankha angangba	Fevers
Manihot escalenta	Euphorbiaceae	U-mangra	Antifungal
Capsicum annum	Solanaceae	U- morok	Antioxidant
Allium odorum	Amaryllidaceae	Maroi nakupi	Hemolytic anemia
Neptunia oleracea	Fabaceae	Esing ekaithabi	Remedy for necrosis in bones
Elaeocarpus floribundus	Elaeocarpaceae	Chorphon	Mouthwash
Alocasia cucullata	Araceae	Palukabi/Singju-pan	Purify blood
Clerodendrum serratum	Verbenaceae	Moirang Khanambi	Dysentery, asthma
Eryngium foetidum	Apiaceae	Awa-phadigom	Arthritis
Ocimum basilicum	Lamiaceae	Naosek-lei	Fever
Oroxylum indicum	Bignoniaceae	Shamba	Tonsil
Polygonum orientale	Polygonaceae	Yellang	Tonic and against headache
Polygonum posumbu	Polygonaceae	Phak-pai	Heartbeat increases
Psophocarpus tetragonolobus	Papilionaceae	Tengnou-manbi	Cough
Rhussemi alata	Anacardiaceae	Heimang	Intestinal worms, hair care
Sesbania grandiflora	Papiplionaceae	Chuchu-rangmei	Diabetes
Zanthoxylum acanthopodium	Rutaceae	Mukthrubi	Fever, cough, and bronchitis
Ludwigia adscendens	Onagraceae	Ising kundo	Skin complaints

TABLE 5.2 *(Continued)*

Name	Family	Local Name	Medicinal Values
Costus speciosus	Zingiberaceae	Khongbam takhelei	Cough
Lobelia nummularia	Campanulaceae	Kihom man	Stomach ache
Lindernia ruellioides	Linderniaceae	Duckbill lindernia	Wound
Mukia maderaspatana	Cucurbitacea	Nom Bil	Aching bodies
Bamboo spp.	Poaceae	Wa	Anti-inflammatory
Glochidion manipurensis	Phyllanthaceae	—	—
Spilanthes paniculata	Asteraceae	Phakphet	Infection of throat and gum
Agrimonia eupatoria	Rosaceae	—	Heal wounds
Cissampelos pareira	Menispermaceae	—	Reduce fever
Fragaria vesca	Rosaceae	—	Tooth cleaner
Sonerila erecta	Melastomataceae	Erect sonerila	—
Nicandra physalodes	Solanaceae	—	Intestinal pain of worms
Utricularia aurea	Lentibulariaceae	—	Heal wound
Oroxylum indicum	Bignoniaceae	Shamba	Tanning and dyeing
Urena lobata	Malvaceae	Sampakpi	Diarrhea
Scutellaria discolor	Lamiaceae	Yenakha	Indigestion
Merremia hederacea	Convolvulaceae	Komolata	Burns and scalds
Dioscorea alata	Dioscoraceae	Haa	Tumors
Plumbago zeylanica	Plumbaginaceae	Telhidak angouba	Dysentery

TABLE 5.3 Distribution of Selected 10 Medicinal Plants in the Urban Areas of Imphal, India

Plant species	Location(s)
Azadirachta indica	Kangla Campus, Manipur University Campus, Lamphel, DM College Campus
Oroxylum indicum (Plate 5.1H)	Khongampat, Lamphel, Dinku Road, Noimizing hill, Singjmai hill
Plumbago zeylanica	IBSD campus, Khongampat, Iroisemba
Centella asiatica	IBSD campus, Porompat, Takyelpat, Lamphel pat
Piper longum	Khongampat, Nomizing hill, Langol hill
Hedychium coronarium	Khongampat, Chingmeirong hill, Nomizing hill
Andrographis paniculata	Khongampat, Langol hill, Chingmeirong hill
Leucas aspera	Iroisemba, Langol hill, Kangla campus
Tinospora cordifolia	Khongampat, Iroisemba, Canchipur MU
Acorus calamus	Iroisemba, Lamphelpat, Takyel

KEYWORDS

- **conservation**
- **Ima Keithel**
- **Imphal**
- **medicinal plants**

REFERENCES

Christian, R. V.; Brigitte, V. L. Tool and method for data collection in ethnobotanical studies of home gardens. *Field Method*, **2004,** *16*(3), 285–306.

Debbarma, M.; Pala, N. A.; Kumar, M.; Bussmann, R. W. Traditional knowledge of medicinal plants in tribes of Tripura in northeast India. *African Journal of Traditional, Complementary and Alternative Medicines*, **2017,** *14*(4), 156–168.

Devi, O. S.; Komar, P.; Das, D. A checklist of traditional edible bio-resources from IMA markets of Imphal valley, Manipur, India. *Journal of Threatened Taxa*, **2010,** *2*(11), 1291–1296.

Devi, O. S.; Komor, P.; Das, D. A checklist of traditional edible bio-resources from IMA market of Imphal Valley, Manipur, India. *Jott. Short Communication*, **2010,** *2*(11), 1291–1296.

Dociu, M.; Dunarintu, A. The socioeconomic impact of urbanization. *International Journal of Academic Research in Accounting, Finance, and Management Science*, **2012,** *2*(1), 47–52.

Imam, A. U. K.; Banerjee, U. K. Urbanization and greening of Indian cities: Problems, practices, and policies. *Ambio*, **2016,** *45*(4), 442–457.

Kadi, A. S.; Nelavigi, K. P. Growth of Urbanization in India. *The International Journal of Science and Technology*, **2015,** *3*(7), 2321–919X.

Kavita, B. D.; Gayathri, N. K. Urbanization in India. *International Journal of Scientific Research and Education*, **2017,** *5*(1), 6166–6168.

Kelly, F. J.; Fussell, J. C. Air pollution and public health: Emerging hazards and improved understanding of risk. *Environ. Geochem. Health*, **2015,** *37*(4), 631–649.

Leishangthem, S.; Sharma, L. D. Study of some important medicinal plants found in Imphal-East district, Manipur, India. *International Journal of Scientific and Research Publication*, **2014,** *4*(9), 1–5.

Lokho, A. The folk medicinal plant of the Mao Naga in Manipur North East India. *International Journal of Scientific and Research Publications*, **2012,** *2*(6), 2250–3153.

Mahanta, P. P. Biodiversity hotspots. *Exotic Northeast*, **2015,** *1*(2), 2455–2135.

Mao, A. A.; Hynniewta, T. H.; Sanjappa, M. Plant wealth of Northeast India with reference to ethnobotany. *Indian Journal of Traditional Knowledge*, **2009,** *8*(1), 96–103.

Mathur, A. A winning strategy for India's North-East. *Jindal Journal of International Affairs*, **2011,** *1*(1), 269–298.

Ningombam, D. S.; Singh, P. K. Ethnobotanical study of *Phologacanthus thyrsformis* need: A conserved medicinal plant of Manipur, Northeast India. *International Journal of Herbal Medicine*, **2014,** *1*(5), 10–14.

Satterthwaite, D.; Granahan, G. M.; Tacoli, C. Urbanization and its implications for food and farming. *Philos Trans R. Soc. Lond. B Biol. Sci.;* **2010,** *365*(1554), 2809–2820.

Singh, A. S. Conservation of biodiversity of Manipur: Sustainable management strategies. *International Journal of Trend in Research and Development*, **2016,** *3*(3), 2394–9333.

Yadav, P. K.; Sharma, K.; Kumar, R. A framework for assessing the impact of urbanization and population on Garo hills landscape of North-East India. *International Journal of Conservation Science*, **2013,** *4*(2), 212–222.

Yumnam, J. Y.; Tripathi, O. P. Traditional knowledge of eating raw plants by the Meitei of Manipur as medicine/nutrient supplement in their diet. *Indian Journal of Traditional Knowledge*, **2012,** *11*(1), 45–50.

CHAPTER 6

Endophytic Fungal Diversity in Selected Medicinal Plants of Western Ghats of India

FAZILATH UZMA[1] and SRINIVAS CHOWDAPPA[2]

[1]Fungal Metabolite Research Laboratory, Department of Microbiology and Biotechnology, Bangalore University, Jnana Bharathi Campus, Bangalore–560056, Karnataka, India, E-mail: faziuzma@gmail.com

[2]Professor, Fungal Metabolite Research Laboratory, Department of Microbiology and Biotechnology, Bangalore University, Jnana Bharathi Campus, Bangalore–560056, Karnataka, India, Tel.: 9972091611/080–22961624; E-mail: srinivasbub@gmail.com

ABSTRACT

Endophytic communities are ubiquitous and colonize the internal tissues of host plants without causing any evident negative effects. The diversity studies the endophytic fungal community helps in discovering new species producing new bioactive compounds with industrial and therapeutic potential and also helps to understand of the role of endophytic fungal communities in the forest ecosystems. In the present study, 131 endophytic fungal isolates were isolated from different plant parts such as leaf, petiole, stem, and root from six medicinal plants used in traditional folklore medicine and procured from the Bisle region in the Hassan District, Western Ghats of Karnataka during the monsoon season of 2013–2014. Among all the endophytes, species of Aspergillus, Cladosporium, Fusarium, Alternaria, and Colletotrichum were the most frequently isolated. Diversity analysis of the fungal isolates was studied using Simpson's dominance index, Simpson's diversity index, Shannon–Weiner index, species richness, and evenness.

6.1 INTRODUCTION

Endophytic fungal communities are ubiquitous, highly diverse and colonize internal plant tissues in a symbiotic association without causing any notable signs of disease to the host plants (Hyde, Soytong, 2008). Approximately one million endophytic species are present in the plant kingdom (Kusari, Lamshöft, & Spiteller, 2009). Endophytic fungi are known to be a storehouse of various novel bioactive compounds having enormous industrial and pharmaceutical prospects (Wang, Dai, 2011). Endophytes improve the host plant's resistance to adversity by secretion of bioactive metabolites which find enormous applications as agrochemicals, antibiotics, antioxidants and anticancer agents (Pinheiro, et al., 2013).

The Western Ghats of India, also known as Sahyadri, starts from the southern tip of Gujarat and extending from the Satpura range in the north and traverses through the states of Maharashtra, Goa, Karnataka, and Kerala, culminating at Kanyakumari, Kerala with an area of 1,600 km. It is a UNESCO world heritage site with an enormous wealth of biological diversity comprising of over 6,000 species of higher plants which also includes an estimated 2,000 endemic species extensively used in traditional medicine practices (Nalini, Sunayana, Prakash, 2014).

From ancient ages, medicinal plants have played an important role in treating and preventing many diseases throughout the world. Several studies have indicated a possible likelihood of medicinal plants hosting endophytic fungi capable of producing pharmacologically important natural products. It can be reasoned that the medicinal properties of plants could be due to the endophytes which reside within them, thereby directing more studies on the plant-endophyte interactions for beneficial compounds. This interaction is controlled by the genes of the host and the endophyte and is modulated by the environment (Moricca, & Ragazzi, 2008). Hence, medicinal plants have been recognized as a repository of fungal endophytes producing novel bioactive metabolites. The ethnomedicinal plants from the Western Ghats of India, widely used in traditional medicinal practices have divulged the presence of diverse endophytic fungal communities (Naik, Shashikala, & Krishnamurthy, 2008). Diversity studies on the endophytic fungal communities from different medicinal plants belonging to Western Ghats of Karnataka have been conducted by a few research groups. This book chapter puts forth some of the diversity work on fungal endophytes for finding novel bioactivities and their colonization pattern during the monsoon season from the Western Ghats of Karnataka. This study is an effort for finding the potential of the endophytic fungi from Western Ghats of Karnataka for their diversity analysis and their importance in the fungal ecosystems.

6.1.1 DIVERSITY OF ENDOPHYTIC FUNGI IN MEDICINAL PLANTS

The endophytic fungi inhabit unique biological niche and are capable of colonizing healthy tissues of plants (Chowdhary, & Kaushik, 2015). Most plant species that have been previously studied host at least one endophytic microbe with the host plants growing in unique environmental settings generally hosting novel endophytic microorganisms (Ryan, et al., 2008). Diversity in endophytic fungal community exists within the host tissues of the selected plant and among the geographically separated individuals of the same host species (Uzma, Konappa, & Chowdappa, 2016). Variation in the fungal diversity may also be associated with the location, climate, and plant age (Petrini, 1991). Diversity analysis of the endophytic fungal assemblages is an emerging challenge, which leads to the discovery of new species producing novel compounds and a better understanding of their role in ecosystems (Rodriguez, et al., 2009).

The study of the diversity of endophytic fungi adds a new dimension to the available literature of diversity of fungal species. It has been reported that the ratio of fungal species: plant species is 33:1 as opposed to six fungi per plant which further necessitates the screening of potential endophytic fungi from diverse habitats (Hawksworth, & Rossman, 1997). Fungal endophytes colonize in the tissues of all plants, regardless of their taxonomical association and environmental preferences. Structure and composition of fungal communities are known to be influenced by many factors such as geographic locations, climatic patterns, seasonality, host plant identity, structure and diversity of surrounding vegetation, physiology and specificity of the colonized tissues. The mechanisms of interactions between the host plant and the endophytic organisms are not completely understood. The abundance of endophytic fungal populations and their composition in host plants could be affected by anatomy and maturity of colonized host tissues apart from the external environmental factors like humidity and temperature. Also, the endophytic community may change in relation to leaf age and type of plant tissue (Fernandes, et al., 2011).

Plants which grow in biodiversity hotspots or with a rich ethnobotanical background have been subjected for the isolation of endophytic fungi, as they are most likely to yield new strains with novel bioactivities. However, an understanding of the diversity is an important pre-requisite before exploring endophytic fungi for industrially important metabolites (Strobel, 2014). Also, the studies on the diversity of endophytes from different plants collected from the same site could also address any discrepancies regarding the presumed nature of endophytes. The endophytic composition of a plant clearly depends on host identity as well as on the geographic location of the host. The composition of endophytes in a monospecific host is affected strongly by local climatic conditions. The species richness of endophytes inhabiting plants in tropical rain forests outnumbers that of temperate forests, although less host specificity is seen (Hoffman, & Arnold, 2010).

6.2 MATERIALS AND METHODS

6.2.1 *SOURCE OF MEDICINAL PLANTS USED FOR THE STUDY*

The medicinal plants selected for the present study are *Tinospora cordifolia* (Willd.) Hook F. and Thomson (*Menispermaceae*), *Piper nigrum* L., *Piper longum* L. (Piperaceae), *Zingiber officinale* Roscoe, *Hedychium coronarium* J. Koenig and *Hedychium flavescens* Carey ex Roscoe (*Zingiberaceae*). Fungal endophytes were obtained from the uninfected/disease-free/aseptic tissues of six wild medicinal plants collected from Bisle region, Hassan district, Western Ghats of Karnataka. Herbarium of plant samples was deposited to National Ayurveda Dietetics Research Institute (Central Council for Research in Ayurveda and Siddha), Department of AYUSH, Ministry of Health and Family Welfare, Govt. of India, (New Delhi) Jayanagar, Bangalore, India. The plant samples were collected during the monsoon season (July-October) in the year 2013–14. The sampling and the subsequent endophytic fungal isolations were repeated thrice.

6.2.2 *ISOLATION AND MAINTENANCE OF ENDOPHYTIC FUNGI*

The collected medicinal plant samples were cut into 0.5 cm^2 segments and were surface sterilized using sodium hypochlorite-ethanol surface sterilization techniques (Nithya, & Muthumary, 2010). The effectiveness of the sterilization procedure was confirmed by the vitality test (Schulz, et al., 1993). The isolates were induced for sporulation by culturing them on Potato Dextrose Agar (PDA). Non-sporulating cultures were grouped under mycelia sterilia and were investigated for colony morphology (Gangadevi, & Muthumary, 2008). In this study, the surface sterilized segments from every single tissue were placed on a 9 cm petri dish containing PDA plates amended with 50 mg/L tetracycline for bacterial growth inhibition and incubated at 24±2°C for 2 to 3 days until the fungal hyphae emerged. The hyphal tips of emergent endophytic fungi from the plant tissues were transferred to fresh PDA plates to obtain pure cultures. After incubation at 24±2°C for 7 to 14 days, the purity of the culture was determined by colony morphology (Uzma, Konappa, & Chowdappa, 2016).

6.2.3 IDENTIFICATION OF THE ENDOPHYTIC FUNGI BASED ON MORPHOLOGICAL FEATURES

The endophytic fungal identification was performed by mounting the fungi on a clean glass slide with lactophenol cotton blue stain and edges were sealed with DPX mountant. The fungi were identified based on their cultural characteristics, the morphology of the fruiting bodies and spores using standard manuals (Barnett, & Hunter, 1998).

6.2.4 FREQUENCY OF ENDOPHYTIC FUNGI

The absolute frequency and relative frequency were calculated according to Ref. (Larran, et al., 2002). The isolation rate and colonization rate of the fungal endophytes from the monsoon season were calculated as per our previous study; wherein similar diversity parameters were used for the endophytic isolates of the winter season (Huang, et al., 2008; Mahapatra, & Banerjee, 2010; Uzma, Konappa, & Chowdappa, 2016).

$$\mathrm{IR} = \frac{\text{number of isolates obtained from tissue segments}}{\text{total number of tissue segments}}$$

$$\mathrm{CR} = \frac{\text{total number of isolates obtained from different tissue segments}}{\text{total number of isolates obtained from overall tissue segments incubated}}$$

6.2.5 STATISTICAL ANALYSIS OF THE FUNGAL ENDOPHYTES

Simpson's diversity index, Simpson's dominance index (D), species richness (S), Shannon-Wiener index (H), and evenness (E) were calculated (Jena, & Tayung, 2013; Uzma, Konappa, & Chowdappa, 2016).

6.3 RESULTS

6.3.1 ENDOPHYTIC FUNGI ISOLATED DURING MONSOON SEASON

About 131 fungal endophytic isolates were obtained from the six selected medicinal plants during the monsoon season. The medicinal plant *T. cordifolia* gave 20 isolates (5 from leaves, 3 from petiole, 6 from stem and 6 from roots). The *P. nigrum* gave 22 isolates (7 from leaves, 6 from petiole, 5 from stem, and 4 from roots). The *P. longum* plant produced 22 endophytic isolates: 6 from leaves, 4 from petiole, 6 from stem, and 6 from roots). A total of 24 isolates were obtained from *Z. officinale*, of which 4 were from leaves, 6 from petiole, 6 from stem (rhizome) and 8 were from the adventitious roots. Twenty-five isolates were obtained from *H. flavescens*: 8 from leaves, 4 from petiole, 6 from stem, and 7 from roots. The *H. coronarium* gave 18 isolates (4 from leaves, 3 from petiole, 3 from stem, and 8 from roots).

6.3.2 DIVERSITY STUDIES OF ENDOPHYTIC FUNGI FROM THE MEDICINAL PLANTS DURING MONSOON SEASON

A total of 131 isolates were recovered from 480 tissue segments of six medicinal plants. They were grouped into 24 fungal taxa, consisting of hyphomycetes (41.66%), ascomycetes (33.33%), coelomycetes (16.16%) and zygomycetes (8.33%) based on their morphology. The extent of colonization varied among the different plant parts studied. The leaves harbored more endophytic fungi than the stem, petiole, and roots. The CR differed among the six plant species. The fungal taxa Aspergillus, Cladosporium, and Trichoderma were the most frequently occurring, followed by Bipolaris, Cylindrocephalum, Lasiodiplodia, Nigrospora, Pestalotiopsis and Phoma which occurred with low frequency. The endophytic fungal isolates demonstrated significant isolation rates (IR) and colonization rates (CR) (Table 6.1). The IR was high for *H. flavescens* than the other studied medicinal plants. The overall colonization rate was high in the leaves of *P. nigrum* (31.81%) and in the roots of *Z. officinale* (33.33%) when compared to the tissue segments of the other selected plants. The absolute and relative frequencies of occurrence of each endophytic fungal species were calculated and are depicted in Table 6.2. Diversity indices of fungal endophytes varied between plant species as well as within tissue fragments. The values of diversity indices suggest that the endophytic colonization in the tissues of the medicinal plants were even, indicating uniform occurrence of various species.

TABLE 6.1 Endophytic Fungal Colonization and Isolation Rates in the Six Medicinal Plants During Monsoon Season

Sl. No.	Medicinal plants selected for the study	Tissue sections	Rate of colonization CR (%)	Rate of isolation (IR)
1	*Tinospora cordifolia* (Willd.) Hook.f & Thomson	Leaf	25	0.25
		Petiole	15	0.15
		Stem	30	0.3
		Roots	30	0.3
2	*Piper nigrum* L.	Leaf	31.81	0.35
		Petiole	27.27	0.3
		Stem	22.72	0.25
		Roots	22.72	0.25
3	*Piper longum* L.	Leaf	27.27	0.3
		Petiole	18.18	0.2
		Stem	27.27	0.3
		Roots	27.27	0.3
4	*Zingiber officinale* Roscoe	Leaf	16.66	0.2
		Petiole	25	0.3
		Stem	25	0.3
		Roots	33.33	0.4
5	*Hedychium flavescens* Carey ex Roscoe	Leaf	32	0.4
		Petiole	16	0.2
		Stem	24	0.3
		Roots	28	0.35
6	*Hedychium coronarium* J. Koenig	Leaf	22.22	0.2
		Petiole	16.66	0.15
		Stem	16.66	0.15
		Roots	44.44	0.4

6.3.2.1 FUNGAL ENDOPHYTES OBTAINED FROM *TINOSPORA CORDIFOLIA*

Out of the 80 segments incubated, we obtained 20 endophytic fungal isolates from *T. cordifolia* (5 from leaves, 3 from petiole, 6 from stem and 6 from roots) were clustered into 9 genera, of which *Aspergillus* sp., *Cladosporium* sp., *Penicillium* sp. and *Trichoderma* sp. occurred with high frequency (2.29%), followed by *Alternaria* sp., *Curvularia* sp., and *Fusarium* sp. (1.52%) which had low frequency of occurrence and found in only one tissue type. The colonization rate is higher in stem and root tissues (30%) than in the other tissues (Table 6.1). The species richness of the fungal isolates was greater in stem and roots (6) compared to the leaf and petiole segments produced five and three endophytic species respectively (Table 6.2). The diversity of the endophytic isolates associated with different tissue segments of *T. cordifolia* was compared using diversity indices. Simpson's diversity indices and Shannon-Wiener index was higher in stem and root tissues. The species evenness revealed minute differences amongst the different tissue types studied (Table 6.3).

6.3.2.2 FUNGAL ENDOPHYTES ISOLATED FROM PIPER NIGRUM

Twenty-two isolates were obtained from 80 tissue segments of *P. nigrum* (7 from leaves, 6 from petiole, 5 from stem, and 4 from roots). They were grouped into 14 genera. The most frequently occurring fungal endophytes were *Cladosporium* sp. and *Rhizopus* sp. (2.29%) and were obtained from more than one tissue type. The CR was high in leaves (31.81%) followed by petiole (27.27%), and the IR was also greater in leaves (Table 6.1). The leaves depicted high species richness (7) followed by petiole and stem (Table 6.2). The species of *Acremonium, Alternaria, Aspergillus, Curvularia, Trichoderma, Fusarium* and *Phoma* were obtained only once. Results of the diversity indices reveal that Simpson's dominance of endophytic fungi is higher in the petiole and roots.The Simpson and Shannon-Wiener's diversity indices were higher in leaf tissues followed by stem tissues (Table 6.3).

6.3.2.3 ENDOPHYTIC FUNGI ISOLATED FROM PIPER LONGUM

More number of isolates (i.e., 22) was obtained from 80 segments of *P. longum;* 6 from leaves, 4 from petiole, 6 from stem and 6 from roots. Thirteen genera were identified from these isolates, amongst which *Alternaria* sp. and *Trichoderma* sp. were the most commonly occurring and were obtained from more than one tissue type. The frequency of occurrence of these endophytes was 2.29%, respectively. The highest colonization rate was observed in leaf, stem and root tissues (27.27%), followed by petiole tissues (18.18%) of *P. longum* (Table 6.1). The endophytic *Acremonium* sp. and *Mucor* sp. had a lower frequency of occurrence (0.76%). The highest number of species (6) was found in the tissue segments of leaves, stem, and roots. The petiole had four endophytic species (Table 6.2). Higher Simpson's dominance of

TABLE 6.2 Species Richness, Absolute (f) and Relative Frequency (fr) of Endophytic Fungi of the Medicinal Plants During Monsoon Season

Sl. No	Fungal endophytes	Tinospora cordifolia				Piper nigrum				Piper longum				Zingiber officinale				Hedychium flavescens				Hedychium coronarium				Absolute frequency (f)	Relative frequency fr (%)
		L	P	S	R	L	P	S	R	L	P	S	R	L	P	S	R	L	P	S	R	L	P	S	R		
1	*Acremonium* sp.						1			1							1						1		1	5	3.816
2	*Alternaria* sp.		1	1				1		1	1		1			1			1			1				9	6.870
3	*Aspergillus* sp.		1	1	1				1	1		1		2		1		2	1	2	1	1		1	1	18	13.740
4	*Bipolaris* sp.																	1								1	0.763
5	*Chaetomium* sp.	1				1		1				1					1								1	6	4.580
6	*Cladosporium* sp.	1	1		1	1		1	1	1		1			2	2	1	1	1	1						16	12.213
7	*Colletotrichum* sp.					1	1						1			1			1	1			1	1		8	6.106
8	*Curvularia* sp.	1		1				1				1	1									1				6	4.580
9	*Cylindrocephalum* sp.																	1								1	0.763
10	*Fusarium* sp.			1	1		1						1		1		1				1			1	1	9	6.870
11	*Lasiodiplodia* sp.																				1					1	0.763
12	*Mucor* sp.				1	1	1				1										1					5	3.816
13	Mycelia sterilia					1				1		1		1	1	1				1	1	1			1	10	7.633
14	*Myrothecium* sp.																1			1						2	1.526
15	*Nigrospora* sp.																								1	1	0.763
16	*Penicillium* sp.	1		1	1								1	1				1								6	4.580
17	*Pestalotiopsis* sp.																	1								1	0.763

TABLE 6.2 *(Continued)*

Sl. No	Fungal endophytes	Tinospora cordifolia				Piper nigrum				Piper longum				Zingiber officinale				Hedychium flavescens				Hedychium coronarium				Absolute frequency (f)	Relative frequency fr (%)
		L	P	S	R	L	P	S	R	L	P	S	R	L	P	S	R	L	P	S	R	L	P	S	R		
18	*Phoma* sp.							1																		1	0.763
19	*Phomopsis* sp.														1		1									2	1.526
20	*Rhizopus* sp.					1	1		1		1		1					1							1	7	5.343
21	*Sordaria* sp.					1	1																			2	1.526
22	*Torula* sp.																1									1	0.763
23	*Trichoderma* sp.	1		1	1				1	1	1	1			1		1				1		1		1	12	9.160
24	*Xylaria* sp.																				1					1	0.763
	Total	5	3	6	6	7	6	5	4	6	4	6	6	4	6	6	8	8	4	6	7	4	3	3	8	**131**	99.989
	Species richness	5	3	6	6	7	6	5	4	6	4	6	6	3	5	5	8	7	4	5	7	4	3	3	8		

TABLE 6.3 Endophytic fungal diversity indices studied from selected medicinal plants during the monsoon season

Sl. No.	Indices	Tinospora cordifolia				Piper nigrum				Piper longum			
		L	P	S	R	L	P	S	R	L	P	S	R
1	Simpson's dominance index	0.200	0.333	0.162	0.162	0.142	0.250	0.200	0.250	0.162	0.250	0.162	0.162
2	Simpson's diversity index	0.800	0.667	0.838	0.838	0.857	0.750	0.800	0.750	0.838	0.750	0.838	0.838
3	Species Richness	5	3	6	6	7	6	5	4	6	4	6	6
4	Shannon-Wiener	1.609	1.098	1.791	1.791	1.945	1.386	1.609	1.386	1.791	1.386	1.791	1.791
5	Evenness	0.999	0.999	0.999	0.999	0.999	0.773	0.999	0.999	0.999	0.999	0.999	0.999

endophytic fungi was observed in the petiole. The diversity indices namely Simpson's and Shannon-Wiener were higher in the tissues of leaves, stem, and roots followed by petiole (Table 6.3).

6.3.2.4 ENDOPHYTIC FUNGI ISOLATED FROM ZINGIBER OFFICINALE

The medicinal plant *Z. officinale* was assessed with 24 isolates (4 from leaves, 6 from petiole, 6 from rhizome, and 8 from adventitious roots which were grouped into 13 genera. The endophytic *Cladosporium* sp. (3.81%) was the commonly isolated endophyte found in three tissue types (Petiole, rhizome and adventitious roots). The other frequently occurring endophytes were *Aspergillus* sp. and *Mycelia sterilia* (2.29%). The colonization rates were higher in adventitious roots (33.33%) followed by petiole and stem (25%) (Table 6.1). The endophytic *Acremonium* sp., *Chaetomium* sp., *Colletotrichum* sp., *Penicillium* sp., *Myrothecium* sp. and *Torula* sp. occurred with low frequency (0.76%) and were obtained only once from each tissue type. Higher species richness was seen in the adventitious roots (8), followed by the petiole and rhizome (Table 6.2). The endophytic abundance varied among the different tissue segments of *Z. officinale.* The adventitious roots presented with a greater number of species (8) whereas the petiole and rhizome had six endophytic species. Simpson's dominance of endophytic fungi was higher in the petiole and rhizome of *Z.officinale*. High Simpson and Shannon-Wiener's diversity indices were observed in adventitious roots followed by petiole and rhizome (Table 6.4). The tissues of *Z. officinale* exhibited uniform evenness.

6.3.2.5 ENDOPHYTIC FUNGI ISOLATED FROM HEDYCHIUM FLAVESCENS

The highest numbers of endophytic fungal isolates (25) were obtained from *H. flavescens* (8 from leaves, 4 from petiole, 6 from stem, and 7 from roots). The roots (28%) and stem (24%) depicted high colonization rates compared to the other tissues of *H. flavescens* (Table 6.1). The species of *Aspergillus* occurred in high frequency of 4.58%. The endophytic fungal abundance varied in different tissues. The different tissues exhibited uniform evenness with a marginal difference. High species richness was depicted in the leaves (Table 6.2). The petiole tissues demonstrated high Simpson's dominance whereas the leaves and roots of *H. flavescens* depicted high Simpson and Shannon-Wiener's diversity indices (Table 6.4).

TABLE 6.4 Endophytic fungal diversity indices studied from selected medicinal plants during the monsoon season

Sl. No	Indices	Zingiber officinale				Hedychium flavescens				Hedychium coronarium			
		L	P	Rh	Ad. R	L	P	S	R	L	P	S	R
1	Simpson's dominance index	0.375	0.219	0.219	0.125	0.156	0.250	0.219	0.142	0.250	0.333	0.333	0.125
2	Simpson's diversity index	0.625	0.781	0.781	0.875	0.844	0.750	0.781	0.857	0.750	0.666	0.666	0.875
3	Species Richness	3	5	5	8	7	4	5	7	4	3	3	8
4	Shannon-Wiener	1.039	1.560	1.560	2.079	1.905	1.386	1.560	1.945	1.386	1.098	1.098	2.079
5	Evenness	0.946	0.969	0.969	0.999	0.979	0.999	0.969	0.999	0.999	0.999	0.999	0.999

L–Leaf, P–Petiole, S–stem, R–root, Rh–Rhizome, Ad. R–Adventitious roots.

6.3.2.6 FUNGAL ENDOPHYTES OBTAINED FROM HEDYCHIUM CORONARIUM

The aromatic medicinal plant *H. coronarium* was used for isolation of endophytic fungi (4 from leaves, 3 from petiole, 3 from stem, and 8 from roots) which were grouped into 11 genera. The colonization rates were highest in roots (44.44%) and leaves (22.22%) of *H. coronarium* (Table 6.1). The number of different endophytic species occurring in the roots was higher than the other parts of *H. coronarium* whereas the evenness was uniform in different tissues. The genus *Aspergillus* was the most frequently isolated (3 isolates) in the leaf, stem and root tissues. The endophytic *Acremonium, Colletotrichum, Fusarium, Mycelia sterilia* and *Trichoderma* also had a good frequency of occurrence (1.52%) when compared to other endophytic fungal species of *H. coronarium* (Table 6.2). The occurrence and abundance of endophytic fungal species were diverse in different tissues. The number of isolates in the roots was high (8 isolates) followed by leaves (4 isolates), petiole and stem (3 isolates). The endophytic isolates of leaves depicted high Simpson's dominance.The Simpson and Shannon-Wiener's diversity indices were higher in roots and leaves when compared to petiole and stem. The species richness was higher in the roots (Table 6.4).

6.4 DISCUSSION

The Western Ghats of India is one of the hot spots of global biodiversity and is reported to have a diverse population of endophytic fungi (Raviraja, 2005; Naik, Shashikala, & Krishnamurthy, 2008; Krishnamurthy, Shashikala, & Naik, 2009). Limited studies on the endophytic fungi of these plants have been conducted. This study explores the endophytic fungal diversity and their colonization pattern in medicinal plants used by the local people for curing various human diseases in the Bisle region of Hassan district, Western Ghats of Karnataka, Southern India. Previously, we had reported the diversity of endophytic fungal isolates from the winter season of the six medicinal plants from the Bisle region, Western Ghats of Karnataka (Uzma, Konappa, & Chowdappa, 2016). This study is an attempt to understand the endophytic fungal assemblages in the members of *Hedychium* (Zingiberaceae), recorded as critically endangered species of India in the Red data book (Manish, 2013) through the different seasons since endophytic fungal isolations from the plants belonging to *Hedychium* sp. have not been reported to the best of our knowledge. Also, the studies on endophytic fungi from these plants have revealed differences in the colonization rates as well as diversity pattern which have not been documented so far. Association of endophytic fungi varies from plant to plant, geographical distribution and also different seasons. The Isolation Rate (IR), Colonization Rate (CR) and Relative Frequency (RF) of the endophytic fungi vary with different medicinal plants (Huang, et al., 2008). In the present study, the IR, CR, and RF of endophytic fungi varied with different tissue segments of the medicinal plants studied during the monsoon season. The difference in colonization by endophytes may be due to the substrate utilization and the physiological state of the host plants. The dominant taxa isolated in the present study *Aspergillus* sp., *Trichoderma* sp., *Alternaria* sp., *Cladosporium* sp., including the sterile forms (*Mycelia sterilia*) were obtained as endophytes in the monsoon season revealing that these fungi are well adapted to survive as endophytes (Kumar, & Hyde, 2004). In our study, the endophytic fungal isolates obtained from the monsoon season were compared with the diversity indices (Shannon-Wiener index and Simpson's diversity index), species richness and evenness. The Simpson's diversity index and their components viz. species richness and evenness were correlated in all the selected medicinal plants. The diversity index increased with a different number of endophytic isolates and species evenness. The colonization and isolation rates also differed considerably between the different tissues of the six medicinal plants. The colonization rate was higher in the leaves and roots of the medicinal plants than the stem and petiole. Our previous study revealed that the rate of colonization was higher in the leaves and stem tissues of these selected medicinal plants during the winter season (Uzma, Konappa, & Chowdappa, 2016). The Simpson's dominance index was high in the petiole tissues of *T. cordifolia, P. nigrum, P. longum, H. flavescens, H. coronarium* except for *Z. officinale* where the leaf tissues demonstrated high dominance index. Also, higher evenness was observed during the monsoon season which may be due to the high number of endophytic isolates obtained during the monsoon season compared to the winter season reported previously (Uzma, Konappa, & Chowdappa, 2016). The present study indicates the significant impact of the monsoon season on the endophytic fungal population. Maximum colonization during monsoon season suggests that environmental factors such as humidity and precipitation are positively correlated with endophytic colonization. Variation in the colonization among the plants may be due to certain site-specific factors. Differences in colonization in plants of a region may be due to the prevailing of diverse host genotypes along with different environmental conditions influencing infection by particular fungal endophytes (Krishnamurthy, Shashikala, & Naik, 2009). During the monsoon season, the colonization frequency (CF) was highest in the roots of *Z. officinale* (33.33%) and lowest in petiole of *T. cordifolia* (15%). The highest relative frequency (RF) was seen in *Aspergillus* sp. (13.74%) and *Cladosporium* sp. (12.21%). The endophytic fungal isolates such as *Colletotrichum, Chaetomium, Phomopsis,* and *Pestalotiopsis* produce slimy conidia that are not forcibly released but dispersed by water, which may account for their isolation in the monsoon season. A high number of endophytic isolates obtained during monsoon season suggests that the colonization of endophytes is associated with the climatic factors. The slimy conidia of fungal spores are dispersed better during wet climates such as rain, thereby accounting for more endophytic fungal isolations during monsoons. Also, the germination of conidia is influenced by climatic factors (Schulthess & Faeth 1998). The *Mycelia sterilia* are sterile forms, often isolated in endophytic studies and cannot be given taxonomic status (Promputtha, et al., 2005). Due to the existence of such non-cultivable endophytes, the real number endophytic species in a sample can be underestimated. Hence, molecular techniques have been used for phylogenetic classification of such sterile forms obtained as *Mycelia sterilia* (Neubert, et al., 2006). In our study, species of *Aspergillus, Cladosporium, Colletotrichum,* and *Fusarium* were the endophytes with high relative frequencies.

The diversity analysis of the monsoon season in the present study reveals that fungal endophytes were obtained in high numbers from the leaf, stem and root tissues of the selected medicinal plants and is supported by the previous work of Huang, et al. (2008), in which the fungal endophytes were more frequent in leaf and stem tissues. In contrast, the species composition of endophytic fungi from *Lippia sidoides* was studied, and the colonization of leaves (50.41%) was higher than that of stems (35.40%) was found (Siqueira, et al., 2011). In Brazil, 95 endophytic fungi from *Bauhinia forficate* were isolated and reported the highest frequency of colonization in the stems (Bezerra, et al., 2012). Species of *Aspergillus, Fusarium, Colletotrichum,* and *Cladosporium* were dominant in the present work, and it may be due to high spore production of these fungi and their cosmopolitan nature, which increases their chance to get established as endophytes (Raviraja, 2005). *Mycelia sterilia*, the fungal taxa which failed to sporulate are also reported in our present work from the three seasons. The *Acremonium, Colletotrichum, Chaetomium, Myrothecium, Phomopsis, Fusarium* and *Pestalotiopsis* spp. as the commonly isolated endophytes from medicinal plants of Western Ghats, Karnataka as per the reports. The reports on the diversity of endophytic fungi obtained from *Z. officinale*, *H. flavescens,* and *H. coronarium* are scarce. Hence, it is essential to screen novel plant species from new habitats for their endophytic assemblages in order to assess their host preferences and their bioactive metabolite production (Nalini, Sunayana, Prakash, 2014). The antagonistic activity of the endophytic actinomycetes of *Z. officinale* against phytopathogenic fungi has been reported earlier (Taechowisan, et al., 2013).

The endophytic fungal isolates were obtained from 15 medicinal shrubs during winter, summer and monsoon seasons. The genera of *Aschersonia, Botryosphaeria, Curvularia, Fusarium, Colletotrichum, Myrothecium, Penicillium, Phyllosticta,* and *Phomopsis* were most commonly isolated. More number of isolates was obtained during the winter season than the monsoon and summer seasons. Also, the high number of isolates obtained during winter and monsoon seasons suggests that the endophytic fungal colonization is correlated with climatic factors (Naik, Shashikala, & Krishnamurthy, 2008). Our study reveals that more number of endophytic isolates was obtained in the monsoon season than in the winter season.

Diversity indices for fungal endophytes analyzed by Shannon-Weiner and Simpson indices indicated differences in relative frequencies, colonization frequencies, and isolation rates. In the monsoon season, high diversity indices were observed for *T. cordifolia* and *P. longum*, whereas low diversity was seen in *H. coronarium*. The frequency of colonization in the tissues of leaves and stems was higher when compared to the roots and petiole. Species richness was predominant in leaves. Our work finds accordance with (Raviraja, 2005), who studied the medicinal plants of Kudremukh region, Western Ghats of Karnataka wherein the number of endophytic species was higher in leaf segments. The Shannon index increases as both the richness and the evenness of the community increase. Simpson's dominance and diversity were analyzed. As the dominance index increases, the diversity decreases. Species evenness refers to the relative closeness of each species in an environment. Species richness is a simple count of species, whereas species evenness quantifies how equal the abundances of the species are. Lesser variation in communities between the species reflects higher species evenness and is independent of species richness.

6.5 CONCLUSION

The present study provides information on colonization of endophytic fungi in six important medicinal plants from the Western Ghats of Karnataka and their diversity analysis. This is the first report on endophytic assemblages from *Hedychium flavescens* and *H. coronarium* during the monsoon season. The diversity analysis of fungal endophytes from medicinal plants helps to comprehend their ecological roles in diverse ecosystems. It provides insight on their biodiversity, plant protection, nutrition and novel bioactive compound production for biotechnological applications. A comparative study of the endophytic population during different seasons would help to understand the host-endophytic interactions, ecological significance, and plant defense mechanisms. Also, the novel strains recovered during the different seasons with novel bioactivities may be useful for agricultural and industrial utilization.

ACKNOWLEDGMENTS

The Financial assistance to Fazilath Uzma (F1–17.1/2012–13/MANF-2012–13-MUS-KAR-11899) granted by Maulana Azad National Fellowship (MANF), University Grants Commission (UGC), New Delhi is gratefully acknowledged.

KEYWORDS

- **diversity analysis**
- **endophytic fungi**
- **medicinal plants**
- **monsoon season**
- **Western Ghats**

REFERENCES

Barnett, H. L.; Hunter, B. B. Illustrated Genera of Imperfect Fungi, APS Press: St. Paul, Minnesota, USA, **1998**.

Bezerra, J. D. P.; Santos, M. G. S.; Svedese, V. M.; Lima, D. M. M.; Fernandes, M. J. S.; Paiva, L. M.; Souza-Motta, C. M. Richness of endophytic fungi isolated from *Opuntia ficus-indica* Mill. (Cactaceae) and preliminary screening for enzyme production. *World J Microbiol Biotechnol.* **2012,** *28*, 1989–1995.

Chowdhary, K.; Kaushik, N. Fungal endophyte diversity and bioactivity in the Indian medicinal plant *Ocimum sanctum* Linn. *Plos One.* **2015,** *10*(11).

Fernandes, G. W.; Oki, Y.; Sanchez–Azofeifa, A.; Faccion, G.; Amaro–Arruda, H. C. Hail impact on leaves and endophytes of the endemic threatened *Coccoloba cereifera* (Polygonaceae). *Plant Ecol.* **2011,** *212*, 1687–1697.

Gangadevi, V.; Muthumary, J. Taxol, an anticancer drug produced by an endophytic fungus *Bartalinia robillardoides* Tassi, isolated from a medicinal plant, *Aegle marmelos* Correa ex Roxb. *World J Microbiol Biotechnol.* **2008,** *4*(5), 717.

Hawksworth, D. L.; Rossman, A. Y. Where are all the undescribed fungi? *Phytopathol.* **1997,** *87*, 888–891.

Hoffman, M. T.; Arnold, A. E. Diverse bacteria inhabit living hyphae of phylogenetically diverse fungal endophytes. Appl Environ Microbiol. **2010,** 76(12), 4063–4075.

Huang, W. Y.; Cai, W. Z.; Hyde, K. D.; Corke, H.; Sun, M. Biodiversity of endophytic fungi associated with 29 traditional Chinese medicinal plants. *Fungal Divers.* **2008,** *33*, 61–75.

Hyde, K. D.; Soytong, K. The fungal endophyte dilemma. *Fungal Divers.* **2008,** *33*, 163–173.

Jena, S. K.; Tayung, K. Endophytic fungal communities associated with two ethnomedicinal plants of Similipal Biosphere Reserve, India, and their antimicrobial prospective. *J Appl Pharm Sci.* **2013,** *3*, S7–S12.

Krishnamurthy, Y. L.; Shashikala, J.; Shankar Naik, B. Diversity and seasonal variation of endophytic fungal communities associated with some medicinal trees of the Western Ghats, Southern India. *Sydowia*, **2009,** *61*(2), 255–266.

Kumar, D. S. S.; Hyde, K. D. Biodiversity and tissue-recurrence of endophytic fungi in *Tripterygium wilfordii. Fungal Divers*, **2004,** *17*, 69–90.

Kusari, S.; Lamshöft, M.; Spiteller, M. *Aspergillus fumigatus* Fresenius, an endophytic fungus from *Juniperus communis* L. Horstmann as a novel source of the anticancer prodrug deoxypodophyllotoxin. *J Appl Microbiol.* **2009,** *107*(3), 1019–30.

Larran, S.; Perelló, A.; Simón, M. R.; Moreno, V. Isolation and analysis of endophytic microorganisms in wheat (*Triticum aestivum* L.) leaves. *World J Microbiol Biotechnol.* **2002,** *18*, 683–686.

Mahapatra, S.; Banerjee, D. Diversity and screening for antimicrobial activity of endophytic fungi from *Alstonia scholaris. Acta Microbiol Immunol Hung.* **2010,** *57*(3), 215–223.

Manish, M. Current status of endangered Medicinal plant *Hedychium coronarium* and causes of Population decline in the natural forests of Anuppur and Dindori districts of Madhya Pradesh, India. *Int Res J Biol Sci-ISCA*. **2013,** *2*(3), 1–6.

Moricca, S.; Ragazzi, A. Fungal endophytes in Mediterranean oak forests: A lesson from *Discula quercina. Phytopathol.* **2008,** *98*, 380–386.

Naik, B. S.; Shashikala, J.; Krishnamurthy, Y. L. Diversity of fungal endophytes in shrubby medicinal plants of Malnad region, Western Ghats, Southern India. *Fungal Ecol.* **2008,** *1*(2), 89–93.

Nalini, M. S.; Sunayana, N.; Prakash, H. S. Endophytic Fungal Diversity in Medicinal Plants of Western Ghats, India. *Int J Biodivers*. **2014**.

Neubert, K.; Mendgen, K.; Brinkmann, H.; Wirsel, S. G. R. Only a few fungal species dominate highly diverse mycofloras associated with the common reed. *Appl. Environ. Microbiol.* **2006,** *72*, 1118–1128.

Nithya, K.; Muthumary, J. Secondary metabolite from *Phomopsis* sp. isolated from *Plumeria acutifolia* Poiret. *Recent Research in Science and Technology.* **2010,** *2*(4).

Petrini, O. Fungal endophytes of tree leaves, In *Microbial Ecology of Leaves* (Eds.) Andrew, I. A., & Hirano S. S., Springer-Verlag, New York, **1991,** 179–197.

Pinheiro, E. A.; Carvalho, J. M.; dos Santos, D. C.; Ade, O. F.; Marinho, P. S.; Guilhon, G. M.; de Souza, A. D.; da Silva, F. M.; Marinho, A. M. Antibacterial activity of alkaloids produced by endophytic fungus *Aspergillus* sp. EJC08 isolated from medicinal plant *Bauhinia guianensis*. *Nat Prod Res*. **2013,** *27*, 1633–1638.

Promputtha, I.; Jeewon, R.; Lumyong, S.; McKenzie, E. H. C.; Hyde, K. D. Ribosomal DNA fingerprinting in the identification of non-sporulating endophytes from *Magnolia liliifera* (Magnoliaceae). *Fungal Divers,* **2005,** *20*, 167–186.

Raviraja, N. S. Fungal endophytes in five medicinal plant species from Kudremukh Range, Western Ghats of India. *J Basic Microbiol.* **2005,** *45*, 230–235.

Rodriguez, R. J.; White, Jr., J. F.; Arnold, A. E.; Redman, R. S. Fungal endophytes: diversity and functional roles. *New Phytol.* **2009,** *182*(2), 314–330.

Ryan, R. P.; Germaine, K.; Franks, A.; Ryan, D. J.; Dowling, D. N. Bacterial endophytes: recent developments and applications. *FEMS Microbiol Lett.* **2008,** *278*(1), 1–9.

Schulthess, F. M.; Faeth S. H. Distribution, abundances, and association of the endophytic fungal community of Arizona fescue (*Festuca arizonica* Vasey). *Mycologia.* **1998,** *90*, 569–578.

Schulz, B.; Wanke, U.; Draeger, S.; Aust, H. J. Endophytes from herbaceous plants and shrubs: effectiveness of surface sterilization methods. *Mycol Res.* **1993,** *97*(12), 1447–1450.

Siqueira, V. M.; Conti, R.; Araújo, J. M.; Souza-Motta, C. M. Endophytic fungi from the medicinal plant *Lippia sidoides* Cham and their antimicrobial activity. *Symbiosis.* **2011,** *53*, 89–95.

Strobel, G. A. Methods of discovery and techniques to study endophytic fungi producing fuel-related hydrocarbons. Nat Prod Rep. **2014,** 31, 259–272.

Taechowisan, T.; Chanaphat, S.; Ruensamran, W.; Phutdhawong, W. S. Antibacterial activity of Decursin from *Streptomyces* sp. GMT-8; an endophyte in *Zingiber officinale* Rosc. *J Appl Pharma Sci.* **2013,** *3*, 074–078.

Uzma, F.; Konappa, N. M.; Chowdappa, S. Diversity and extracellular enzyme activities of fungal endophytes isolated from medicinal plants of Western Ghats, Karnataka. *Egyptian Journal of Basic and Applied Sciences,* **2016,** *3*(4), 335–342.

Wang, Y.; Dai, C. C. Endophytes: a potential resource for biosynthesis, biotransformation, and biodegradation. *Ann Microbiol.* **2011,** *61*, 207–215.

CHAPTER 7

Araceae of Agasthyamala Biosphere Reserve, South Western Ghats, India

JOSE MATHEW,[1] P. M. SALIM,[2] P. M. RADHAMANY,[1] and KADAKASSERIL VARGHESE GEORGE[3]

[1]Department of Botany, Kerala University, Kariavattom, Thiruvananthapuram–695581, Kerala, India, E-mail: polachirayan@yahoo.co.in, radhamanym@rediffmail.com

[2]M.S. Swaminathan Research Foundation, Puthoorvayal, Kalpetta, Wayanad–673 577, Kerala, India, E-mail: salimpichan@yahoo.com

[3]Department of Botany, St. Berchmans' College, Changanassery, Kottayam–686101, Kerala, India, E-mail: kvgeorge58@yahoo.in

ABSTRACT

The forests along the Southern Western Ghats are the most species-rich ecological regions in peninsular India with respect to species diversity and endemism. Agasthyamala Biosphere Reserve (ABR) in the southern Western Ghats is a major genetic estate with an enormous biodiversity of ancient lineage. The members of Araceae (Arum family) have a significant role in ethnopharmacognostic and biodiversity conservation point of view. A review of the family Araceae in the context of the proposed project "Kerala part of South Western Ghats endemic plants" has yielded several interesting specimens. This study has brought to light some novelties with special reference to new distributional record of *Arisaema madhuanum* and *Alocasia longiloba* to flora of Kerala. In the standard taxonomic treatment of Araceae of ABR, identification keys of all genera, their species characterization along with correct name, distribution, phenology, and conservation status are included.

7.1 INTRODUCTION

With respect to species diversity and endemism, the forests along the southern Western Ghats are the most species-rich ecological regions in peninsular India. About 80% of the flowering plant species of the entire Western Ghats are distributed in this ecological region. Studies conducted by earlier researchers have brought out more than 50 new flowering plant taxa from the Agasthyamala Biosphere Reserve of southern Western Ghats during the last three decades (Henry et al., 1984; Sivarajan, 1985; Sivadasan and Kumar, 1987; Kumar, 1989; Sivadasan, 1989; Sivadasan et al., 1989; Sivadasan and Mohanan, 1991; Mohanan and Henry, 1991; Kumar and Manilal, 1994; Sivadasan et al., 1994; Sivadasan et al., 1997; Sivadasan and Mohanan, 1999; Kumar et al., 2000, 2001, 2004, 2011; Mohanan and Sivadasan, 2002; Remadevi and Binojkumar, 2003; Sabu et al., 2013; Robi et al., 2013; Sujanapal et al., 2013; Robi et al., 2014; Sivu et al., 2014). In spite of these studies, many areas in the southern part of the Western Ghats, especially that of the Montane hill forest area still remains unexplored or underexplored.

This study aims to document the Aroid flora of the seasonally inundated tropical montane forests and grasslands of ABR part of southern Western Ghats based on extensive field studies. It is expected that intensive floristic studies giving due emphasis to correct identity, distribution, present status and extent of threat if any, endemism and the dynamism have tremendous significance to safeguard biodiversity of this fragile area.

7.2 STUDY AREA

The study area selected for the present investigation is Agasthyamala Biosphere Reserve, which comes under Kerala and Tamil Nadu states of India. The proposed study area that includes reserve forest, vested forest, revenue land, and private land, extend over 3500.36 sq.kms (1828 km^2 in Kerala and 1672.36 km^2 in Tamil Nadu) and it comprises 10 forest divisions/sanctuaries viz., Konni, Achankovil, Thenmala, Punalur, Schenduruny, Trivandrum, Neyaar, Peppara, Agasthyavanam Biological Park and Kalakkad Mundanthurai (Figure 7.1). The whole area is hilly, undulating, and highly rugged. The lofty main Ghats with elevations varying from 85 m to 1868 m. Cliffs and higher slopes of the eastern side protect the forest vegetation from the adverse effects of violent dry northeastern winds. The highest peak is Agasthyarkoodam with an altitude of 1868 m, which is located on the eastern boundary of the tract. The major rock types of this tract are Magmatic Gneissic, Charnockite, and Khondalite of an Archaean complex. Quartz, Garnet, Hornblende, Feldspars, and Black Mica are also found as constituents in these formations. These rocks have suffered intensive deformation like faulting and folding during the different phases of orogeny, most probably due to tectonic disturbances. The general foliation trend shows NW - SE direction, with a steep dip towards SW (Rajesh et al., 2001). The heavy rainfall and high temperature, causing alternate cycles of wetting and drying phenomena favor the process of laterization. The major soil types met within this tract are red loamy soil, laterite soil, alluvial soil, sandy loam, and clayey soil. Generally, the climate in this area is moderately hot and humid. The low-lying area enjoys a healthy and fairly moderate climate, with not much appreciable variation in either seasonal or diurnal temperatures. However, the interior areas experience a little more climatic variation. The three distinct seasons noted in this tract are cold, hot and wet seasons. The hottest season is noted during February to May and the coldest from December to January. In the upland area, the temperature declines towards elevated regions. Variations in the radiant energy of the sun with respect to seasons, cloudiness, altitudes, latitudes and diurnal changes were also noted. The temperature varies from 20°C to 36°C in the lower stretches and 17°C to 30°C at higher altitudes. Mist is common on the higher slopes during November to January. This forests region gets heavy rain showers from both south-west monsoon (June to mid-August) and northeast monsoon (mid-September to mid-November). Bulk of the precipitation is from the south-west monsoon. The average rainfall received during the last ten years is 2800.10 mm and the average number of rainy days in a year is 131 (Radhakrishna Pillai and Muhammad, 2007). Maximum rainfall is observed in June, July, and October and lowest during December, January and February months. The tract also receives pre-monsoon showers, preceded by thunderstorms, during April-May. There are two prevailing winds blow on in this tract, following the monsoons. From March-April onwards, the tract experiences a light wind which will gradually develop into south-west monsoon round about the beginning of June. Westerly winds that blow during the south-west monsoon are mild and harmless. But, the easterly winds in months of January and February are much violent and strong. They cause much havoc and damage to the forest crops growing on the hilltops. The desiccating effect of these winds cause much damage to the forests and may act as the driving force to spread the accidental forest fires. The profound rainfall and bright sunshine cause a humid and warm climate. Humidity varies from 65 to 98% in different localities in accordance with time and season. The highest relative humidity is noticed during the months of June, July, and August (south-west monsoon) and the lowest in February, when precipitation is kept minimum. General relative humidity is lower in the afternoon and highest during the early morning hours, when the atmospheric temperature will be the minimum.

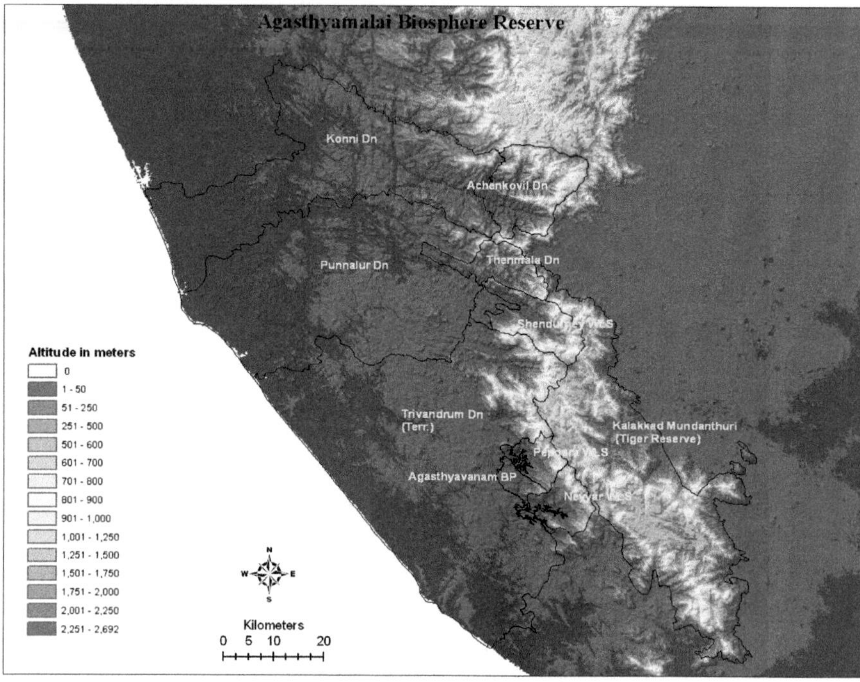

FIGURE 7.1 **(See color insert.)** Map of Agasthyamala Biosphere Reserve.

7.2.1 *FOREST TYPES*

Agasthyamala Biosphere Reserve has a vast stretch of natural forests. The forest has spread an area of 3500.36 sq.kms, comprising evergreen, semi-evergreen, and deciduous type of forests. As per the classification of the forest types of India (Champion and Seth, 1968; FSI, 2011) the main forest types met within this tract are: West coast tropical evergreen forests, Southern hilltop tropical evergreen forests, West coast semi-evergreen forests, Southern moist mixed deciduous forests, Wet reed brakes, Bamboo brakes, Canebrakes, Myristica swamps and Grasslands with Shola patches. Besides these natural forest types, plantations of *Teak, Matchwood, Acacia, Rosewood,* and *Pepper* were also located in the forest area.

7.3 METHODOLOGY IN BRIEF

Intensive floristic explorations and ethnobotanical studies were conducted during the period 2009–2015 by frequent field visits and specimen collections. Plant specimens were taxonomically identified using the standard publications and different herbaria. The herbarium specimens were prepared as per the standard specifications (Fosberg and Sachet, 1965; Bridson and Forman, 1991). The occurrence and distribution of endemic species were verified and analyzed with the help of pertinent literature (Ahammedullah and Nayar, 1987; IUCN, 2011; Sasidharan, 2013). The systematic documentation of the identified species was done according to the classification system of Bentham and Hooker (1862–1883) with necessary alterations as suggested by Hutchinson (1926, 1934, 1956, 1973) and Brummit and Powell (1992). Artificial diagnostic (Dichotomous parallel) keys were prepared for the identification of families, genera, and species. Name of the authors and citation of the original publication of the generic name is given based ON Farr et al. (1979) and IPNI (2015).

7.4 SYSTEMATIC TREATMENT AND FLORISTIC ANALYSIS

The plants collected from the study area during the period 2009–2015 were subjected to systematic treatment, and the appropriate keys were prepared to identify the genus.

7.4.1 KEY TO THE GENERA

1.	Climbers with nodular adventitious roots	2
1:	Erect rhizomatous herbs or shrubs	3
2.	All leaves entire; perianth 6	Pothos
2:	Leaves both dissented and entire; perianth absent	Rhaphidophora
3.	Leaves palmately or pinnately foliolate or dissected	4
3:	Leaves entire	6
4.	Spathe spirally twisted; rootstock creeping	Anaphyllum
4:	Spathe not spirally twisted; rootstock not creeping	5
5.	Lamina pedatisect; plants leafy during flowering	Arisaema

5:	Lamina triparty compound; plants leafless during flowering	Amorphophallus
6.	Epiphytes with bulbiferous shoots; leafless during flowering	Remusatia
6:	Terrestrial or epiphytic without bulbiferous shoots; leafy during flowering	7
7.	Rootstock creeping, not tuberous	Lagenandra
7:	Rootstock tuberous, not creeping	8
8.	Leaves broadly ovate, to 50 cm long	9
8:	Leaves ovate-hastate or orbicular, to 10 cm	10
9.	Rootstock up to 2 m long. Basal lobes of leaves acute	Alocasia
9:	Corm up to 15 cm broad. Basal lobes of leaves obtuse	Colocasia
10.	Leaves orbicular, peltate; male flower zone fertile to apex, a hemispherical-dome	Ariopsis
10:	Leaves triangular-ovate; male flower zone sterile to apex, tapering	11
11.	Spadix included in the spathe; ovules basal and apical	Theriophonum
11:	Spadix exerted from the spathe; ovules basal	Typhonium

7.4.2 TAXONOMIC TREATMENT OF THE SPECIES

7.4.2.1 ALOCASIA LONGILOBA MIQ.

Alocasia longiloba Miq., Fl. Ned. Ind. II. 207. 1855; *A. amabilis* W. Bull; A. argyrea Sander; *A. cochinchinensis* Pieree ex Engl. & K. Krause; *A. curtisii* N.E.Br.; *A. cuspidata* Engl.; *A. denudata* Engl.; *A. denudata* var. *elongata* Engl.; *A. eminens* N.E.Br.; *A. korthalsii* Schott; *A. lowii* Hook.f.; *A. lucianii* Pucii ex Rodigas; *A. pucciana* Andr.; *A. putzeysii* N.E.Br.; *A. singaporensis* Linden; *A. spectabilis* Engl. & K. Krause; *A. thibantiana* Mast.; *A. veitchii* (Lindl.) Schott; *A. watsoniana* Sander; *Caladium lowii* Lem.; *C. veitchii* Lind. (Figure 7.2a–c).

Herbs, terrestrial to 1500 mm tall. Stem rhizomatous, elongate, 80–600 x 20–80 cm. Petiole terete. Leaves 1–3, subtended by lanceolate papery-membranous cataphylls; blade pendent, usually dark green, with venation grey-green adaxially, narrowly hastate-sagittate, 27–85 x 14–40 cm. Pointed posterior lobes. Spathe 7–17 cm, abruptly thin 1.5–3.5 cm from base, stipitate, female zone 1.5–2.5 cm, pistil green, subglobose, stigma white in color, sterile zone 7–10 mm long and narrower than fertile zone, rhombic hexagonal synandrodes, top flattened, male zone ivory in color, subcylindric. Fruit ripening orange-red. Fl. & Fr.: June–October.

Habitat and Distribution: In rain forests and regrowth understory and on rocks of Cambodia, Laos, Vietnam, China, Malaysia, Borneo, Java, Sulawesi and India (World). In India, only found in South Western Ghats.

Materials Examined: India, Kerala: southern Western Ghats, Kollam, Achankovil, Rainforest near Kumbrauvatty waterfall, 17 July 2014, *J. Mathew 4711*(MGUH; MG university herbarium, Kottayam, Kerala); Wayanad, 900 Forests, 10 October 2011, *P.M. Salim 5118* (MSSRF).

Note: It is a new distributional record in Kerala.

7.4.2.2 AMORPHOPHALLUS BULBIFER (ROXB.) BLUME

Amorphophallus bulbifer (Roxb.) Blume, Rumphia 1: 148. 1837; *A. aculatum* Hook.f.; *A. taccoides* Hook.f.; *A. tuberculiger* (Schott) Engl.; *Arum bulbiferum* Zipp. Ex Kunth; *A. spectabile* Zipp ex Kunth; *Conophallus bulbifer* (Roxb.) Schott; *C. tuberculiger* Schott; *Pythonium bulbiferum*(Roxb.)Schott. (Figure 7.2d).

Corm globose, to 150 mm across Petiole to 1 m long, 20–30 mm thick, brown color spotted; lamina 300–500 mm across, pinnatisect, bulbiferous at the base. Lobes to 20 mm, obovate, lanceolate. Spadix sessile; spathe to 400 x 100 mm, at the broadest part, pale pink; female flowers covering 45 mm of the spike; male flowers to appendages cylindrical. Fl. & Fr.: June–September.

Habitat and Distribution: In deciduous forests of India and Malaysia (World). In Kerala, found in central and south Kerala.

Materials Examined: India, Kerala: southern Western Ghats, Kollam, Achankovil, Thura, 11 June 2011, *J. Mathew 5177, 5178* (MGUH).

7.4.2.3 AMORPHOPHALLUS BONACCORDENSIS SIVAD. & N. MOHANAN

Amorphophallus bonaccordensis Sivad. & N.Mohanan, Blumea 39: 295. 1994; Mohanan & Sivad., Fl. Agasthyamala 750, 2002 (Figure 7.2e).

Herbs, corms subglobose. Leaf trichotomously decompound 300–750 mm long, smooth, greenish brown mottles; rachis of the segments 150–200 cm long, shallowly channeled above and with decurrent leaf bases; leaflets sessile, Stolons cylindric, 40–50 x 4–7 mm. Spadix stipitate. Pistillate flowers subspirally arranged; ovary sessile, subglobose, 18–2 x 2–3 mm, greenish, 2/ 3 loculed, style very short, cylindric, 0.5–0.8 x 0.8–1 mm; stigma 2/3-lobed, covered with short unicellular papillae; stamen 1–13 mm high, Fl. & Fr.: March–April.

Habitat and Distribution: Evergreen forests of South Western Ghats (World). Endemic to Agasthyavam Biological Park. Rare

Materials Examined: India, Kerala: southern Western Ghats, Thiruvananthapuram, Bonacord, March 2011, *J. Mathew 4714,4718*(MGUH).

7.4.4.4 AMORPHOPHALLUS COMMUTATUS (SCHOTT) ENGL.

Amorphophallus commutatus (Schott) Engl., Monogr. Phan. 2: 319. 1879; *Conophalus commutatus* Schott, *Kattuchena*(Mal).

Herbs. Leaves tripartitely compound, leaflets elliptic, base acute, glabrous, membranous; petiole to 500 mm long, terete; peduncle 80–100 mm long and 7–9 mm diameter, pale yellowish in color, 4 cataphylls. Spathe ovate-acute, limb expanded, tube purple in color. Spadix 230–250 mm long. Female zone ca. 15 mm long. Male zone cylindrical, ca. 30 mm long, appendix elongate, narrowly conical with rounded apex, 180–200 cm long, 13–15 mm diameter at base and tapering. Female flowers ca. 2.5 mm long, style very short; stigma ca. 1.5 mm diameter. 3-lobed. Male flowers pale yellowish, sessile; 2-lobed; dehiscence by apical slit. Fl. & Fr.: May–September.

Habitat and Distribution: In moist deciduous forests of Western Ghats (World). Endemic.

Material Examined: India, Kerala: southern Western Ghats, Kollam, Achankovil, 11 August 2011, *J. Mathew 3785* (MGUH).

7.4.4.5 AMORPHOPHALLUS NICOLSONIANUS SIVAD.

Amorphophallus nicolsonianus Sivad., Pl. Syst. Evol. 153: 165, 1986 (Figure 7.2 (f–g).

Herbs with corm Leaves 250–300 mm long; leaflets 130–150 x 40–50 mm, ovate or oblong, acuminate, base acute or obtuse, glabrous. Peduncle terete, 170–200 mm long, spathe ovate-lanceolate, acuminate, 80–100 x 25–30 mm, greenish-brown, spadix sessile, slightly longer than the spathe, flowers confined towards the base; appendix narrowly cylindric and sterile. Fl. & Fr.: January-March.

FIGURE 7.2 (See color insert.) (a–c) *Alocasia longiloba* Miq.; (d) *Amorphophallus bulbifer* (Roxb.) Blume; (e) *Amorphophallus bonaccordensis* Sivad. & N. Mohanan; (f–g) *Amorphophallus nicolsonianus* Sivad.; and (h) *Amorphophallus paeoniifolius* (Dennst.) Nicolson.

Habitat and Distribution: In evergreen forests of South Western Ghats (World). Endemic.

Material Examined: India, Kerala: southern Western Ghats, Kollam, Achankovil, Kumbaruvatty, 11 August 2011, *J. Mathew 2712* [Sterile] (MGUH).

7.4.4.6 AMORPHOPHALLUS PAEONIIFOLIUS (DENNST.) NICOLSON

Amorphophallus paeoniifolius (Dennst.) Nicolson, Taxon 26: 337. 1977; *A. campanulatus* Decne.; *A. chatty* Andrews; *A. decurrens* (Blanco) Kunth; *A. dixenii* K.Larsen & S.S.Larsen; *A. dubius* Blume; *A. giganteus* Blume; *A. gigantiflorus* Hayata; *A. malaccensis*; *A. microappendiculatus* Engl.; *A. paeoniifolius* (Dennst.) Nicolson var. *campanulatus* (Decne.) Sivad.; *A. rex* Prain; *A. sativus* Blume; *A. virosus* N.E.Br.; *Arum campanulatus* Roxb.; *A. decurrens* Blanco; *A. phalliferum* Oken; *A. rumphii* Oken; *Candarum hookeri* Schott; *C. roxburghii* Schott; *C. rumphii* Schott; *Conophallus giganteus* Schott ex Miq.; *C. sativus* (Blume) Schott; *Dracontium paeonifolium* Dennst.; *D. polyphyllum* Dennst.; *D. polyphyllum* G. Forst.; *Hydrosme gigantiflora* (Hayata) S.S.Ying; *Kunda verrucosa* Raf.; *Plesmonium nobile* Schott; *Pythion campanulatum* Mart. *Kattuchena & Chena*(Mal.). (Figure 7.2h).

Herb with corm Petiole to 500 mm long, 30–40 mm thick, green with brown moltings; lamina 700–1200 cm across; lobes to 14 x 5.5 mm, ovate, acuminate, decurrent at base into a wing to petiole. Spathe 380 mm across, campanulate, undulate; spadix as long as spathe; appendage obovoid, 200–250 mm across, rugose, deep pink. Flowers on the lower half of the spadix. Fl. & Fr.: May-August.

Habitat and Distribution: In moist deciduous forests of India, Sri Lanka and Pacific Islands (World). In India, South Western Ghats.

Materials Examined: India, Kerala: southern Western Ghats, Kollam, Achankovil, Kalingamala, 12 August 2011, *J. Mathew 3251*, 3252 (MGUH).

7.4.4.7 ANAPHYLLUM BEDDOMEI ENGL.

Anaphyllum beddomei Engl., Planzenreich. Arac. Lasioid. 26. 1911. (Figure 7.3d–f).

Cormous herbs. Leaf lamina 3-sected, middle lobe pinnately lobed. Petioles 500–1000 mm high. Spathe green, 50–80 x 25–50 mm, broadeacute at apex, base rounded. Spadix blackish purple, 30–50 mm long, 3–5 mm diam. Perianth 3–4. Stamens 4–6. Ovary 1-celled; ovule 1; style 0; stigma discoid. Berries ovoid, 0.6–0.8 cm long, turn to orange redFl. & Fr.: February–March.

Habitat and Distribution: Evergreen forests of South Western Ghats (World). Endemic to Agasthyamala Biosphere Reserve. Threatened

Materials Examined: India, Kerala: southern Western Ghats, Kollam, Vellakkaltheri, 11 March 2011, *J. Mathew 3413, 3414*(MGUH).

7.4.4.8 ANAPHYLLUM WIGHTII SCHOTT

Anaphyllum wightii Schott, Gen. Aroid. t. 83. 1858. *Keerikkizhangu & Sulli.* Figure 7.3 (a -c).

Tall herbs with a creeping rhizome. Leaf lobes 3–8 pairs, 230–285 x 80–100 mm, acute; petiole 600–1500 mm long, 5–10 cm thick, brownish. Spadix 150–300 mm long, 50–80 mm broad, spathe open, deep brown, twisted; spikes 45–50 mm x 8–9 mm; flowers bisexual, densely arranged, perianth lobes 4, obovate, truncate; stamens 6, ovary 1-celled, ovule solitary; style short, stigma discoid. Fruit a globose achene. Fl. & Fr.: January-February.

Habitat and Distribution: Evergreen forests of South Western Ghats (World). Endemic to Western Ghats. Rare.

Materials Examined: India, Kerala: southern Western Ghats, Kollam, Vellakkaltheri, 11 March 2011, *J. Mathew 3433, 3434*(MGUH).

7.4.4.9 ARIOPSIS PELTATA NIMMO

Ariopsis peltata Nimmo, Cat. Pl. Bombay 252. 1839. (Figure 7.3g– h).

Epiphytic or terrestrial annuals, rootstock globose, ca. 1 cm across. Leaves 1 or 2, peltate, lamina 5–9 cm across, orbicular, entire, membranous, glabrous; petiole 7–8 cm long, slender. Peduncle 2–5 cm long, solitary; spathe yellow or white, 3–4 cm long, cymbiform, open, tube absent; spike 2–2.5 cm long, cylindrical; female flowers 5–7, on one side of the spike towards the lower part; ovary ovoid, 4-ridged, 1-celled, ovules many, on 5-parietal placentas; stigma 4-lobed; male flowers sunken in upper fleshy part of spadix; stamens 3; anthers divaricate, pubescent. Fl. & Fr.: June–August.

Habitat and Distribution: Lithophytes and epiphytes in evergreen forests of India and western Malesia (World): Throughout in Kerala.

Materials Examined: India, Kerala: southern Western Ghats, Kollam, Thoovalmala, 14 July 2011, *J. Mathew 2997, 2998*(MGUH).

FIGURE 7.3 (See color insert.) (a–c) *Anaphyllum wightii* Schott; (d–f) *Anaphyllum beddomei* Engl.; (g–h) *Ariopsis peltata* Nimmo; (i–j) *Ariopsis peltata* var. *brevifolia* J. Mathew & Kad. V. George.; and (k–m) *Arisaema agasthyanum* Sivad. & Sathish.

7.4.4.10 ARIOPSIS PELTATA VAR. BREVIFOLIA J. MATHEW & KAD. V. GEORGE

Ariopsis peltata var. ***brevifolia*** J.Mathew & Kad.V.George, Telopea 18: 152. 2015. (Figure 7.3i–j).

Small herb to 2 cm tall; lamina peltate, 10–15 mm long, 10–15 mm wide, Inflorescence 1 or 2 in each floral sympodium, stigma with 3 lobes, white in color, male flower zone fertile to apex, hemispherical-dome, synandria all connate apically, forming continuous surface punctured by 2 rings of cavities with somewhat prominent margin into which pollen is shed from the 3(or 4) surrounding thecae of which each pair of thecae belongs to a different synandrium. Fruits a 3-sided berry. Fl. & Fr.: May–July.

Habitat and Distribution: Lithophytes in evergreen forests of South Western Ghats (World). Endemic to Agasthyamala Biosphere Reserve. Rare.

Materials Examined: India, Kerala: southern Western Ghats, Kollam, Kottavasal, 14 July 2011, *J. Mathew 3074, 3075* (MGUH).

7.4.4.11 ARISAEMA AGASTHYANUM SIVAD. & SATHISH

Arisaema agasthyanum Sivad. & Sathish, Aroideana 10: 18. 1987. (Figure 7.3k–m).

Herbs. Leaflets 7–8, 60–100 x 20–30 mm. Spathe 60–150 x 40–50 mm, convolute at base; black. Spadix 45–60 mm long. Ovary 1-celled; ovules 4–6; stigma sessile. Spadix appendix thick blunt at tip with a lower stripe having subulate sterile flowers; sterile flowers 1–2 in male spadix, many in female.

Habitat and Distribution: In evergreen forests of South Western Ghats (World). Endemic to Agasthyamala Biosphere Reserve. Rare.

Materials Examined: India, Kerala: southern Western Ghats, Thiruvananthapuram, Pongalappara, 15 August 2013, *J. Mathew 4758, 4759*(MGUH).

7.4.4.12 ARISAEMA ATTENUATUM BARNES & C.E.C. FISCH.

Arisaema attenuatum Barnes & C.E.C. Fisch., Bull. Misc. Inform. Kew 1936: 275, 1936.

Herb. Leaflets 7–12, 150–250 x 50–70 cm, lanceolate, apex acuminate; petiole to 600 mm long, 10–30 mm thick, brownish. Peduncle thick; spathe 100–220 mm long, tube 50–150 mm long, 10–32 mm broad, with thick greenish moltings; limb ovate, apex curved down, green with brown streaks. Spadix 100–150 mcm long, clavate at apex, tapering and acuminate, female flowers many in lower 10–30 cm long; neuter flowers filiform, to 10 mm long.

Habitat and Distribution: In evergreen forests of India (World). In Kerala, found in Idukki, Kollam, and Thiruvananthapuram.

Materials Examined: India, Kerala: southern Western Ghats, Kollam, Kottavasal, 14 July 2011, *J. Mathew 3966, 3967*(MGUH).

Note: In recent treatments of *Arisaema* (Gusman and Gusman, 2006), *A. attenuatum* was reduced to synonymy of *A. leschenaultii* without proper investigation of the type material and protologue. A critical examination of types and other materials of *A. attenuatum* and *A.leschenaultii* coupled with the materials collected from Kollam (*J. Mathew 3966*) and field observations showed that the former has a distinct warrant to be a good and individual taxon. Tapering spadix appendix of *A. attenuatum* is the distinct character *vs.* clavate cylindrical appendix of *A. leschenaultii*. Thus, we state that Barnes & C.E.C. Fisch. was correct in treating *A. attenuatum* as a distinctive species and here it is reinstating as a specific taxa.

7.4.4.13 ARISAEMA BARNESII C.E.C. FISCH.

Arisaema barnesii C.E.C. Fisch., Bull. Misc. Inform. Kew 342. 1933. (Figure 7.4a–c).

Herb with corm 10–20 mm across. Leaf solitary, pentafoliate; petiole 15–25 cm long, greenish-brown with inconspicuous white spots. Leaflets elliptic-lanceolate, 10–15 cm long and 3–5 cm wide, distinctly petiolulate, lateral veins 9–11 on each side, leaflets covered with silvery trichomes abaxially; petiolules c. 2 cm long. Peduncle 10–15 cm long, silvery-white in color. Spathe tube cylindric, 4–5 cm long, *c.* 1 cm in diam., green to pale green with number of white vertical bands; throat marging galeate-auriculate, slightly recurved; limb greenish with a distinct white colored quindent marking, ovate-lanceolate, apex acuminate with a greenish-purple tail up to 3 cm long, recurved upwards terminating into a knob. Male spadix *c.* 4 cm long with basal fertile portion 1–1.5 cm long, cylindrical, densely flowered; appendix obclavate, *c.* 3 cm long, purplish basally and greenish distally tapering into a small warded apical knob. Fl. & Fr.: April- June

Habitat and Distribution: In evergreen forests and sholas of Western Ghats (World); endemic. Found throughout in Kerala. Threatened.

Materials Examined: India, Kerala: southern Western Ghats, Kollam, Achankovil, Vellakkaltheri, 10 June 2013, *J. Mathew 4910, 4911* (MGUH).

FIGURE 7.4 (See color insert.) (a–c) *Arisaema barnesii* C.E.C. Fisch.

7.4.4.14 ARISAEMA JACQUEMONTII BLUME

Arisaema jacquemontii Blume, Rumphia 1: 95. 1836; *Arisaema wightii* Schott.

Paradioecious herb. Leaves 1, rarely2. Leaflets 5–7, elliptic-ovate to lanceolate, 45–135 x 20–40 mm in size. Spadix appendage rather stout, curved, truncate in male and cuneate in female; in male, spadix 30–54 mm long, in female, 58–62 mm Stamens stipitate, anthers subglobose, ca. 1 mm broad, dehiscing by longitudinal slits. Ovary subglobose, 1.5–2 mm long, style sessile, stigma disciform. Berries globose. Fl.& Fr.: June–December.

Habitat and Distribution: Margins of evergreen forests and sholas of South India and Himalayas (World). In Kerala, found in Idukki and Kollam.

Material Examined: India, Kerala: southern Western Ghats, Kollam, Achankovil, Vellakkaltheri, 10 June 2013, *J. Mathew 2910* (MGUH).

7.4.4.15 ARISAEMA LESCHENAULTII BLUME

Arisaema leschenaultii Blume, Rumph. 1: 93. 1836; *A. pulchrum* N. E. Br., *Arisaema tuberculatum* C.E.C. Fisch.; *A. convolutum* C.E.C. Fisch.; *A. longicaudatum* Blatt.; *A. filicaudatum* N.E.Br. *Pambucholam* (Mal.). (Figure 7.5a–e).

FIGURE 7.5 (See color insert.) (a–e)*Arisaema leschenaultii* Blume; (f–h) *Arisaema sarracenioides* Barnes & C.E.C. Fisch.; and (i–l) *Arisaema madhuanum* Nampy & Manudev.

HerbLeaf one; leaflets 7–12, to 25 x 7 cm, lanceolate, acuminate; nerves many, close, parallel, forming an intramarginal vein, prominent, glabrous; petiole to 70 cm long, 1–3 cm thick, brownish. Spadix peduncled, below the level of leaf; spathe 10–22 cm long, tube 5–15 cm long, 1–3.5 cm broad, with thick greenish-brown streaks; limb ovate, apex curved down, finely acuminate, green with brown streaks. Spadix 110–150 mm long, apex clavated, female flowers many neuters filiform, simple or bifurcate. Fl.& Fr.: June–December.

Habitat and Distribution: Margins of evergreen forests, sholas, and grasslands of India, Sri Lanka and Japan (World); found throughout in Kerala.

Materials Examined: India, Kerala: southern Western Ghats, Kollam, Achankovil, Vellakkaltheri, 10 June 2013, *J.* Mathew *3414–3421*, (MGUH).

7.4.4.16 ARISAEMA MADHUANUM NAMPY & MANUDEV

Arisaema madhuanum Nampy & Manudev, Edinburgh J. Bot. 71(2): 269. 2014. (Figure 7.5i–l).

Paradioecious herb, tuberous perennial herbs, to 75 cm tall. Cataphylls 1–3, 2.5–2.2 cm long, obtuse, membranous. Pseudostem 40–60 cm long, cylindrical, light brownish petiole streaked with black mottling, ligule not prominent. Leaf solitary; petiole 40–60 cm long, 0.5–2 cm thick, brownish green with black mottling. Leaflets 8–12, sessile, subequal, lamina usually spathulate, rarely obovate, 100–500 × 50–120 mm, cuneate base and acuminate apex, margins entire. Peduncle to 25 cm long, 0.5–1.5 cm thick, deep purple, exserted by 6–15 cm from the pseudostem. Spathe 18–32 cm long including the limb portion; spathe tube cylindric but opened towards distal end, V-shaped towards mouth, 8–10 cm long, 1.5–2.5 cm wide, deep purple basally, pale purple distally, dark purplish within, with white longitudinal stripes along veins externally; spathe limb oblong-lanceolate, 100–150 mm long, 20–25 mm wide, purple, ovate-lanceolate, vaulted, abruptly tapering. Male spadix sessile, just reaching the mouth of the spathe tube, 8–9 cm long, slender, terete, base slightly thickened, ca3–5 mm wide, gradually tapering to a distal portion, distal end purple, fertile portion 25–35 mm long at the base; appendix sessile, ca.3 mm thick basally, narrowly tapering, basally purplish, distally greenish; neuters absent. Male flowers sessile or, towards the distal end of the fertile zone, shortly stipitate, each consisting of 2–7 purple-pink anthers; anthers reniform, 0.6–0.8 × 0.5–0.7 mm. Female spadix with basal portion cylindrical, ca. 22–30 x 8 mm, densely flowered; appendix filiform at the apex, 6–10 cm sized purple filiform, ovaries green, ovoid; stigmas subsessile, discoid. Fruiting spadix cylindrical dimensions. Frui ts 2-seeded berries. Fl.& Fr.: June–July.

Habitat and Distribution: In understory of semi-evergreen forests of South Western Ghats (World); endemic. In Kerala, found in Kollam.

Materials Examined: India, Kerala: southern Western Ghats, Kollam, Achankovil, Vellakkaltheri, 10 June 2013, *J. Mathew 4418–4423* (MGUH).

Notes. It is a new distributional record in Kerala with much more plasticity from the type materials. Certain variations have been noted in this new collection and are enlisted in the Table 7.1.

TABLE 7.1 Morphological Variations Noted in the New Population of *Arisaema madhuanum*

Character	New population	Type
Form	Para-dioecious	Dioecious
Height	30 to 75 cm	115 cm
Color of petiole	Light brownish petiole streaked with black mottling	Pale green with brown mottling
Length of petiole	40–60 cm	35.5–50 cm
Leaflets	8–12	7–10
Shape and size of leaf lamina	Lamina usually spathulate, very rarely obovate; 10–50 × 5–12 cm	Obovate-spathulate; 13–40 × 2.2–11 cm
Color of peduncle	Deep purple	Pale green with brown mottling
Color of spathe	Deep purple basally, pale purple distally, dark purplish within, with white longitudinal stripes along the veins	Pale to purple basally, greenish distally, dark purplish within, with white longitudinal stripes along the veins
Shape of the spathe	Cylindrical in base, V-shaped towards mouth	Cylindrical, funnel-shaped towards mouth
Male spadix	Neuters absent	Neuters present

7.4.4.17 ARISAEMA SARRACENIOIDES BARNES & C.E.C. FISCH.

Arisaema sarracenioides Barnes & C.E.C. Fisch. in Hook., Ic. Pl. ser. 5, 4: t. 3307. 1936. (Figure 7.5f–h).

Dioecious herbs Pseudostem 30–45 cm long, cylindrical, green with purple spots. Cataphylls 3, the last one usually degraded. Leaf solitary; petiole 200–350 mm long, 5–20 mm thick, pale green; leaflets 7–9, lamina size 150–280 × 25–45 mm in size., Inflorescence dioecious, emerging after the leaf; peduncle to 18 cm long, 5–7 mm thick, mottled as in petiole, pale purple towards the distal end, exerted by 6–15 cm from the pseudostem. Spathe 120–170 mm long excluding the limb; tube cylindric, funnel-shaped towards the mouth, with a small bend at the mid-point of spathe, mouth of tube obscured by a lid-like limb tipped by 10–14 cm long tail, limb green to purple. Male spadix sessile, 80–90 mm long, slender, terete, base slightly thickened, ca. 3–5 mm wide, gradually tapering to a terminal appendage, white with purple mottling. Neuters absent. Male flowers sessile or towards the distal end of the fertile zone; stipe to 1 mm long; anthers reniform, 0.6–0.8 × 0.5–0.7 mm, dehiscing through an apical slit. Female spadix with basal portion cylindrical, ovaries green, unilocular, ovoid; stigmas sub-sessile, discoid. Fruiting spadix cylindrical. Fruits fleshy, 1–2 seeded berry. Fl. & Fr.: May–June.

Habitat and Distribution: In understory of semi-evergreen forests of South Western Ghats (World); endemic. In Kerala, found in Idukki and Kollam.

Materials Examined: India, Kerala: southern Western Ghats, Kollam, Achankovil, Thoovalmala, 10 June 2013, *J. Mathew 3061, 4014* (MGUH).

7.4.4.18 ARISAEMA TORTUOSUM VAR. TORTUOSUM HOOK. F.

Arisaema tortuosum (Wall.) Schott in Schott & Endl., Melet. Bot. 1: 17. 1832, var. *tortuosum* Hook. f., Fl. Brit. India 6: 502. 1893;. *Arum tortuosum* Wall. (Figure 7.6a–d).

Herb with corms leaves paired, pedatisect; leaflets 8–10, 10–12 x 3–4 cm, oblanceolate, abruptly acuminate, base acute, sparsely hispid below, petiolulate; petiole 32–35 cm long, 2-together. Spadix solitary; spathe 8–12 cm long, green; limb ovate, acuminate, with white and purple streaks; tube 4–6 cm long; spadix 12–18 cm long, sigmoidally curved, appendage narrow. Berry orange-red, ca. 6 mm across, obovoid; seeds 1–2. Fl. & Fr.: May–June.

Habitat and Distribution: In understory of semi-evergreen forests and grasslands of Himalayas, south India and Sri Lanka (World). Found throughout in Kerala.

Materials Examined: India, Kerala: southern Western Ghats, Kollam, Achankovil, Thoovalmala, 10 June 2013, *J. Mathew 3326, 3327*(MGUH).

7.4.4.19 ARISAEMA TYLOPHORUM C.E.C. FISCH.

Arisaema tylophorum C.E.C. Fisch., Bull. Misc. Inform. Kew 346. 1933. *Arisaema wightii* sensu C.E.C. Fisch. In Gamble, Fl. Pres. Madras 1891(1105). 1936, non-Schott 1859. (Figure 7.6e–g).

FIGURE 7.6 (See color insert.) (a–d) *Arisaema tortuosum* var. *tortuosum* Hook. f.; and (e–g) *Arisaema tylophorum* C.E.C. Fisch.

Corm 1–2 cm across, globose. Leaflets 7, to 16 x 4 cm, elliptic-oblong, caudate-acuminate, base attenuate, glabrous; petiolule 1–2 cm long; petiole to 40 cm long, 0.8 cm thick, brown spotted. Spadix solitary, at the level of the leaves; peduncle from the petiole 10 cm long; spathe to 10 cm long; tube deep brown; limb ovate, horizontal, green with white lines, finely acuminate, acumen 4–5 cm long; spadix to 6 cm long, 3 mm thick, narrowed above; curved near the apex and terminated by a small knot of 3 mm thick. Fl. & Fr.: May-August.

Habitat and Distribution: In understory of evergreen forests of Western Ghats (World); endemic. In Kerala, found in Palakkad, Idukki, and Kollam.

Materials Examined: India, Kerala: southern Western Ghats, Kollam, Achankovil, Aramba, 17 May 2013, *J. Mathew 3786, 3788*(MGUH).

Note: In recent treatments of *Arisaema* (Gusman & Gusman, 2006), *A. tylophorum*was reduced to synonymy of *A. barnesii* without proper investigation of the type material. A critical examination of types and other materials of *A. tylophorum* and *A. barnesii* coupled with the materials collected from Thiruvananthapuram (*J. Mathew 3786*) and field observations showed that the former has a distinct warrant to be a good and individual taxa. Plants to 300 mm tall; spathe limb ovate, green without a white patch at the base; spadix appendix terminating in a distinct, minutely warted knob is the identical character of *A. tylophorum* vs. tapering spadix appendix seen in *A. barnseii*. Thus, we argue that C.E.C. Fisch. was correct in treating *A. tylophorum* as a distinctive species and here it is reinstating as a specific taxa.

7.4.4.20 COLOCASIA ESCULENTA (L.) SCHOTT

Colocasia esculenta (L.) Schott, Melet. Bot. 1: 18. 1832; *Alocasia dussii* Dammer; *A. illustris* W.Bull; *Aron colocasium* (L.) St.- Lag.; *Arum chinense* L.; *A. colocasia* L.; *A. colocasioides* Desf.; *A. esculentum* L.; *A. lividum* Salisb.; *A. nymphaeifolium*(Vent.) Roxb.; *A. peltatum* Lam., *Caladium acre* R.Br.; *C. colocasia* (L.) W.Wight; *C. colocasioides* (Desf.) Brongn.; *C. esculentum* (L.) Vent.; *C. glycyrrhizum* Fraser; *C. nymphaeifolium* Vent.; *C. violaceum* Desf.; *Calla gaby* Blanco; *C. virosa* Roxb.; *Colocasia acris* (R.Br.) Schott; *C. aegyptica* Samp.; *C. antiquorum* R.Br.;*C. colocasia* (L.) Huth; *C. euchlora* K.Koch & Linden; *C. formosana* Hayata; *C. gracillis* Engl.; *C. himaensis* Royle; *C. konishii* Hayata; *C. neocaledonica* Van Houtte; *C. nymphaeifolia* (Vent.) Kunth; *C. peltata* (Lam.) Samp.; *C. vera* Hassk.; *C. violacea* (Desf.) auct.; *C. virosa* (Roxb.) Kunth; *C. vulgaris* Raf.; *Leucocasia esculenta* (L.) Nakai; *Steudnera virosa* (Roxb.) Prain; *Zantedeschia virosa* (Roxb.) K.Koch. *Chembu, Kaattuchembu & Seppankizhangu,* (Mal.)

Stout herbs. Leaves many, 230–280 x 160–180 mm, ovate, sagittate at base, glabrous; nerves six pairs. Peduncle solitary / few together, 100–200 mm long, stout; spathe 150–200 mm long, yellow, constricted above the base; limb acuminate; spadix 80–100 mm long, cylindrical, terete appendages, and obtuse; male flowers above, 5–6 cm of the spadix, female flowers on lower, 2 cm of the spadix; ovary 1-celled, ovules many on 2–4 parietal placentas. Fruit an aggregated berries, globose in shape. Fl. & Fr.: Janury-March.

Habitat and Distribution: In marshy areas of Pantropical countries (World): Throughout in Kerala.

Materials Examined: India, Kerala: southern Western Ghats, Kollam, Achankovil, Thura, 10 March 2011, *J. Mathew 3547, 3548*(MGUH).

7.4.4.21 LAGENANDRA MEEBOLDII (ENGL.) C.E.C. FISCH.

Lagenandra meeboldii (Engl.) C.E.C. Fisch. in Gamble, Fl. Pres. Madras 1576(1099). 1931; *Cryptocoryne meeboldii* Engl. (Figure 7.7a).

Rhizome 5–10 mm thick, creeping. Leaves clustered, 50–80 x 20–40 mm, ovate, apex acute, base rounded; petiole 50–100 mm. Peduncle to 25 mm long, concealed in the leaf sheath. Spathe to 100 x 15, mm, acuminate at apex, twisted Spadix 20 mm long, produced above into a cusp. Female flowers in a clusters of 5 mm across; male flowers in a globose clusters of 3 mm across. Fl. & Fr.: January–March.

Habitat and Distribution: Marshy areas in evergreen forests of Western Ghats (World); endemic. Found throughout Kerala.

Materials Examined: India, Kerala: southern Western Ghats, Kollam, Kallar, 11 Feb. 2011, *J. Mathew 3787, 3788* (MGUH).

7.4.4.22 LAGENANDRA OVATA (L.) THW.

Lagenandra ovata (L.) Thw., Enum. Pl. Zeyl. 334. 1864; *Arisarum ovatum* (L.) Raf.; *Arum ovatum* L.; *Caladium ovatum* (L.) Vent.; *Cryptocoryne ovata* (L.) Schott; *Lagenandra insignis* Trimen. *Aandavazha*
Karin-pola (Mal.). (Figure 7.7b).

Perennial herbs; rhizome 4 cm thick and cylindrical. Leaves entire, to 400 x 130 mm in size, ovate-oblong, acuteat apex; petiole 300–700 mm long. Peduncle axillary and erect, 150–200 mm long; spathe 150–200 x 50–100 mm, broadly ellipsoid, brown, acuminate at apex; spadix 30–50 mm long. Male flowers many, in terminal globose head; stamens 1 / 2; female flowers many, basal, subglobose; ovary obovoid; ovule 1. Fruitlets obovoid and ridged; seed 1, brown in color. Fl. & Fr.: August–February.

Habitat and Distribution: In along banks of streams of India and Sri Lanka (World). Found in south Kerala.

Materials Examined: India, Kerala: southern Western Ghats, Thiruvananthapuram, way to Athirumala, 11 Jan. 2014, *J. Mathew 4516, 4583* (MGUH).

7.4.4.23 LAGENANDRA TOXICARIA DALZ.

Lagenandra toxicaria Dalz. in Hook.'s J. Bot. Kew. Gard. Misc. 4:289.1852. *Andavazha & Neerchengazhi* (Mal.).

Aquatic herbs, 40–50 cm tall; rhizome slender, 3–3.5 cm across. Leaves 23–26 x 7–10 cm, elliptic to ovate, lanceolate, acuminate, cuneate to narrowly cordate at base; petiole 12–14 cm long. Spathe 10–14 cm long, greenish, purple outside; limb much longer than tube, caudate acuminate, smooth; spadix with a basal pistillate portion, a staminate portion and a terminal naked appendix; pistillate flowers subtruncate; staminate flowers form a subcylindric mass. Berries 3–5 seeded. Fl. & Fr.: Throughout the year.

Habitat and Distribution: In along banks of streams of southern Western Ghats (World); endemic. Found throughout in Kerala

Materials Examined: India, Kerala: southern Western Ghats, Kollam, Pulikkayam, 11 Feb. 2011, *J. Mathew 3414, 3415* (MGUH).

7.4.4.24 POTHOS ARMATUS FISCHER

Pothos armatus Fischer, Kew Bull. 126. 1929.

Armed climbers, spines 2–4 mm long, a few at nodes. Leaves ovate-orbicular with cordate base in lower, while upper leaves elliptic-oblong, acuminate at apex and acute at base; nerves many, irregular and prominent at below; petiole 10 mm long. Spadix on short lateral branchlets; peduncles 30–50 mm long with 3–4 cataphylls; spathe 15–20 mm long, oblong, and acute; spadix 50 mm long, slender. Flowers 1.5 mm across, in 3–5 together; orbicular bracts with pubescence; stamens 6; ovary obovoid, truncate above. Fl. & Fr.: January–March.

Habitat and Distribution: In evergreen forests of Southern Western Ghats (World); endemic. Found in Central and south Kerala. Rare.

Materials Examined: India, Kerala: southern Western Ghats, Kollam, Kottavasal, 11 Feb. 2011, *J. Mathew 3416, 3417* (MGUH).

7.4.4.25 POTHOS CRASSIPEDUNCULATUS SIVAD. & N. MOHANAN

Pothos crassipedunculatus Sivad. & N.Mohanan, Pl. Syst. Evol. 168: 221. 1989. (Figure 7.7c–d).

Climbers. Leaves to 200 x 70 mm, elliptic-lanceolate, apex acuminate, prominent intramarginal veins at the margin, base cuneate or cordate; petiole to 20 mm long and pulvinate. Inflorescence lateral, ca. 10 mm long, cataphylls 3 - 4; peduncle to 40 mm long; spathe 20–25 x 15 mm, ovate-acute, apiculate, white in color; spadix sessile, 20 x 4 mm in size, sub-cylindric. Flowers to 2 mm diameter, bisexual, trimerous; tepals 6, fleshy, apex flat; filaments short, 3 x 0.6 mm; anthers 4lobed; stigma sessile, convex with tuft of papillae at center; ovary trilocular. Fl. & Fr.: December–May.

Habitat and Distribution: In evergreen forests of Southern Western Ghats (World); endemic. Found in Palakkad, Thrissur, Kollam, and Thiruvananthapuram. Rare.

Materials Examined: India, Kerala: southern Western Ghats, Kollam, Vellakkaltheri, 11 Feb. 2011, *J. Mathew 3445, 3446* (MGUH).

7.4.4.26 POTHOS SCANDENS L.

Pothos scandens L., Sp. Pl. 968. 1753; *Batis hermaphrodita* Blanco; *Podospadix angustifolia* Raf.; *Pothos angustifolius* (Raf.)C. Presl; *P. chapelieri* Schott; *P. cognatus* Schott; *P. decipiens* Schott; *P. exiguiflorus* Schott; *P. fallax* Schott; *P. hermaphrodites* (Blanco) Merr.; *P. horsfieldii* Miq.; *P. leptospadix* de Vriese; *P. longifolius* C. Presl; *P. zollingeri* Engl.; *P. zollingerianus* Schoot; *Tapanava indica* Raf.; *T. rheedei* Hassk. *Anapparuva, Paruvakodi & Paruval* (Mal.).

Large climbers; stem angled, 2–4 mm thick. Leaf to 90 x 30 mm, lanceolate in shape, acuminate, nerves many, glabrous; petiole 30–60 mm long, broadly winged. Spadix axillary; peduncle to 50 mm long, spathe 5 mm across, orbicular, brown in color; spadix 3–5 mm across, shape globose. Flowers densely packed; bracts 3–5, orbicular; stamens 6, free; ovary obovoid, stigma 3-toothed. Fl. & Fr.: October–November.

Habitat and Distribution: Common in all forest types of Paleotropics (World); throughout in Kerala.

Materials Examined: India, Kerala: southern Western Ghats, Kollam, Pulikkayam, 10 July 2011, *J. Mathew 3233, 3234* (MGUH).

7.4.4.27 REMUSATIA VIVIPARA (ROXB.) SCHOTT

Remusatia vivipara (Roxb.) Schott in Schott & Endl., Melet. Bot. 18.1832; *Arum viviparum* Roxb., Hort. Bengal 65. 1814. *Marachembu & Maratthaali* (Mal.). (Figure 7.7e–f).

Herbs, leaves 200–300 mm across, ovate, acute apex and cordate base, glabrous and glossy above; petiole 350–400 mm long; spadix solitary with large cataphylls; spathes obovate or rhomboid in shape, spreading, yellowish white; tube short, ca. 5 cm long, ellipsoid, limb 80–100 x 60–70 mm in size, acute; spadix 5–6 cm long; female flowers many, ovary globose, 1-celled; ovules many, stigma 3-lobed, globose; male flowers many, on clavate above part of the spadix; stamens 6 and united.

Habitat and Distribution: In dry to moist deciduous forests of Paleotropics (World); throughout in Kerala.

Materials Examined: India, Kerala: southern Western Ghats, Kollam, Vellakkaltheri, 10 July 2011, *J. Mathew 3258, 3259*(MGUH).

7.4.4.28 RHAPHIDOPHORA PERTUSA (ROXB.) SCHOTT

Rhaphidophora pertusa (Roxb.) Schott, Bonplandia 5: 45. 1857; *Pothos pertusus* Roxb.; *Monstera pertusa* (Roxb.) Schott; *Rhaphidophora lacera* Hassk.; *Scindapsus peepla* Thwaites; *S. pertusus* (Roxb.) Schott. *Scindapsus pertusus* (Roxb.) Schott in Schott & Endl., Melet. Bot. 21. 1832. *Elitthadi, Aanachurukki, Aanamakudam, Aanathippali, Athithippali, Gajathippiali, Ilathimaravazha, Pudayavu & Teyaarvalli* [Sasidharan, 2013] (Mal.).

Climbers, cylindrical stem, 40 mm thick and fleshy. Leaves size 250–350 x 250–300 mm, widely ovate, pinnatisect or sometimes entire, acute apex and rounded base; petiole to 200–350 mm long, channeled. Inflorescence solitary and axillary; spathe to 200 x 80 mm long, ovate to oblong, acute, creamy white in color; spadix 120 mm long and 20 mm thick. Flowers unisexual, thickly packed. Female flowers basal; ovary obconical and truncate; ovule solitary; stigma knob-like. Male flowers above; stamens 4–6, free. Fl. & Fr.: August–September.

Habitat and Distribution: In evergreen forests of India and Sri Lanka (World). Throughout in Kerala.

Materials Examined: India, Kerala: southern Western Ghats, Kollam, Thura, 10 July 2011, *J. Mathew 3352,3376*(MGUH).

FIGURE 7.7 (See color insert.) (a) *Lagenandra meeboldii* (Engl.) C.E.C. Fisch.; (b) *Lagenandra ovata* (L.) Thw.; (c–d) *Pothos crassipedunculatus* Sivad. & N. Mohanan; and (e– f) *Remusatia vivipara* (Roxb.) Schott.

7.4.4.29 THERIOPHONUM INFAUSTUM N.E. BR.

Theriophonum infaustum N.E. Br., J. Linn. Soc. Bot. 18: 260. 1881; *Calyptrocoryne minuta* Schott; *C. wightii* Schott; *Theriophonum wightii* (Schott) Engl.

Herbs, corm globose in shape. Leaves radical, to 80 x 60 mm, ovate in shape, base hastate, apex acute; petiole to 150 mm long, slender. Inflorescence axillary, spadix to 4 cm long; spathe globose in the lower part, constriction above, elliptic limb with greenish-yellow color. Male flowers many, stamens 1 or 2; sessile anthers. Female flowers few, ovary conical in shape, 1-celled; ovules few, basal. Filiform neuters. Fruit a group of globose berries. Fl. & Fr.: July–September.

Habitat and Distribution: Moist deciduous forests, also in marshy areas of south Western Ghats (World); endemic. Throughout in Kerala.

Materials Examined: India, Kerala: southern Western Ghats, Kollam, Thura, 10 July 2011, *J. Mathew 1754, 1757*(MGUH).

7.4.4.30 TYPHONIUM FLAGELLIFORME (LODD.) BLUME

Typhonium flagelliforme (Lodd.) Blume, Rumphia 1: 134. 1835; *Arum angustifolium* Griff.; *A. cuspidatum* Blume; *A. divaicatum* L.; *A. flagelliforme*Lodd.; *A. ptychiurum* Zipp. Ex Kunth; *Heterostalis flagelliformis* (Lodd.) Schott; *Typhonium cuspidatum* (Blume) Decne.; *T. cuspidatum* var *ptychiurum* Blume; *T. flagelliferum* Griff.; *T. hastiferum* Miq.; *T. incurvatum* Blatt. & McCann; *T. reinwarditianum* de Vriese & Miq. Ex. Miq.; *T. sylvaticum* Voigt. *Karinthakara* (Mal.).

Cormous herbs. Leaves simple, rarely divided, ovate, 3–15 by 1–6 cm, base cordate, apex acute, chartaceous. Spathe 6–15 cm long, tubular below, mouth constricted, expanded above into an apically twisted limb; spadix exerted; male and female flowers widely separated, neuter flowers just above the female, heteromorphic; ovary 1-locular; ovules 1–2, basal. Berry ovoid. Seeds 1 or 2, globose. Fl. & Fr.: July–September.

Habitat and Distribution: Plaines of Indo-Malesia (World). Throughout in Kerala.

Materials Examined: India, Kerala: southern Western Ghats, Kollam, Thura, 14 July 2011, *J. Mathew 1844, 1846*(MGUH).

7.5 RESULTS AND DISCUSSION

During the present study, 30 Aroids belonging to 12 genera were recorded. *Arisaema* was the largest genus in the family with 9 taxa followed by

Amorphophallus (5 spp.), *Lagenandra* (3 spp.) and *Pothos* (3 spp.). According to Sasidharan (2013), 45 taxa (except the ornamentals) of Aracaeae elements grown in the forests of Kerala. Thus, ABR holds 66% of the total Araceae flora of Kerala. However, *Amorphophallus smithsonianus* Sivad., an earlier reported endemic taxa from the Agasthyamala region could not be relocated. The floristic analysis resulted in the enumeration of 16 plant taxa (53%) are considered as endemic to the Western Ghats. Out of which 8 taxa (26%) are coming under rare and threatened category. As according to Mathew & George (2015), a high percentage of the endemicity and rarity of the plant species in the Agasthyamala Biosphere Reserves throws light on the significance of phytogeography of the south Western Ghats.

7.5.1 FLORISTIC NOVELTIES

The present investigation helped to enrich the available literature on a floristic wealth of Kerala in terms of additions and reinstatement. The population of *Alocasia longiloba* and *Arisaema madhuanum* from the lower bed of Achankovil River forms new reports of the taxa from the state. Besides that, *Arisaema attenuatum* and *A. tylophorum* were reinstated.

7.5.2 ETHNOBOTANY

The study throws light on the food habits of *Malampandarm* and *Kani* tribes inhabited in the study area. Consumption of wild edible plants constitutes an integral part of their nomadic life. Based on the field investigation, the leaves and pseudostems of *Amorphophallus bulbifer* and *Amorphophallus paeoniifolius* are their integral food items. These food items were pre-treated in limewater and roasted in pots or specially processed bamboo reapers. *Anaphyllum* (*Keerikizhang*), is used against snake bite. For that its fresh rhizome is crushed in a cleaned black stone and is given twice or thrice a day. The leaves of the *Pothos scandens* are ground into a paste with water and is given in morning and evening for diabetic treatment,

7.5.3 THREAT TO THE BIODIVERSITY

Agasthyamala Biosphere Reserve, situated in a highly fragile area, is prone to threat due to various man made interventions (changes in the land use, habitat loss, and fragmentation, expansion of plantation area, irresponsible tourism, road construction, invasion of alien species), climatic/weather changes and wild fires. Hence, stringent measures should be adopted to safe-guard the ecology and biodiversity of this forestland.

7.6 FUTURE PERSPECTIVES

It is expected that the data generated from this study would provide a guideline for future conservation and bioprospecting studies. Based on the observations on the high percentage of the endemism, adequate measures should be adopted to ensure the protection of the members of Araceae in its natural habitat. Besides, ex-situ conservation methods should be encouraged.

7.7 CONCLUSION

The Western Ghats is a rich repository of native and endemic species and is recognized as one of the globally important biodiversity hotspots. Agasthyamala Biosphere Reserve in the southern Western Ghats is the most species rich ecological regions in peninsular India with respect to species diversity and endemism. The floristic analysis of the study area has brought to light the rich diversity of the Arum plants with 30 taxa including 16 endemics, 8 rare/threatened and few species with signs of adaptive evolution. The study also shows the ethnobotanical importance of these plants.

KEYWORDS

- **Agasthyamala Biosphere Reserve**
- **Araceae**
- **conservation**
- **endemism**
- **Western Ghats**

REFERENCES

Ahmedullah, M.; Nayar, M. P. *Endemic Plants of the Indian Region.* I. *Peninsular India.* Botanical Survey of India: Calcutta, **1987**.

Bentham, G.; Hooker, J. D. *Genera Plantarum,* Vols. 1–3. L. Reeve & Co.: London, **1862–1883**.

Bridson, D. M.; Forman, L. *The Herbarium Handbook.* Royal Botanic Gardens: Kew, **1991**.

Brummitt, K. R.; Powell, C. E. *Authors of Plant Names.* Royal Botanic Gardens: Kew, **1992**.

Champion, H.G.; Seth, S. K. *Revised survey of the forest types of India.* Manager of Publications: New Delhi, **1968**.

Farr, E. R.; Leussink, J. A. Stafleu, F. A. *Index Nominum Genericorum,* 3 Vols. Utrecht: The Netherlands, **1979**.

Fosberg, F. R.; Sachet, M. M. *Manual for Tropical Herbaria* (Reg. Veg. 39). IAPT: Utrecht, **1965**.

FSI. *Atlas Forest Type of India.* Forest Survey of India: Dehradun, **2011**.

Gusman, G.; Gusman, I. *The Genus Arisaema: A Monograph for Botanists and Nature Lovers.* Edition 2. A.R. Gantner Verlag, Ruggell: Leichstenstein, **2006**.

Henry, A. N.; Gopalan, R.; Swaminathan, M. S. A new *Symplocos* Jacq. (Symplocaceae) from Southern India, *J. Bombay Nat. Hist. Soc.* **1984**, *81*, 169–171.

Hutchinson, J. *The Families of Flowering Plants*–Vol. I. Macmillan & Co. Ltd.: London, **1926**.

Hutchinson, J. *The Families of Flowering Plants*–Vol. II Monocotyledons (ed.3). Clarendon Press: Oxford, **1973**.

Hutchinson, J. *The Families of Flowering Plants*–Vol. II. Macmillan & Co. Ltd.: London, **1934**.

Hutchinson, J. *The Families of Flowering Plants*–Vols. I & II (ed. 2). Clarendon Press: Oxford, **1956**.

IPNI. *International Plant Names Index*. Published on the internet: http://www.ipni.org, **2015**.

IUCN. *2011 IUCN Red List of threatened species*: http://www.iucnredlist.org, **2011**

Kumar, C. S. Two novelties in the genus *Trias* Lindl. (Orchidaceae). *Blumea* **1989**, *34*, 103–109.

Kumar, C. S.; Manilal, K.S. *A catalogue of Indian Orchids.* Bishen Singh Mahendra Pal Singh: Dehra Dun, **1994**. 162 p.

Kumar, E. S. S.; Geetha Kumary, M. P.; Pandurangan, A. G. *Cinnamomum alexei*Kosterm. (Lauraceae): a new record for India. *Bangladesh J. Plant Taxon.* **2011,** *18(2),* 199–201.

Kumar, E. S. S.; Khan, A. E. S.; Binu, S. A new species of *Thottea* Rottb. (Aristalochiaceae) from Kerala, South India. *Rheedea* **2000,** *10*, 117–120.

Kumar, E. S. S.; Khan, A. E. S.; Binu, S.; Almeida, S. M. *Grewia palodensis* (Tiliaceae)–a new species from Kerala, India. *Rheeda* **2001**, *11*, 41–43.

Kumar, E. S. S.; Nair, G. M.; Yeragi, S. S. *Andrographis chendurunii*–a new species of Acanthaceae from India. *Nord. J. Bot.* **2004,** *22*, 683–685.

Manudev, K. M.; Nampy, S. *Arisaema madhuanum*, a new species of Araceae from India. *Edinburgh J. Bot.* **2014,** *71*, 269–273.

Mathew, J.; George, K.V. Checklist of Orchids of Kottavasal Hills in Achancoil Forests, Southern Western Ghats, (Kollam, Kerala), India. *Journal of Threatened Taxa* **2015**, *7*(10), 7691–7696

Mohanan, M.; Henry, A. N. *Cinnamomum chemungianum* (Lauraceae): a new species from Kerala, southern India. *J. Bombay Nat. Hist. Soc.* **1991**, *88*, 97–99.

Mohanan, N.; Sivadasan, M. *Flora of Agasthyamala.* Bishen Singh Mahendra Pal Singh: Dehra Dun, 2002.

Radhakrishna Pillai, P.; Muhammad, P. P. *Working Plan, Achancovil Forest Division*. Forest and Wildlife Department: Government of Kerala, 2007.

Rajesh, V. J.; Arima, M.; Santosh, M. Geology of the Achankovil Shear zone, southern India. *Gondwana Research* **2001,** *4*, 744–745.

Remadevi, S.; Binojkumar, M. S. *Ecbolium ligustrinum* (Vahl) Vollesen var. *aryankavensis*–a new variety from Kerala, India. *J. Econ. Taxon. Bot.* **2003**, *27*, 1189–1901.

Robi, A. J.; Sujanapal, P.; Udayan, P. S. *Cinnamomum agasthyamalayanum* sp. nov. (Lauraceae) from Kerala, India. *Int. J. Adv. Research* **2014**, *2 (10*), 1012–1016.

Robi, A. J.; Udayan, P. S. *Actinodaphne shendurunii* (Lauraceae): a new species from the southern Western Ghats, India. *Int. J. of Plant, Animal, and Envt. Sc.* **2013,** *3 (3)*, 185–188.

Sabu, T.; Mohanan, N.; Krishnaraj, M.V.; Shareef, S. M.; Shameer, P. S.; Roy, P. E. *Garcinia pushpangadaniana* (Clusiaceae), a new species from the southern Western Ghats, India. *Phytotaxa* **2013,** *116 (2),* 51–56.

Sasidharan, N. *Flowering plants of Kerala* (CD) 2.0. Kerala Forest Research Institute: Thrissur, 2013.

Sivadasan, M. *Amorphophallus smithsonianus* (Araceae)–a new species from India, and a note on A. sect. Synantherias. *Willdenowia* **1989**, *18*, 435–440.

Sivadasan, M.; Kumar, C. S. A new species of *Arisaema* (Araceae) from India with a note on variation and evolution of staminate flowers. *Aroideana* **1987**, *10 (4)*, 18–21.

Sivadasan, M.; Mohanan, N. *Ixora agasthyamalayana:* a new species of Rubiaceae from India. *Bot. Bull. Acad. Sin.* **1991**, *32*, 307–311.

Sivadasan, M.; Mohanan, N. *Pavetta bourdillonii* (Rubiaceae): a new species from India. *Bot. Bull. Acad. Sin.* **1999,** *40*, 61–63.

Sivadasan, M.; Mohanan, N.; Kumar, C. S. *Pothos crassipedunculatus*–a new species of Pothos Sect. Allopothos (Araceae) from India. *Pl. Syst. Evol.* **1989,** *168*, 221–225.

Sivadasan, M.; Mohanan, N.; Rajkumar, G. A new subspecies of *Symplocos macrophylla* Wall.x DC (Symplocaceae) from India. *Rheedea* **1997,** *7*, 89–92.

Sivadasan, M.; Mohanan, N.; Rajkumar, G. *Amorphophallus bonacordensis*-a new species of Araceae from India. *Blumea* **1994,** *39,* 295–299.

Sivarajan, V. V. A new species of *Thottea* (Aristolochiaceae) from India. *Pl. Syst. Evol.* **1985**, *150*, 201–204.

Sivu, A. R.; Ratheesh, N. M. K.; Pradeep, N. S.; Kumar, E. S. S.; Pandurangan, A. G. A new species of *Memecylon* (Melastomataceae) from the Western Ghats, India. *Phytotaxa* **2014**, *162 (1),* 044–050.

Sujanapal, P.; Robi, A. J.; Udayan, P. S.; Dantus, K. J. *Syzygium sasidharanii* sp. nov. (Myrtaceae)–A new species with edible fruits from Agasthyamala Hills of Western Ghats, India. *Int. J. Adv. Research.* **2013**, *1 (5)*, 44–48.

CHAPTER 8

Flowering Plant Diversity in the Alpine Regions of Eastern Himalaya

DIPANKAR BORAH*, A. P. DAS, SUMPAM TANGJANG, and TONLONG WANGPAN

Department of Botany, Rajiv Gandhi University, Rono Hills, Doimukh 791 112, Arunachal Pradesh, India,
**E-mail: dipankarborah085@gmail.com*

ABSTRACT

The Himalayas form a graceful and vast abode of floristic and faunal elements and also represent diverse human cultures spreading through its length and breadth. Eastern Himalayas, a biodiversity hotspot is not only a home to the world's highest mountains but is also amongst the highest diversity rich areas of the world. The easternmost part of the Indian Himalayas harbors many special vegetation types, depending upon altitudinal and climatological stratification. Arunachal Pradesh is one of the richest states in the region in terms of biodiversity, owing to its unique geographical position and altitudinal gradients. The article ventures preliminary account of flowering plant diversity of Nagula wetland complex of Arunachal Pradesh, which has more than 100 alpine freshwater lakes fed by melting snow. The altitude ranges from 3,500–4,500 meters above mean sea level. The study recorded a total of 106 species, falling under 68 genera and 32 families. Asteraceae is the most dominant family followed by Orobanchaceae, Gentianaceae, etc. If the area's natural vegetation is conserved without any disturbance that will maintain not only the pristine beauty but also the rich and original biological elements of the area. A high proportion of angiosperms of this area can be adopted for ornamental gardens, and some others can be tested for their medicinal properties.

8.1 INTRODUCTION

Mountainous regions exhibit the most vulnerable and challenging environments of the world. They cover both the extremities hot and cold with a high amount of biological and other resources. The longest mountain belt of the world, the great Himalayas covers most of the states on the northern boundary of India and some other countries. The entire region is now declared by IUCN as the 'Himalaya Biodiversity Hotspot' (Myers et al., 2000; Mittermier et al., 2005; Sharma et al., 2010). The Eastern Himalaya, included under the 'global 200 Ecoregions', falls in the 'Crisis Eco-Regions' where nearly 90% of the natural habitat is highly degraded (Brooks et al., 2006; Chettri et al., 2010). The region is characterized by a wide range of attitudinal gradients and climatic shifts. The region is also considered to be a meeting place for several biogeographical regions including Indo-Malayan, Palaearctic and Sino-Japanese. And, even the European elements slowly migrated towards the east and all these allowed the large-scale exchange of genes and formation of extremely rich biodiversity in the Eastern Himalaya (Das, 2004; Chettri et al., 2010; Sharma et al., 2015). Due to the wide diversity of altitude from 150 m to over 8000 m AMSL and the composition of the extreme topography have led to the assemblage of innumerable forms of vegetation right from the tropical lowlands to alpine meadows and to permafrost areas (Das et al., 2008; Chettri et al., 2010; Das, 2013). This complexity of topography and the climatic drifts created different microclimatic niches, where a high degree of endemism flourishes. The Eastern Himalaya, as a part of India covers an area of 93,984 km^2 that covers through the states of Arunachal Pradesh, Sikkim and the Darjeeling district of West Bengal (Das, 2016), whereas the Lesser Himalayas also extends through the other Northeastern states.

Arunachal Pradesh, the land of dawn-lit mountains, is an multilingual ethnic region which is situated between 26° 30' N and 29° 30 ' N and 91° 30' E and 97° 30' E with an area of 83,743 km^2 (Gurung et al., 2003; Debajit et al., 2015; Bharali et al., 2012), it shares 160 km long border with Bhutan in the west, 1030 km with China in the North, 440 km with Myanmar and other Northeastern states (Assam & Nagaland) in the South, East, and South East. It stretches from the snow-capped mountains in the north to the plains of Brahmaputra valley in the South. The state is gifted with wide topographical variations, vegetation, and wildlife. Among the districts, Tawang is the smallest with an area of 2172 km^2 and with the central location of 27°45′N latitude and 90°15′E longitude at the Northwest extremity of South Tibet. The district was carved out from the West Kameng district and hence the border exhibit considerable diversity of climate and topography.

Great Himalayas act as the prized source of numerous important rivers and innumerable streams which nourishes wide areas of the Indian subcontinent in its south and help to maintain the diverse form of rich vegetation there. On the other hand, with its overall all-round conditions, the Himalayas itself represent a unique ecosystem. High altitude wetlands (HAW) are the water bodies found above 3000 m AMSL and are fed by glaciers or snow (Panigrahy et al., 2012) and change in their structure and blasts could affect millions beneath them, hampering the hydrological cycle, flood cycles, etc.

The high altitude Arunachal Himalayas is with about 1672 lakes of different sizes that cover a total area of 11,863 ha. *i.e.*, 7.6 % of the total wetland area of the state which ranks second after Jammu and Kashmir (Debajit et al., 2015). Of these, the Tawang district itself is accommodating 108 of these lakes, some of which can be approached through man-made roads, and are mostly falling along the international boundaries (Debajit et al., 2015). These lakes are generally occurring in clusters and often referred to as 'Wetland Complexes.' Bhagajang, Nagula, Thembang Bapu CCA and Pangchen Lumpo are such important wetland complexes situated above 3000 m altitude enjoying alpine and tundra climate and experiencing diurnal and seasonal temperature variation causing freezing and melting cycle which influence the physiology of these lakes (Panigrahy et al., 2012).

The Nagula Wetland Complex is situated north of the Tawang Township in Western Arunachal Pradesh within the altitudinal range of 3,500–4,500 m AMSL. It comprises more than 100 alpine freshwater lakes that are fed mostly by snow-melt water (Panigrahy et al., 2012).

The Alpine zone, where trees no longer tend to grow (above tree line), the vegetation is mostly in stunted form due to the extreme cold and are dominated by low herbs and prostrate shrubs. The alpine flora flourished in severe environmental stresses which is characterized by a short, cold, unpredictable growing season, low nutrient availability, low pressures of CO_2, high UV irradiance, limited water availability, strong winds and widely fluctuating temperature of air and soil surface (Rundel et al., 2016). The alpine plants have acquired some adaptive features operating in numerous ways to counteract the stresses of such an environment. Floral preformation is one of the most successful adaptations seen in the Himalayan plants, where flowers with floral parts, inflorescence or whole aerial parts are pre-formed as miniatures inside winter buds, and in some cases, this is done through the formation of pseudobulbs (Rawat & Gaur 2004). Several workers studied the preformation of alpine flowers in other parts of the world and also in the Himalayas (Bliss, 1971; Smith et al., 1987; Korner 1999, 2003; Tsukaya et al., 2001). However, the alpine vegetation of the Arunachal Himalaya has received very limited attention regardless of being a part of the

Eastern Himalaya and of a global biodiversity hotspot. The article reports a preliminary account of flowering plant diversity around some of the lakes of Nagula Wetland Complex of Arunachal Pradesh.

8.2 MATERIALS AND METHODS

Seven lakes of the Nagula Wetland Complex [Penga-teng Tso, Kyo-Tso, Sungetser-Tso, and other four unnamed lakes] in the Tawang district of Arunachal Pradesh and the areas around those lakes were visited twice, during the summer (August) of 2017 when the temperature remains high, lake water remain in melted form and majority of the species remain in their flowering phenophase. The plants were collected at random, and field characters were noted in a Field Note Book, and most of the plants were also photographed. Specimens were then processed into mounted Herbarium sheets following conventional methods (Jain and Rao, 1977) in the Department of Botany of the Rajiv Gandhi University. Specimens were then identified using different floras including Flora of Assam, Materials for the flora of Arunachal, Flora of British India. For the updated nomenclature and family delimitation, www.theplantlist.org was extensively consulted. The voucher specimens are stored at ARUN (Herbarium of BSI, Itanagar) (Plates 8.1–8.5).

To record the uses of these plants, primarily, local people of different age groups were consulted. In addition, different published literature was also consulted to understand the range of usefulness of the recorded species of plants (Das et al., 1990; Rai et al., 1998).

8.3 ENUMERATION OF FLOWERING PLANTS

The survey of seven lakes and their margins in the Tawang district of Arunachal Pradesh, revealed a total of 106 flowering plant species, representing 68 genera and 32 families of angiosperms (Table 8.1, Figure 8.1). These are mostly herbs which dominate the alpine zone and shrubs which forms the permanent vegetation in the study area. Rhododendrons are the most prominent ones. Asteraceae appeared as the most dominant family represented by 20 species, and it was followed by Orobanchaceae (8 spp.). Plant family Gentianaceae, Ranunculaceae, and Orchidaceae are represented with six species each. No clear delimitations were observed on their range of distribution probably due to the narrow altitudinal range and of almost uniform environmental conditions. However, some plants like *Senecio raphanifolius, Pedicularis siphonantha, Parnassia nubicola,* etc. were common throughout, almost covering every corner of these lakes. On the other hand, some plants were restricted to specific lakes such as *Galearis roborovskyi, Pegaeophyton scapiflorum,* etc. Most of the plants were growing in the open ground except for few which were present under the rhododendron as like parasitic *Boschniakia himalaica.* Again, not a single species of aquatic angiosperms was found growing in these lakes.

PLATE 8.1 **A.** Penga-teng Tso Lake; **B.** Alpine vegetation; **C.** *Fragaria daltoniana*; **D.** *Ligusticopsis wallichiana*; **E.** *Geranium collinum*; **F.** *Anaphalis nubigena.*

PLATE 8.2 (See color insert.) **G.** *Galearis roborovskyi*; **H.** *Pegaeophyton scapiflorum*; **I.** *Rhodiola sedoides*; **J.** *Cirsium eriophoroides*; **K.** *Corydalis polygalina*; **L.** *Codonopsis foetens*; **M.** *Aconitum fletcherianum*; **N.** *Geranium polyanthes*; **O.** *Impatiens chungtienensis.*

PLATE 8.3 P. *Corydalis cashmeriana*; **Q.** *Primula glomerata*; **R.** *Ligularia fischeri*; **S.** *Euphrasia bhutanica*; **T.** *Corydalis jigmei*; **U.** *Saussurea obvallata.*

PLATE 8.4 **V.** *Potentilla bryoides*; **W.** *Cyananthus lobatus*; **X.** *Taraxacum campylodes*; **Y.** *Impatiens radiata*; **Z.** *Pedicularis trichoglossa*; **A1.** *Persicaria vivipara*; **B1.** *Caltha palustris*.

PLATE 8.5 **C1.** *Utricularia brachiata*; **D1.** *Arisaema jacquemontii*; **E1.** *Platanthera latilabris*; **F1.** *Onosma hookeri*; **G1.** *Cyananthus flavus*.

8.4 DIVERSITY OF FLOWERING PLANTS

8.4.1 HABIT GROUPS

Though the vegetation is sparse, but is dominated by herbaceous plants (Table 8.1). There are only six species that can be recognized as shrubs, and one species is a subshrub. Again, most of the herbs (64 species) are perennial which can survive under snow or in a very cold environment. These chaemephytes are nicely adapted for such climatic conditions. After that annuals (therophytes) are prevailing in this region and are represented by 34 species. Also, there is one biennial plant (*Verbascum thapsus*). Such type of vegetation can be formed only under highly stressed environment and where there is enough moisture in the habitat during the growth season.

TABLE 8.1 Enumeration of Plants Recorded from Nagula Wetland Complex in the Alpine Region of Tawang District of Arunachal Pradesh and Their Uses

	Name	Family	Habit	Potentiality
1.	*Aconitum fletcherianum* G.Taylor	Ranunculaceae	PH	M
2.	*Aconitum heterophyllum* Wall. ex Royle	Ranunculaceae	AH	M
3.	*Allium wallichii* Kunth	Amaryllidaceae	PH	M, E
4.	*Anaphalis busua* (Buch.-Ham.) DC.	Asteraceae	AH	O
5.	*Anaphalis nubigena* DC.	Asteraceae	AH	O
6.	*Anaphalis triplinervis* (Sims) Sims ex C.B.Clarke	Asteraceae	AH	M, O
7.	*Anemone rupestris* Wall. ex Hook.f. & Thomson	Ranunculaceae	AH	O
8.	*Arisaema jacquemontii* Blume	Araceae	AH	O
9.	*Aster sikkimmensis* Hook.f. & Thomson	Asteraceae	SS	O
10.	*Bistorta calostachya* (Diels) Soják	Polygonaceae	PH	O
11.	*Boschniakia himalaica* Hook.f. & Thomson	Orobanchaceae	AH, Parasite	SI
12.	*Calceolaria mexicana* Benth.	Scrophulariaceae	AH	O
13.	*Caltha palustris* L.	Ranunculaceae	PH	M
14.	*Carduus edelbergii* Rech.f.	Asteraceae	AH	O
15.	*Cassiope fastigiata* (Wall.) D.Don	Ericaceae	S	M, O
16.	*Cirsium eriophoroides* (Hook.f.) Petr.	Asteraceae	PH	SI-Calcium indicator
17.	*Cirsium falconeri* (Hook.f.) Petr.	Asteraceae	PH	M

TABLE 8.1 *(Continued)*

	Name	Family	Habit	Potentiality
18.	*Cirsium verutum* (D.Don) Spreng.	Asteraceae	PH	M
19.	*Codonopsis foetens* Hook.f. & Thomson	Campanulaceae	AH	M
20.	*Corydalis cashmeriana* Royle	Fumariaceae	PH	M, O
21.	*Corydalis ecristata* (Prain) D.G.Long	Fumariaceae	PH	O
22.	*Corydalis jigmei* C.E.C.Fisch. & Kaul	Fumariaceae	PH	O
23.	*Corydalis polygalina* Hook.f. & Thomson	Fumariaceae	PH	O
24.	*Cremanthodium reniforme* (DC.) Benth.	Asteraceae	PH	O
25.	*Cyananthus flavus* C.Marquand	Campanulaceae	PH	O
26.	*Cyananthus lobatus* Wall. ex Benth.	Campanulaceae	PH	O
27.	*Dasiphora fruticosa* (L.) Rydb.	Rosaceae	S	M, O
28.	*Euphorbia wallichii* Hook.f.	Euphorbiaceae	PH	M
29.	*Euphrasia bhutanica* Pugsley	Orobanchaceae	AH	M
30.	*Fragaria daltoniana* J.Gay	Rosaceae	PH	M, E
31.	*Galearis roborovskyi* (Maxim.) S.C.Chen, P.J.Cribb & S.W.Gale	Orchidaceae	AH	O [CITES]
32.	*Gaultheria nummularioides* D.Don	Ericaceae	S	E, M, O
33.	*Gentiana elwesii* C.B.Clarke	Gentianaceae	PH	M
34.	*Gentiana leucomelaena* Maxim.	Gentianaceae	AH	-
35.	*Gentiana pedicellata* (D.Don) Wall.	Gentianaceae	AH	-
36.	*Geranium collinum* Stephan ex Willd.	Geraniaceae	PH	M, O
37.	*Geranium polyanthes* Edgew. & Hook.f.	Geraniaceae	PH	M
38.	*Geranium wallichianum* D.Don ex Sweet	Geraniaceae	PH	M
39.	*Halenia elliptica* D.Don	Gentianaceae	AH	O
40.	*Herminium lanceum* (Thunb. ex Sw.) Vuijk	Orchidaceae	AH	M, O
41.	*Impatiens chungtienensis* Y.L.Chen	Balsaminaceae	AH	M, O
42.	*Impatiens racemosa* DC.	Balsaminaceae	AH	O
43.	*Impatiens radiata* Hook.f.	Balsaminaceae	AH	O
44.	*Impatiens scabrida* DC.	Balsaminaceae	AH	O
45.	*Lactuca macrorhiza* (Royle) Hook.f.	Asteraceae	PH	M
46.	*Leycesteria formosa* Wall.	Caprifoliaceae	S	O
47.	*Ligularia amplexicaulis* DC.	Asteraceae	PH	M
48.	*Ligularia discoidea* S.W.Liu	Asteraceae	PH	-
49.	*Ligularia fischeri* (Ledeb.) Turcz.	Asteraceae	PH	-
50.	*Ligusticopsis wallichiana* (DC.) Pimenov & Kljuykov	Apiaceae	PH	M

TABLE 8.1 *(Continued)*

	Name	Family	Habit	Potentiality
51.	*Meconopsis grandis* Prain	Papaveraceae	PH	M, O
52.	*Meconopsis paniculata* (D. Don) Prain	Papaveraceae	PH	M, O
53.	*Myriactis wightii* DC.	Asteraceae	AH	M
54.	*Onosma hookeri* C.B.Clarke	Boraginaceae	PH	M
55.	*Paris polyphylla* Sm.	Melanthiaceae	PH	M
56.	*Parnassia nubicola* Wall. ex Royle	Celastraceae	PH	M
57.	*Parochetus communis* D.Don	Leguminosae	AH	O
58.	*Pedicularis longiflora* Rudolph	Orobanchaceae	AH	M, O
59.	*Pedicularis megalantha* D.Don	Orobanchaceae	AH	M, O
60.	*Pedicularis siphonantha* D.Don	Orobanchaceae	AH	M, O
61.	*Pedicularis denudata* Hook.f.	Orobanchaceae	AH	O
62.	*Pedicularis pantlingii* Prain	Orobanchaceae	AH	O
63.	*Pedicularis trichoglossa* Hook.f.	Orobanchaceae	PH	M, O
64.	*Pegaeophyton scapiflorum* (Hook.f. & Thomson) C.Marquand & Airy Shaw	Brassicaceae	PH	M
65.	*Persicaria vivipara* (L.) Ronse Decr.	Polygonaceae	PH	E
66.	*Platanthera clavigera* Lindl.	Orchidaceae	PH	O
67.	*Platanthera latilabris* Lindl.	Orchidaceae	PH	O
68.	*Pleurospermum album* C.B.Clarke ex H.Wolff	Apiaceae	PH	M
69.	*Ponerorchis chusua* (D.Don) Soó	Orchidaceae	PH	M
70.	*Potentilla bryoides* Soj k	Rosaceae	PH	-
71.	*Potentilla multifida* L.	Rosaceae	PH	O
72.	*Potentilla peduncularis* D.Don	Rosaceae	PH	O
73.	*Primula glomerata* Pax	Primulaceae	PH	M, O
74.	*Primula prolifera* Wall.	Primulaceae	PH	M, O
75.	*Primula sikkimensis* Hook.	Primulaceae	PH	M, O
76.	*Primula hopeana* Balf.f. & R.E.Cooper	Primulaceae	PH	M, O
77.	*Prunella vulgaris* L.	Lamiaceae	PH	M, O
78.	*Rheum nobile* Hook.f. & Thomson	Polygonaceae	PH	M, O
79.	*Rhodiola sedoides* Liden &P.Bharali	Crassulaceae	PH	-
80.	*Rhododendron campanulatum* D. Don	Ericaceae	S	M, O
81.	*Rhododendron lepidotum* Wall. ex G. Don	Ericaceae	S	M, O, P
82.	*Roscoea purpurea* Sm.	Zingiberaceae	PH	E, M, O
83.	*Rumex maritimus* L.	Polygonaceae	AH	E, M

TABLE 8.1 *(Continued)*

	Name	Family	Habit	Potentiality
84.	*Salvia campanulata* Wall. ex Benth.	Lamiaceae	PH	M, O
85.	*Satyrium nepalense* D.Don	Orchidaceae	AH	M, O
86.	*Sauromatum diversifolium* (Wall. ex Schott) Cusimano & Hett.	Araceae	AH	M
87.	*Saussurea costus* (Falc.) Lipsch.	Asteraceae	PH	M, O, R
88.	*Saussurea obvallata* (DC.) Edgew.	Asteraceae	PH	M, R
89.	*Saxifraga sphaeradena* Harry Sm.	Saxifragaceae	AH	M
90.	*Saxifraga moorcroftiana* (Ser.) Wall. ex Sternb.	Saxifragaceae	PH	M
91.	*Saxifraga pallida* Wall. ex Ser.	Saxifragaceae	AH	M
92.	*Saxifraga hispidula* D.Don	Saxifragaceae	PH	-
93.	*Sedum multicaule* Wall. ex Lindl.	Crassulaceae	PH	M, O
94.	*Senecio graciliflorus* (Wall.) DC.	Asteraceae	PH	M
95.	*Senecio raphanifolius* Wall. ex DC.	Asteraceae	PH	M
96.	*Silene birgittae* Bocquet	Caryophyllaceae	AH	-
97.	*Soroseris hookeriana* (C.B.Clarke) Stebbins	Asteraceae	PH	M
98.	*Swertia grandiflora* Harry Sm.	Gentianaceae	PH	M, O
99.	*Swertia hookeri* C.B.Clarke	Gentianaceae	PH	-
100.	*Taraxacum campylodes* G.E.Haglund	Asteraceae	PH	M, O
101.	*Thalictrum cultratum* Wall.	Ranunculaceae	PH	M, O
102.	*Thalictrum reniforme* Wall.	Ranunculaceae	PH	M
103.	*Trifolium repens* L.	Fabaceae	PH	O
104.	*Utricularia brachiata* Oliv.	Lentibulariaceae	AH	SI -Carnivorous
105.	*Valeriana hardwickii* Wall.	Caprifoliaceae	PH	M
106.	*Verbascum thapsus* L.	Scrophulariaceae	BH	M, O

Abbreviations: **Habit: AH** = Annual Herb; **BH** = Biennial Herb; **PH** = Perennial Herb; **S** = Shrub; **SS** = Subshrubs. **Potentiality: E** = Edible; **M** = Medicinal; **O** = Ornamental; **P** = Packing; **R** = Religious; **SI** = scientifically interesting.

8.4.2 PLANTS WITH USEFUL PURPOSES

Our own observation and scanning through wide range of literature (including: Das et al., 1990; Rouf et al., 2003; Bonjar 2004; Turker et al., 2005; Kunwar et al., 2005; Zi-yan et al., 2006; Ballabh et al., 2008; Khan et al., 2008; Rai et al., 2008; Singh et al., 2008; Rajbhandari et al., 2009; Joshi et al., 2010; Negi et al., 2011; Rana et al., 2011; Agnihotri et al., 2013; Aslam et al., 2014; Lone et al., 2014; Sharmila et al., 2014; Gupta et al., 2015; Jabeen et al., 2015; Jin Kang et al., 2016; Joshi, 2016; and Tsering et al., 2016) revealed the usefulness of most of these plants. Out of the recorded 106 species, 65 plants were medicinally important, 58 were ornamentals, 5 were edible, 2 have socio-cultural importance, and the leaves of 1 species are known to use for packing different materials.

8.4.2.1 POTENTIAL ORNAMENTAL SPECIES

Among the enumerated flowering plants, 58 species has immense potential to use as ornamentals. Species like *Pleurospermum album, Geranium collinum, Geranium polyanthes, Geranium wallichianum, Bistorta calostachya,* etc. are also growing naturally very well in nearby town areas. Instead of filling our temperate gardens with exotic plants, numerous of these native species with decorative flowers and foliage also can be equally attractive and beautiful.

8.4.2.2 POTENTIAL MEDICINAL PLANTS

The region is also well known for its medicinal plant's diversity such as *Valeriana hardwickii, Swertia grandiflora, Swertia hookeri, Aconitum fletchetarum, Meconopsis paniculata, Verbascum Thapsus,* etc. These medicinal plants were reported to be used in curing several diseases including high blood pressure, nervous overstrain, lung disorders, free radical overloads, and high altitude sickness. Hence, the vegetation of this entire area needs to be surveyed immediately, and some areas should be recognized to conserve for the conservation of medicinal plants and their cultivation. Interestingly, a traditional healer of Zimithang village has cultivated a wide variety of medicinal plants in his greenhouse [ca. 2500 m AMSL], which attracts many research scholars from different parts of the country. Thus, contributing to the development of scientific knowledge across the region and also validating the properties and medicinal values of these plants and thus the traditional knowledge of the indigenous community is being shared with the scientific community.

In this region, the majority of the population follow Buddhism, and these lakes are considered as an ethical symbol of their faith. Moreover, the strict ethical sense among the native people prohibits them from hunting and other associated activities that maintain the fragility and serenity of these lakes. However, the growing demand for traditional medicines is leading to haphazard and unregulated exploitation of these resources. For instance, the growing market of the fungus-insect combination, *Cordyceps sinensis,* has led to its massive unregulated exploitation.

8.4.3 FACTORS FOR VEGETATION CHANGES

Any change in the structure and function of alpine vegetation is considered as an important indicator of climate change (Körner 2003; Das et al., 2008; Chettri et al., 2015). Effects of global warming is prominent in the Himalayas and that has imparted several undesirable changes in alpine vegetation, which visibly altered the structure and overall function of such an ecosystem. The occurrence of a species in any vegetation is its positive response to the total of the habitat conditions. To save and to conserve these ecosystems, the first requirement is the continuation of the original prevailing physical climate. Changes in the climate of an area are the effect of any natural and man-made modification in and/or in nearby areas. Civilization related (so-called) developmental activities are highly important factors that are influencingthe climate change. Different areas, like the most part in the Himalayas, temperate regions in Arunachal Pradesh are facing such changes. Extension of settlements and communication networks, intensive tourism, increase in vehicular traffic, etc. are some such important factors. It is now almost impossible to avoid these changes and, in the long run, that will seriously affect the environment, which will adversely affect the ecosystems. Living in such high up on the mountains is obviously a challenging task. Different aspects of tourism like offering local food on sales, homestays, guides, travel-related activities, etc. are improving the life of local people. But this aspect also needed to be strictly regulated and monitored.

8.5 CONCLUSION

The High altitude wetlands of Tawang district have the potential to shelter and maintain rich and important floral wealth. In order to understand the present status of the area's biological wealth, a detailed inventory is the urgent immediate need. In addition to the declaration of in situ Protected

Area, some ex situ conservatories are to be established under the supervision of biodiversity experts, basically taxonomists.

KEYWORDS

- **Arunachal Pradesh**
- **Flowering Plant Diversity**
- **Himalayas**
- **Nagula Wetland Complex**

REFERENCES

Agnihotri, P.; Hussain, T. An Overview of genus *Saxifraga* L. in Indian Himalayas. *Ann. Pl. Sci.* **2013**, *2*(08), 278–283.

Aslam, K.; Nawchoo, I. A.; Bhat, M. A.; Ganie, A. H.; Aslam, N. Ethno-pharmacological review of genus Primula. *Intn. J. Adv. Res.* **2014**, *2*(4), 29–34.

Atarod, M.; Nasrollahzadeh, M.; Sajadi, S. M. Green synthesis of a Cu/reduced graphene oxide/Fe_3O_4 nanocomposite using *Euphorbia wallichii* leaf extract and its application as a recyclable and heterogeneous catalyst for the reduction of 4-nitrophenol and rhodamine B.*RSC Advances* **2015**, *5*, 91532–91543. doi 10.1039/C5RA17269A.

Ballabh, B.; Chaurasia, O. P.; Ahmed, Zakwan; Singh, Shashi Bala. Traditional medicinal plants of cold desert Ladakh—Used against kidney and urinary disorders. *J. Ethnopharm.* **2008**, *118* (2), 331–339.

Bharali, S.; Khan, M. L. Rhododendrons in Arunachal Pradesh. **2012.** *New Zeal. Rhododendron Assn. Bull.* No. *100*, 81–85.

Bliss, L. C., Arctic and Alpine Plant life Cycles. *Ann. Rev. Ecol. System.* **1971**. *2*. doi, 10.1146/annurev.es.02.110171.002201.

Bonjar, Shahidi. Evaluation of antibacterial properties of some medicinal plants used in Iran. *J. Ethnopharm.* **2004**, *94* (2–3), 301–305.

Chettri, N.; Sharma, E.; Shakya, B.; Thapa, R.; Bajracharya, B.; Uddin, K.; Oli, K. P.; Choudhury, D. *Biodiversity in the Eastern Himalayas: Status, Trends, and Vulnerability to Climate Change*. Climate Change Impact and Vulnerability in the Eastern Himalayas–Technical Report **2010,** ICIMOD, Kathmandu.

Das, A. P. & Chanda, S. Potential ornamentals from the flora of Darjeeling Hills, West Bengal (India). *J. Econ. Tax. Bot.* **1990,** *14*(3), 675–687.

Das, A. P. Conservation efforts for East Himalayan Biodiversity and need for the establishment of corridors. In Ghosh, C. & Das, A. P., *Recent Studies in Biodiversity and Traditional Knowledge in India* **2011**, Sarat Book House, Kolkata. pp. 329–346.

Das, A. P. Floristic studies in Darjiling hills. *Bull. Bot. Surv. India* **2004**, 43(1–4), 1–18.

Das, A. P. The present status of the flowering plants of Darjiling and Sikkim. In: Gupta, Asha (ed.), *Biodiversity Conservation and Utilization* **2013**, Pointer Publishers, Jaipur. Pp. 83–96.

Das, A. P.; Bhujel, R. B. & Lama, D. Plant Resources in the Protected Areas and Proposed Corridors of Darjeeling, India. In *Biodiversity Conservation in the Kangchanjunga Landscape* **2008**, Eds. N. Chettri, B. Shakya & E. Sharma. ICIMOD, Kathmandu. Pp. 57–79.

Das, S. K. Floristic study of Algae under the ice covers in the alpine lakes of Arunachal Pradesh India (Eastern Himalaya). *Cryptog. Biodiv. Assessm.* **2016**, *1*(1), 75–83.

Gupta, V.; Bansal, P.; Bansal, R.; Mittal, P.; Kumar, S. Folklore Herbal Remedies Used in Dental Care in Northern India and Their Pharmacological Potential. *Am. J. Ethnomed.* **2015**, *2*(6), 365–372.

Gurung D. J.; Pant. R. M. Cultural tourism in Arunachal Pradesh. *A J. Busi. Stud.* **2003**, *7* (2003), 1–7.

Hajra, P. K.; Verma, D. M.; Giri, G. S. (1996) Materials for the flora of Arunachal Pradesh, Botanical Survey of India

Hooket J. D. (1875) Flora of British India Vol-I, State for India in Council.

https://en.wikipedia.org/wiki/Geography_of_Arunachal_Pradesh (accessed on 20 March 2019).

https://en.wikipedia.org/wiki/Tawang_district (accessed on 20 March 2019).

Jabeen, N.; Ajaib, M.; Siddiqui, M. F.; Ulfat, M.; Khan, Babar. A survey of ethnobotanically important plants of District Ghizer, Gilgit-Baltistan. *Fuuast J. Biol.* **2015,** *5*(1), 153–160.

Jain, S. K.; Rao, R. R. (1977). Handbook of Field and Herbarium Methods. Today and Tomorrow's Printers and Publishers, New Delhi, 157 pp.

Jin Kang; Yongxiang Kang; Xiaolian Ji; Quanping Guo; Guillaume Jacques; Marcin Pietras; Nasim Łuczaj; Dengwu Li; Łukasz Łuczaj. Wild food plants and fungi used in the mycophilous Tibetan community of Zhagana (Tewo County, Gansu, China). *J. Ethnobiol. Ethnomed.* **2016**, *12*, 21. doi.org/10.1186/s13002-016-0094-y.

Joshi (Shrestha), Sudha. Pyrrolizidine alkaloids in some species of Senecio Linnaeus (Senecioneae: Asteraceae). *The Pl. Innov.* **2016**, *5*(9), 106–109.

Joshi, M.; Dhar, U. *In vitro* propagation of *Saussurea obvallata* (DC.) Edgew.—an endangered ethnoreligious medicinal herb of Himalaya. *Pl. Cell Rep.* **2010**, 21(10), 933–939. https://doi.org/10.1007/s00299-003-0601-1.

Kanjilal, U. N.; Das, A.; Kanjilal, P. C.; De, R. N. (1939) Flora of Assam. Volume 3. Government of Assam.

Khan, S. W.; Khatoon, S. Ethnobotanical studies on some useful herbs of Haramosh and Bugrote valleys in Gilgit, northern areas of Pakistan. *Pak.J. Bot.* **2008**, 40(1), 43–58.

Körner, C., *Alpine Plant Life: Functional Plant Ecology of High Mountain Ecosystems*. **2003**. Springer, Berlin.

Korner, C., *Alpine Plant Life*. **1999.** Springer, Berlin.

Kunwar, R. M.; Adhikar, N. Ethnomedicine of Dolpa district, Nepal: the plants, their vernacular names and uses. *Lyonia* **2005**, *8*(1), 43–49.

Lone, S. H.; Bhat, K. A.; Bhat, H. M.; Majeed, R.; Anand, R.; Hamid, A.; Khuroo; M. A. Essential oil composition of *Senecio graciliflorus* DC: Comparative analysis of different parts and evaluation of antioxidant and cytotoxic activities. *Phytomedicine* **2014**, 21(6), 919–925. doi.org/10.1016/j.phymed.2014.01.012.

Mittermeier, R. A.; Gil, P. R.; Hoffmann, M.; Pilgrim, J.; Brooks T.; Mittermeier, C. G.; Lamoreux, J.; Da Fonseca, G. A. B. Hot*spots Revisited: Earth's Biologically Richest and Most Endangered Terrestrial Ecoregions*. Conservation International. **2005**.

Myers, N.; Mittermeier, R. A.; Mittermeier, C. G.; da Fonseca, G. A. B.; Kent, J. Biodiversity Hotspots Conservation Priorities. *Nature* **2000**, *403*, 853–858.

Negi, V. S.; Maikhuri, R. K.; Vashishtha, D. P. Traditional healthcare practices among the villages of Rawain valley, Uttarkashi, Uttarakhand, India. *Indian J. Trad. Knowl.* **2011**, *10*(3), 533–537.

Panigrahy, S.; Patel, J. G.; Parihar, J. S. *High Altitude Lakes on India. National Wetland Atlas* **2012,** Space Applications Centre, ISRO, Ahmedabad, India.

Rai, P. C. & Das, A. P. Analysis of the flora of Neora Valley National Park in Darjeeling District of West Bengal, India. In: Das, A. P. (ed.), *Perspectives of Plant Biodiversity* **2002**, Bishen Singh Mahendrapal Singh, Dehradun. 135–150.

Rai, P. C.; Sarkar, A.; Bhujel, R. B. & Das, A. P. Ethnobotanical studies in some fringe areas of Sikkim and Darjeeling Himalayas. *J. Hill Res.* **1998**, *11*(1), 12–21.

Rajbhandari, M. R.; Mentel; Jha, P. K.; Chaudhary, R. P.; Bhattarai, S.; Gewali, M. B.; Karmacharya, N.; Hipper, M.; Lindequist. U. Antiviral Activity of Some Plants Used in Nepalese Traditional Medicine. *Evidence-Based Complem. Alt. Med.* **2009**, *6*(4), 517–522.

Rana, M. S.; Samant, S. S. Diversity, indigenous uses, and conservation status of medicinal plants in Manali wildlife sanctuary, Northwestern Himalaya. *Indian J. Trad. Knowl.* **2011**, *10*(3), 439–459.

Rawat, D. S.; Gaur, R. D. Floral preformation in alpine plants of the Himalaya. *Curr.Sci.* **2004**, *86*(11), 1481–1482.

Rouf, A. S. S.; Islam, M. S.; Rahman, M. T. Evaluation of antidiarrhoeal activity *Rumex maritimus* root. *J. Ethnopharm.* **2003**, *84* (2–3), 307–310. doi.org/10.1016/S0378–8741(02)00326–4.

Rundel, P. W.; Millar, C. I. Alpine ecosystems. In: *Ecosystems of California.* Chapter 29. **2016**: University of California Press, Berkeley, California. Pp. 613–634.

Sarma, D.; Mallik, S. K.; Nabam, J.; Dutta, T.; Das, P.; Singh, A. K. Mountain Lakes of Tawang, Arunachal Pradesh. *Fishing Chimes* **2015**, *35*(6), 54–58.

Sharma, E.; Chettri, N.; Oli, K. P. Mountain biodiversity conservation and management: a paradigm shift in policies and practices in the Hindu Kush-Himalayas. *Ecol. Res.* **2010**, *25*, 905–923.

Sharmila, S.; Kalaichelvi, K.; Abirami, P. Ethnopharmacobotanical information of some herbaceous medicinal plants used by Toda tribes of Thiashola, Manjoor, Nilgiris, Western Ghats, Tamilnadu, India. *Intn. J. Pharmac. Sci. Res.* **2014**, *6*(1), 315–320. DOI: http://dx.doi.org/10.13040/IJPSR.0975–8232.

Singh, K. N.; Lal, Brij. Ethnomedicines used against four common ailments by the tribal communities of Lahaul-Spiti in western Himalaya. *J. Ethnopharm.* **2008**, *115* (1), 4 147–159. https://doi.org/10.1016/j.jep.2007.09.017.

Smith, A.; Young, T. P. Tropical Alpine Plant Ecology. *Ann. Rev. Ecol. System*. **1987**, 18. doi, 10.1146/annurev.ecolsys.18.1.137.

Tsering, J.; Tag, H.; Gogoi, B. J. Vijay Veer Traditional Anti-poison Plants Used by the Monpa Tribe of Arunachal Pradesh. Herbal Insecticides, In: Vijay, Veer; Gopalakrishnan, R. (eds.) *Herbal Insecticides, Repellents and Biomedicines: Effectiveness and Commercialization* **2016**, Springer, New Delhi. doi.org/10.1007/978–81–322–2704–5_10.

Tsukaya, H.; Tsuge, T. Morphological Adaptation of Inflorescences in Plants that Develop at Low Temperatures in Early Spring: The Convergent Evolution of "Downy Plants." *Pl. Biol.* **2001**, *3*(5), 536–543. doi, 10.1055/s-2001–17727.

Turker, A. U.; Gure, E. Common mullein (*Verbascum thapsus* L.). *Recent Adv. Res. Phytoth. Res.* **2005**, *19*(9), DOI, 10.1002/ptr.1653.

www.theplantlist.org (accessed on 20 March 2019).

Zi-Yan Li Xiao-Dong; Yang Zhi-Rong; MaJing-Feng; Zhao Hong-bin; Zhang-Liang Li. Flavonoids from Pegaeophyton scapiflorum. *Chem. Nat. Comp.* **2006**, *42*(6), 732–733.

PART II

Ethnopharmacology and Medicinal Plants

CHAPTER 9

Ethno-Bioprospection in Northeast India

AMRITESH C. SHUKLA,[1] NURPEN M. THANGJAM,[2] LALDINGNGHETI BAWITLUNG,[2] DEBASHIS MANDAL,[2] and BERNADETTE MONTANARI[3]

[1]Department of Botany, University of Lucknow, Lucknow–226007, India E-mail: amriteshcshukla@gmail.com

[2]Department of Horticulture, Aromatic and Medicinal Plants, School of Earth Sciences and Natural Resources Management, Mizoram University, Aizawl–796004, India

[3]International Institute of Social Studies, Erasmus University Rotterdam, The Hague, Netherlands

ABSTRACT

Large variety of tribes and communities lives in the northeastern part of India. These communities firmly believe in traditional systems of medicine and herbal medicine, and tend to rely heavily on popular traditional methods for the treatment of diseases prescribed by traditional healers. The present article contains collection of data for the traditional methods in the treatment of common ailments. The data gathered in this chapter are based on some primary literatures, as well as their confirmation from various secondary and tertiary literatures. The results of the present findings shows that the local people tend to use decoction as the most popular method of herbal medicine treatment followed by paste, juice, infusion, and powder.

The study also reveals that these traditional methods are widely used for the treatment of common ailments and diseases in various parts of India. The study makes an important contribution to the documentation of the traditional herbal healing practices to prevent knowledge erosion and the development of curative pharmaceutical products.

9.1 INTRODUCTION

9.1.1 NORTHEAST INDIA

Northeast India comprises the adjacent Seven Sister States (Arunachal Pradesh, Assam, Manipur, Meghalaya, Mizoram, Nagaland, and Tripura), along with the Himalayan state of Sikkim. It is connected to eastern part of India by a narrow corridor squeezed between the Bhutan and Bangladesh (Plate 9.1).

PLATE 9.1 Map of India and part of the northeastern region.

The northeastern part of India constitutes 8% of the total size of the country with a approximate population of 40 million, according to the 2011 census) which represent 3.1% of the total population. Geographically, the northeastern region is categorized into the Eastern Himalayas, Northeast Hills (Lushai Hills and Patkai-Naga Hills), and the Brahmaputra and Barak Valley Plains. The climatic condition of the NE region is humid sub-tropical climate with hot, humid summer, rigorous monsoons, and mild winters. Altitude varies from sea level to over 7,000 mt (23,000 ft) above mean sea level (MSL). The region is subject to heavy rainfall, an average of 10,000 mm (390 in).

9.1.2 TRIBES AND COMMUNITIES OF THE NORTHEASTERN REGION OF INDIA

The NE region of India have more than 220 ethnic groups, and equal number of dialects. The hilly states of the region (Arunachal Pradesh, Mizoram, Meghalaya, Manipur, and Nagaland) are mainly inhabited by tribal people who have diversity within the tribal groups. The important tribes and communities in these states are: Anal, Assamese, Bhutia, Bodo, Biate, Chakma, Chhetri, Dimasa, Garo, Gangte, Gurung, Hmar, Hajong, Hrankhwl, Jamatia, Khasi, Karbi, Khampti, Kom, Koch, Kuki, Lepcha, Limbu, Lushai, Mao, Mizo, Maram, Meitei, Mishing, Monsang, Naga, Noatia, Nepali, Paite, Pnar, Poumai, Purvottar Maithili, Rabha, Reang, Rongmei, Simte, Singpho, Sylheti, Teddim, Tangkhul, Vaiphei, Zeme Naga, and Zou.

9.1.3 TRADITIONAL ETHNO-MEDICINAL PRACTICES OF THE NER

Various plant parts that are used for medicinal purpose have been documented in ancient literatures. The historical development of traditional pharmaceutical formulae is traced back to *Charak Samhita*, the first systematic documentation of *Ayurveda. Ayurveda* also recommended comprehensive *Materia Medica* which includes medicinal plants, minerals; metals and animal-based products; although people have overtime prioritized the use of medicinal plants.

The tribal people and ethnic races throughout the world have develop their own cultures, customs, cults, religious rites, taboos, legends and myths, folktales, medicinal plants, etc. (Rout et al., 2010). The tribal practice of folk medicine includes a number of medicinal plants that are consumed daily as tubers, bulbs, flowers and various plant species and fruits which are consumed in times of food shortages and famine. Because most tribal communities are scattered and live in isolation and are deprived of efficient means of communication and transportation, they rely heavily on traditional systems of medicine, the most affordable and accessible form of treatment (Yineger and Yewhalaw, 2007).

Similarly, the northeastern part of India as one of the most important flora and fauna biodiversity hotspot in the world has its own traditional cultural practices whereby the local people have been using traditional means of treatment, using medicinal plants (Fig.9-1).

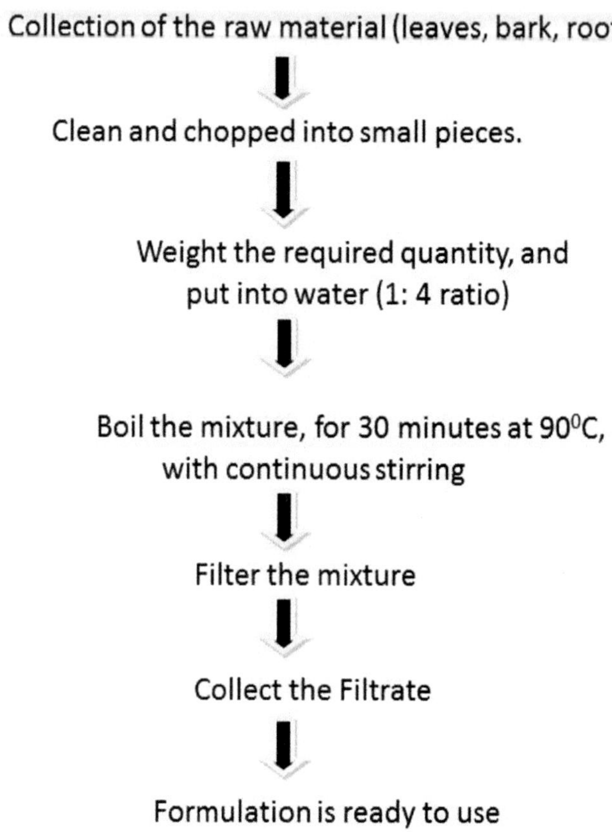

FIGURE 9.1 Schematic representation for developing formulation under decoction method.

Literature also reveals that in northeastern region of India, number of researches have already been made. In Assam (Hajra & Baishya, 1981; Das et al., 2008; Sajem and Gosai 2006); Meghalaya (Rao, 1981; Chetri 1994); Nagaland (Rao & Jamir 1982); Arunachal Pradesh (Dam & Hajra, 1981; Tiwari et al., 1980); Manipur (Lokho 2012; Gurumayum et al., 2014; Ningombam et al., 2014; Singh et al., 2014); Tripura (Das et al., 2012; Bhowmik et al., 2013; Debnath et al., 2014); Sikkim (Bharati et al., 2010);

however from Mizoram (Fischer, 1938; Mahanti 1994, Singh, 1996; Lalramnghinglova & Jha, 2000; Sharma et al., 2000; Lalramnghinglova, 2003, Bharadwaj & Gakhar, 2005, Rai et al., 2007 Lalfakzuala, 2007; Rai & Lalramnghinglova, 2010; Shukla et al., 2011; Shukla et al., 2012; Lalsangluaii et al., 2013(a); Lalsangluaii et al.,., 2013(b); Shukla et al., 2014(a); Shukla et al., 2014(b); Singh et al., 2014; Vermfa et al., 2014 and Shukla et al., 2015), have already been recorded.

9.2 DATA COLLECTION

The information were collected with the help of various primary and secondary literatures as well as through the interviews among different tribes and communities and the traditional practitioners, residing in northeastern region. The interviewed pertaining to names of the ethnomedicinal formulae, plant parts used, methods for combining and preparing, dosages and duration of treatment were recorded, using a set of questioner, and categorized scientifically. The information thus collected were identified with the help of herbariums and the expertise available at the State Forest Department, Government of Mizoram, and finally submitted to the Department of Horticulture, Aromatic and Medicinal Plants, School of Earth Sciences and Natural Resources Management, Mizoram University, Aizawl.

9.3 OBSERVATIONS

The present chapter highlights the various traditional herbal formulae used to treat common diseases, widely used by the tribes and communities in northeastern region. The traditional healers administer locally prepared folk remedies for which they use various methods of preparations, i.e., decoction, infusion, paste, juices, and powder. The study further shows that decoction is the most commonly used method followed by paste, juice, infusion, and powder. Decoctions are made in similar processes as those used in Ayurvedic preparations (*quath* or *kawath*). Infusions are made according to two traditional means: Cold infusion or *Hima* where the dried plant material is coarsely powdered and soaked in plain water and then filtered; and hot infusion or *Phanta* whereby boiled water is used for a hot infusion. Paste, juice, and powder follow standard preparation processes [Plate 2 (A-F)].

PLATE 9.2 (See color insert.) (A–F): An overview of traditional formulae: (A&B) decoction, (C) infusion, (D) juice, (E) paste, and (F) powder.

Further, an overall 213 plant species belonging to 144 families which are traditionally elaborated as formulae were recorded. These are used for treating ailments among 29 northeastern tribal communities. A total of 62 plant species belonging to 49 families are prepared in *decoction* by the traditional healers, 46 plant species which belong to 27 families are made into *paste*, 44 plant species in 27 families are *juiced,* and 40 plant species in 26 families are typically *infused* and 21 plant species from 15 families are powdered.

It was further recorded that the traditional formulae were used to treat sexual asthenia, arthritis, bleeding, bronchitis, high blood pressure, cholera, cold and coughs, cuts and wounds, diarrhea, diabetes, dysentery, fractured bones, fever, gastritis, jaundice, malaria, piles, ringworms, skin infection, snake bite, stomach-ache, typhoid, urinary problems. The study further recorded that Adi-minyong, Apatani, Apong, Assamese, Barpeta, Bhutia, Boro, Chakma, Chungtia, Dhimanji, Dimasa, Idu, Jaintia, Kamrup, Khasi, Lepcha, Limbus, Madahi, Meitei, Mishing, Mizo, Mog, Mongpa, Monpas, Naga, Nyishis, Padam, Rangia, Singphao were the important tribal communities living in Arunachal Pradesh, Assam, Manipur, Meghalaya, Mizoram, Nagaland, Sikkim and Tripura regions. Decoction is the most popular method for 18 tribal communities; paste (16), juice (13), powder (12) and infusion (6) (Figures 9.1–9.8).

9.4 DATA ANALYSIS

Literature reveals that numbers of researches have already been recorded on ethnomedicinal plants, traditionally using for various ailments in north-eastern part of India but there is no such report containing different methods of formulation, practiced among different tribes and communities, at a single place; as recorded in the present investigation.

Borah et al., (2012); Buragohain, (2011); Choudhury et al., (2010); Das et al., (2006); Gogoi et al., (2001); Kalita et al., (2015); Rout et al., (2010), Rout et al., (2012); Saikai et al., (2010); Sajem et al., (2010); Sarma et al., (2002); Sarma et al., (2008); Shankar et al., (2012); Sharma et al., (2011); Taid et al., (2014) have reported that *Achyranthes aspera* L., *Acorus calamus* L., *Adhatoda vasica* Nees, *Adhatoda zeylanica* Medic, *Ageratum conizoides* L., *Alocasia fornicate* (Kunth), *Alocasia indica (L.)* Schott, *Amaranthus spinosus* L., *Amorphophallus paeonifolius* (Dennst.) Nicolson, *Bixa orellana* L., *Clerodendrum Viscosum* L, *Colocasia esculenta* L. Schott., *Desmodium laxiflorum* L, *Homalomena Aromatic* (Spreng.) Schott, *Pistia stratiotes* L., *Plumbago zeylanica* Linn., *Pothos scandens* L., *Psidium guajava* L., *Rhus semi alata* Murr.; *Rhus semialata Murr*, *Rungia parviflora* Nees; *Thunbergia coccinea* L.; *Zanthoxylum armatum* DC are the plants traditionally using among tribal communities in Assam.

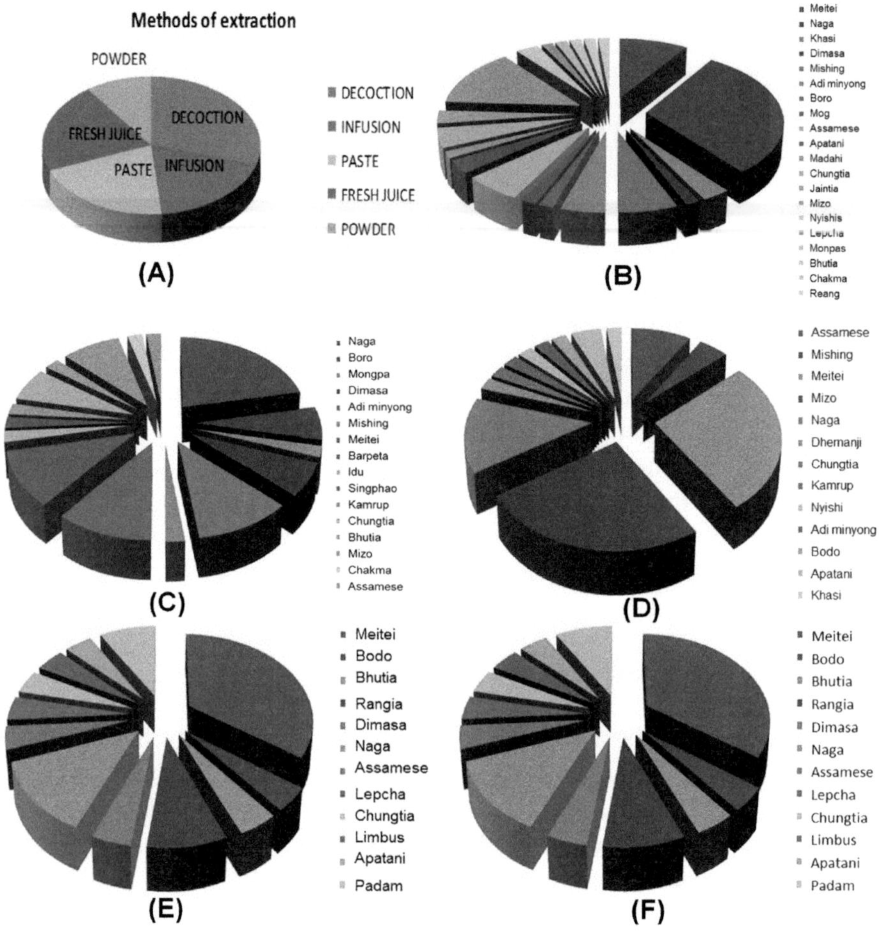

FIGURE 9.2 Various forms of formulae popularized among various tribes and communities. (A): Popular traditional formulae used in northeastern India; (B): Decoction formulae; (C): Paste formulae; (D): Juice formulae; (E): Infusion formulae; (F): Powder formulae.

Similarly, Lalfakzuala et al., (2007); Lalhmingliani et al., (2015); Lalmuanpuii et al., (2013); Lalsangluaii et al., (2013); Lalsangluaii et al (2014); Rai & Lalramnghinglova (2009); Shukla et al (2014); Shukla et al., (2015); have reported the plants *Amomum dealbatum* Roxb, *Ananas comosus* (L.) Merr., *Arenga pinnata* (W.) Merr., *Artocarpus chama* Lam, *Camellia sinensis* Var., *Cassia alata* L., *Curcuma longa* L., *Dysoxylum gobara* (Buch.–Ham.) Merr., *Elaeagnus conferta* Roxb, *Ficus semicordata* Buch–Ham. Ex Sm., *Lindernia ruelloides* (Colsm.), *Mangifera indica* L., *Parkia*

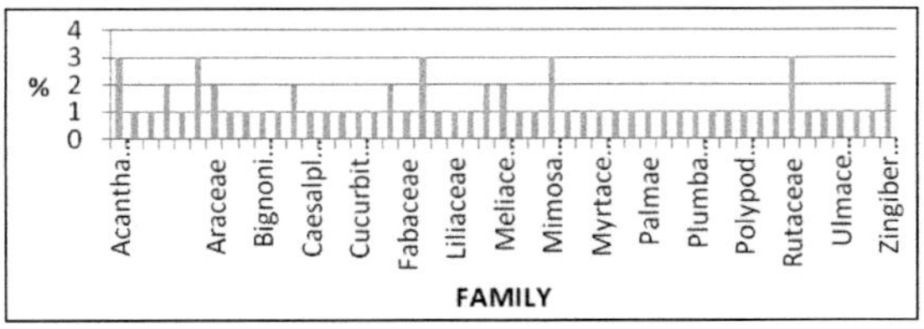

FIGURE 9.3 Percentage of plant families used in decoction formulae.

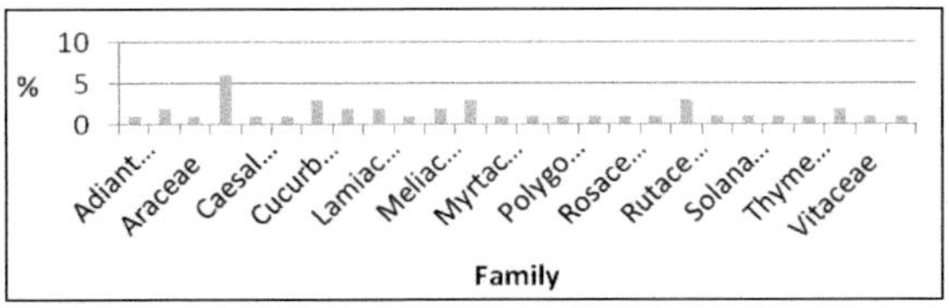

FIGURE 9.4 Percentage of plant families used in infusion formulae.

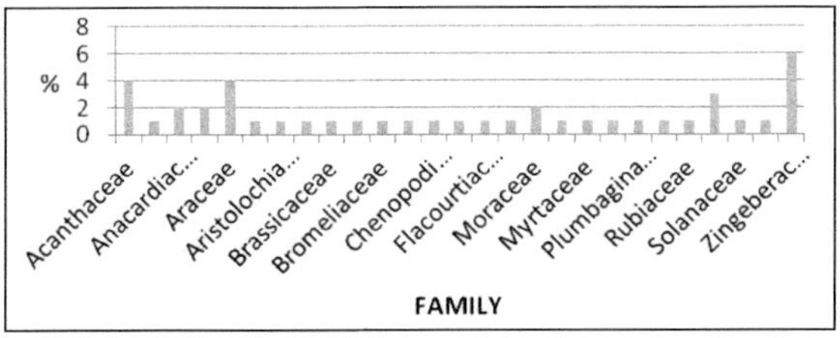

FIGURE 9.5 Percentage of plant families used in juice formulae.

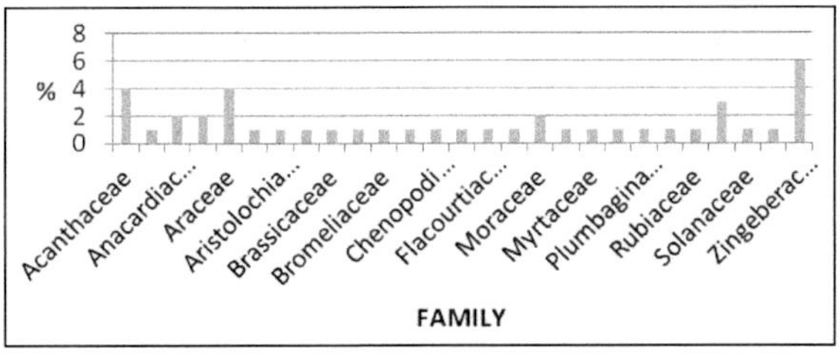

FIGURE 9.6 Percentage of plant families used in powder formulae.

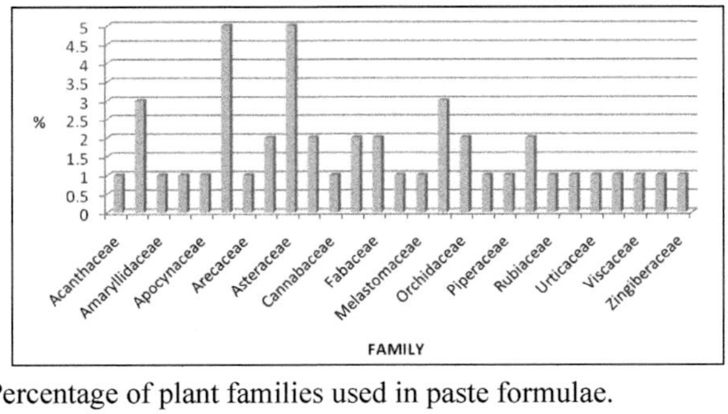

FIGURE 9.7 Percentage of plant families used in paste formulae.

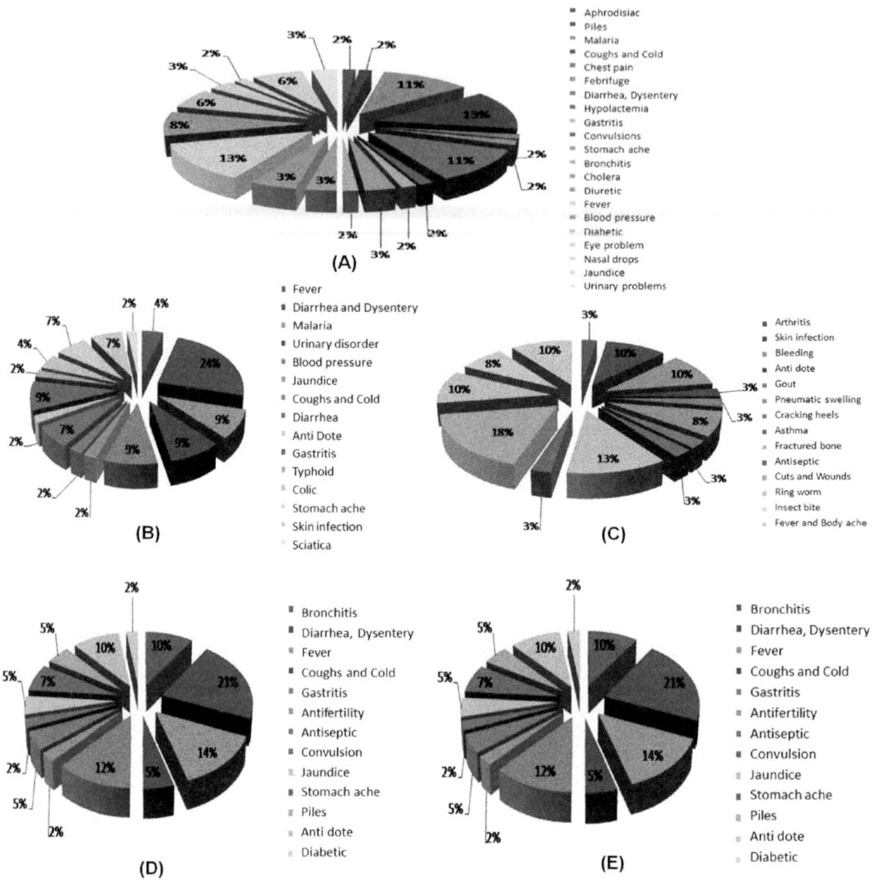

FIGURE 9.8 Common ailments treated by the various traditional preparation methods (a) decoction; (b) infusion; (c) paste; (d) juice; and (e) powdered formulae.

roxburghii (Syn. P. javaniica), *Zanthoxylum rhetsa* (Roxb.) DC; traditionally using among tribal communities in Mizoram; Bhowmik et al., (2013); Das et al., (2012); Debnath et al., (2014); have reported that *Adhatoda vasica* Nees, *Kaempferia rotunda* Linn., *Kalanchoe pinnata* pers., *Leucus aspara* Spreng are the plants traditionally using among tribal communities in Tripura; Bharati et al., (2010); reported that *Alstonia scholaris* R. Br., *Datura metel* L., *Mallotus philippinensis* Muel. Arg., *Melastoma malabathricum* L., are the plants traditionally used among tribal and communities in Sikkim. Laloo et al., (2006) also reported *Ilex khasiana* Purk. Plants are used in Meghalaya.

Baruah et al., (2013); Benniamin, (2011), Khongsai et al., (2011); Shankar et al., (2012) have reported *Adiantum lunulatum* Burm, *Aesculus assamica* Griffith, *Calicarpa arborea* Roxb., *Centilla asiatica* L., *Clerodendron colebrookianum* walp, *Dendrobium nobile* Lindl., *Drynaria quercifolia L., Musa paradisiacal* L., *Nephrolepis auriculata L., Ocimum sanctum* L., *Piper mullesua* Ham. Ex D.Don, *Piper nigrum* L. *Piper peepuloides* Roxb., *Psidium guajava* L., *Rauvolfia densiflora* Benth, *Spilanthus acmella* L., *Zanthoxylum armatum* Skeels DC Roxb, are the plants traditionally using among tribal in Arunachal Pradesh, India.

However, Gurumayum et al., (2014); Lokho (2012); Ningombam et al., (2014); Singh et al., (2014); have reported that *Abelmoschus esculentus* (L.) Moench, *Aegle marmelos* (Linn.) Correa, *Benincasa hispida* (Thunb.) Cogn., *Cannabis sativa* L., *Chenopodium ambrosioides* L., *Coriandrum sativum* L., *Elsholtzia blanda* (Benth.), *Eupatorium birmanicum* DC., *Glycosmis arborea* (Roxb.) DC., *Gynura bicolor* DC, *Hibiscus Sabdariffa* L., *Melia azaderach* L., *Sida rhombifolia* L. were the plants traditionally using among tribal communities in Manipur. Similarly; Jamir et al., (2012); Kichu et al., (2015); have reported that *Alnus nepalensis* D.Don, *Alstonia scholaris* (Linn.), *Artemisia vulgaris* Linn, *Catharanthus roseus* (L.) G.Don, *Cymbopogon citratus* Nees., *Gossypium herbaceum* Linn, *Lantana camara* Linn., *Ocimum basilicum* Linn., *Psidium guajava* L., *Sonerila maculata* Roxb, *Zanthoxylum rhetsa* (Roxb.)DC was the plants traditionally using among tribal communities in Nagaland.

Further, the finding of the current investigation shows that among all the traditional methods of formulations (decoction, infusion, paste, juice, and powder); 'decoction' was the method most frequently used among the tribes and communities of the entire northeastern region. Similar, observations have also been recorded in the studies of paste, juice, infusion, powder respectively; which also favors the current observation.

9.5 CONCLUSION

The results of the study reveal the mixed range methods of plant preparation and practices in the tribal communities of the region. It includes a detailed description of their various applications. It is hopeful that these can serve as a valuable database to record the traditional use of these plants for the prevention of traditional plant use knowledge against erosion and for pharmaceutical development.

KEYWORDS

- **ethno-bioprospecting**
- **northeast India**
- **traditional knowledge**
- **tribal communities**

REFERENCES

Anon. http://www.pnas.Org/content/109/39/15835.short (accessed on 20 march 2019).

Baruah, S.; Borthakur, S. K.; Gogoi, P.; Ahmed, M. Ethnomedicinal plants used by Adi–Minyong tribe of Arunachal Pradesh, Eastern Himalaya. *Indian Journal of Natural Products and Resources*, **2013,** *4*(3), 278–282.

Benniamin, A. Medicinal ferns of Northeastern India with special reference to Arunachal Pradesh. *Indian Journal of Traditional Knowledge*, **2011,** *10*(3), 516–522.

Bharati, K. A.; Sharma, B. L. Some ethnoveterinary plant records for Sikkim Himalaya. *Indian Journal of Traditional Knowledge*, **2010,** *9*(2), 344–346.

Bhardwaj, S.; Ghakar, S. K. Ethnomedicinal plants used by the tribals of Mizoram to cure cut and wound. *Indian Journal of Traditional Knowledge,* **2005,** *4*(1), 75–80.

Bhowmik, S.; Datta, B. K.; Mandal, N. C. Traditional usage of medicinal plants among the Mog community people and their chemical justification. *Acta Biologica Indica*, **2013,** *2*(1), 361–366.

Borah, S. M.; Borah, L.; Nath, S. C. Ethnomedicinal plants from Disoi Valley reserve forest of Jorhat district, Assam. *Plant Sciences Feed*, **2012,** *2*, 59–63.

Buragohain, J. Ethnomedicinal plants used by the ethnic communities of Tinsukia District of Assam, India. *Recent research in Science and Technology*, **2011,** *3*(9), 31–42.

Chhetri, R. B. Further observation on Ethnomedicobotany of Khasi Hills in Meghalaya, India. *Ethnobotany*, **1994,** *6*, 33.

Choudhury, M. D.; Bawari, M.; Singh, L. S. Some antipyretic ethno medicinal plants of Manipuri community of Barak valley, Assam, India. *Ethnobotanical Leaflets*, **2010,** *1*, 4.

Cluis, Corinne. "Bioprospecting: A New Western Blockbuster, After the Gold Rush, the Gene Rush." *The Science Creative Quarterly* (University of British Columbia), **2013**, 8.

Dam, D. P.; Hajra, P. K. Observation on Ethnobotany of the Monpas of Kameng district Arunachal Pradesh. In S. K. Jain (ed). *Glimpses of Indian Ethnobotany,* **1981,** pp. 107–114.

Das, A. K.; Dutta, B. K.; Sharma, G. D. Medicinal plants used by tribes of Cachar district, Assam, *Indian Journal of Traditional Knowledge*, **2008,** *7* (3), 446–454.

Das, N. J.; Saikia, S. P.; Sarkar, S.; Devi, K. Medicinal plants of north kamrup district of Assam used in primary health care system. *International Journal of Traditional Knowledge*, **2006,** *5*(22), 489–493.

Das, S.; Choudhury, M. D. Ethnomedicinal uses of some traditional medicinal plants found in Tripura, India. *J Med Plant Res*, **2012,** *6*, 4908–4914.

Debnath, B.; Debnath, A.; Shilsharma, A.; Paul, C. Ethnomedicinal knowledge of Mog and Reang communities of south district of Tripura, India. *Indian J Adv Plant Res*, **2014,** *1*(5), 49–54.

Fanai, L.; Kumar, A.; Mishra, R. K.; Pandey, A.; Shukla, A. C.; Dikshit, A. Ethnomedicinal concept for modern medicine of Homalomena aromatica Schott ProcInt Conf cum Exhibition on Drugs Discovery & Development from Natural Resources (Eds Lalhlenmawia H), **2014,** pp. 224–236 (ISBN, 978–81–923224–1-4).

Fischer, C. E. C. *The Flora of Lushai Hills Records of the Botanical Survey of India*, **1938,** *12*, 75–161.

Gogoi, R.; Borthakur, S. K. Notes on herbal recipes of Boro tribe in Kamrup district, Assam Ethnobotany, **2001,** *13*, 15–23.

Gurumayum, S.; Soram, J. S. International Journal of Pure & Applied Bioscience. *Int J Pure App Biosci*, **2014,** *2*(1), 147–155.

Hajra, P. K.; Baishya, A. K. Ethnobotanical notes on the Miris (Mishings) of Assam plains. S. K. Jain (eds). *Glimpses of Indian Ethnobotany*, 1981, pp. 161–169.

Jamir, N. S.; Lanusunep, N. P. Medico-Herbal medicine practiced by the Naga tribes in the state of Nagaland (India) *Indian J Fund App Life Sci*, **2012,** *2*(2), 328–333.

Kalita, D.; Boissya, C. L. Some folk uses of plants by Missing tribal of Assam, *Vasundhara*, **2000**, *5*, 79–84.

Kalita, G. J.; Rout, S.; Mishra, R. K.; Sarma, P. Traditionally used medicinal Plants of Bajali Sub-division, Barpeta District, Assam. *Journal of Medicinal Plants*, **2015,** *3*(2), 08–17.

Khongsai, M.; Saikia, S. P.; Kayang, H. Ethnomedicinal plants used by different tribes of Arunachal Pradesh. *Indian Journal of Traditional Knowledge*, **2011,** *10*(3), 541–546.

Kichu, M.; Malewska, T.; Akter, K.; Imchen, I.; Harrington, D.; Kohen, J.; Jamie, J. F. An ethnobotanical study of medicinal plants of Chungtia village, Nagaland, India. *Journal of Ethnopharmacology*, **2015,** *166*, 5–17.

Lalfakzuala, R.; Lalramnghinglova, H.; Kayang, H. Ethnobotanical usages of plants in western Mizoram. *Indian J Tradit Knowl*, **2007,** *6*, 486–493.

Lalhmingliani, E.; Gurusubramanian, G.; Kumar, N. S.; Lalfelpuii, R.; Lalremsanga, H. T. Ethnomedicinal Uses of Host Plants of Wild Silk Moths in Mizoram. *J Environ Soc Sci.* **2015,** *2*(1), 114.

Lalmuanpuii, J.; Rosangkima, G.; Lamin, H. Ethno-medicinal practices among the Mizo ethnic group in Lunglei district, Mizoram. *Science Vision*, **2013,** *13*(1), 24–34.

Laloo, R. C.; Kharlukhi, L.; Jeeva, S.; Mishra, B. P. Status of medicinal plants in the disturbed and the undisturbed sacred forests of Meghalaya, northeast India: population structure and regeneration efficacy of some important species. *Current Science*, **2006,** *90*(2), 225–232.

Lalramnghinglova, H.; Jha, L. K. Ethnobotany: A Review. *J Econ Taxon Bot* **1999,** 23, 1–27.

Lalramnghinglova, H. *Ethnomedicinal Plants of Mizoram.* Bishensingh Mahendrapal Singh, Dehradun, **2003,** pp. 3- 95.

Lalsangluaii, F.; Chinlampianga, M.; Shukla, A. C. Efficacy and Potency of Paris polyphylla Smith, an Ethno-medicinal Plant of Mizoram. *Science and Technology Journal,* **2013,** *1*(1), 36–40.

Lalsangluaii, F.; Kumar, A.; Shukla, A. C.; Dikshit, A. Tradition to Technology: an Approach for Drug Development against Human. *Pathogenic Fungi Science Vision*, **2013,** 13(2), 49–57.

Lokho, A. The folk medicinal plants of the Mao Naga in Manipur, North East India. *Int J Sci Res Publ*, **2012,** *2*(6), 1–8.

Mahanti, N. *Tribal Ethnobotany of Mizoram.* Inter-India Publications, New Delhi, **1994,** pp. 5–94.

Ningombam, D. S.; Devi, S. P.; Singh, P. K.; Pinokiyo, A.; Thongam, B. Documentation and assessment on knowledge of ethnomedicinal practitioners: A case study on local Meetei healers of Manipur. *Int J Pharma Biosci*, **2014,** *9*, 53–70.

Rai, P. K., Lalramnghinglova, H. Ethnomedicinal plant resources of Mizoram, India: Implication of traditional knowledge in health care system. *Ethnobotanical Leaflets*, **2010,** *14*, 274–305.

Rai, P. K.; Lalramnghinglova, H. a Lesser known ethnomedicinal plants of Mizoram, North East India: An Indo-Burma hotspot region. *Journal of Medicinal Plants Research,* **2010,** *4*(13), 1301–1307.

Rai, P. K.; Lalramnghinglova, H. Threatened and less known ethnomedicinal plants of an Indo-Burma hotspot region: Conservation implications. *Environmental Monitoring and Assessment*, **2011,** 178, 53–62.

Rao, R. R.; Jamir, N. S. Ethnobotanical studies in Nagaland I medicinal plants, *Economic Botany,* **1982,** *36*(2), 176–181.

Rao, R. R. Ethnobotany of Meghalaya: Medicinal Plants Used by Khasi and Garo Tribes. *Economic Botany*, **1981,** 35, 4–9.

Route, J.; Sajem, A. L.; Nath, M. Medicinal plants of North Cachar Hills district of Assam used by Dimasa tribe. *IJTK*, **2012,** *1,* 520–527.

Rout, J.; Sajem, A. L.; Nath, M. Traditional medicinal knowledge of the Zeme (Naga) tribe of North Cachar Hills district, Assam on the treatment of diarrhea, Assam University. *Journal of Science and Technology*, **2010,** *5*(1), 63–69.

Saikia, B.; Borthakur, S. K.; Saikia, N. Medico-ethnobotany of Bodo tribals in Gohpur of Sonitpur district, Assam. *Indian Journal of Traditional Knowledge*, **2010,** *9*(1), 52–54.

Sajem, A. L.; Rout, J.; Nath, M. Traditional tribal knowledge and status of some rare and endemic medicinal plants of North Cachar Hills District of Assam, Northeast India. *Ethnobotanical Leaflets,* **2008,** *12*, 261–275.

Sajem, A. L.; Gosai, K. Traditional use of medicinal plants by the Jaintia tribes in North Cachar Hills district of Assam, northeast India. *Journal of Ethnobiology and Ethnomedicine*, **2006,** *2*(1), 1.

Sajem, L.; Gosai, K. Ethnobotanical investigations among the Lushai tribe in North Cachar Hill District of Assam, Northeast. *Ind. J.Trad. Knowl.* **2010,** *9*(1), 108–113.

Sarma, H.; Sarma, A. M.; Sarma, C. M. Traditional knowledge of weeds: a study of herbal medicines and vegetables used by the Assamese people [India]. *Herba Polonica*, **2008,** *54*(2), 80–87.

Sarma, S. K.; Saikia, M. Utilization of wetland resources by the rural people of Nagaon district, Assam. *Indian J Traditional Knowledge*, **2010,** *9*, 145–151.

Sarma, S. K.; Bhattacharjya, D. K.; Devi, B. Traditional use of herbal medicines by Madahi tribe of Nalbari district of Assam. *Ethnobotany;* **2002,** *14*, 103–111.

Shankar, R.; Rawat, M. S.; Deb, S.; Sharma, B. K. Jaundice and its traditional cure in Arunachal Pradesh. *Journal of Pharmaceutical and Scientific Innovation*, **2012,** *1*(3), 93–97.

Sharma, H. K.; Changta, L.; Dolui, A. K. Traditional medicinal plants in Mizoram, India. *Fitoteapia*, **2001,** *72*, 146–161.

Sharma, J.; Gaur, R. D.; Pauli, R. M. Conservation status and diversity of some important plant in the Shivalik Himalaya of Uttarakhand, India. *Int J Med Aron Plants* **2011,** *1, 2,* 75–78.

Sharma, P. P.; Mujumdar, A. M. Traditional knowledge on plants from Toranmal Plateau of Maharashtra. *Indian Journal of Traditional Knowledge,* **2003,** *2*, 292–296.

Sharma, U. K.; Pegu, S. Ethnobotany of religious and supernatural beliefs of the Missing tribes of Assam with special reference to the 'Dobur Uie.' *Journal of Ethnobiology and Ethnomedicine.* **2011,** *7*, 16.

Shukla, A. C.; Chinlampianga, N.; Lalsangluaii, F.; Gupta, R.; Verma, A.; Lalramnghinglova, H. Traditional use of medicinal plants among the tribal communities in Mizoram, *Northeast India Ethnobotany*, **2014,** *26* (1&2), 1–9.

Shukla, A. C.; Kumar, A.; Verma, A.; Mishra, R. K.; Pandey, A.; Dikshit, A. Eucalyptus citriodora can be a potential source of an Antidermatophytic agent. *Proceeding of National Seminar on Recent Advances In Natural Product Research, India,* **2012,** pp. 36–44.

Shukla, A. C.; Lalsangluaii1 F, Singh, B.; Kumar, A.; Lalramnghinglova, H.; Dikshit, A. Homalomena aromatica: an ethnomedicinal plant can be a potential source of antimicrobial drug development. *European Journal of Environmental Ecology,* **2015,** *2*(2), 96–104.

Shukla, A. C.; Pandey, K. P.; Mishra, R. K.; Dikshit, A.; Shukla, N. Broad spectrum antimycotic plant as a potential source of therapeutic agent. *J of Natural Products,* **2011,** *4*, 42–50.

Shukla, A. C.; Zosiamliana, J. H.; Khiangte, L.; Singh, B.; Kumar, A.; Dikshit, A. Botanical Antimicrobials: an Approach from Traditional to Modern System of Drug Development. *Science and Technology Journal*, **2014,** *2*(1), 67–76.

Singh, B.; Shukla, A. C.; Singh, R. Traditional Medicinal Plants and their Prospects for New Drug Development. *Science and Technology J.* **2014,** *2*(1), 45–52.

Singh, T. T.; Sharma, H. R.; Devi, A. R.; Sharma, H. M. Plants used in the treatment of fever by the scheduled caste community of Andro village in Imphal East District, Manipur (India). *Tren Life Sci*, **2014,** *3*(3), 23–29.

Taid, T. C.; Rajkhowa, R. C.; Kalita JC. A study on the medicinal plants used by the local traditional healers of Dhemaji district, Assam, India for curing reproductive health-related disorders Adv *Appl Sci Res*, **2014,** *5*(1), 296–301.

Tiwari, K. C.; Majumdar, R.; Bhattacharjee, S. Folklore medicine from Assam and Arunachal Pradesh (District Tirap). *J Crude Drug Res.* **1979,** *17*, 61–67.

Verma, A.; Bharati, V. K.; Shukla, A. C. Ethno-medicinal Usage of Plants by Mizo tribes in Skin Disease Treatment. In "Environment, Biodiversity and Traditional System" (eds) Khanna, D. R., Solanki, G. S., Pathak, S. K.; Publisher Biotech Books, New Delhi, **2014,** pp. 263–272.

Yineger, H.; Yewhalaw, D. Traditional medicinal plant knowledge and use by local healers in Sekoru District, Jimma Zone, Southwestern Ethiopia. *Journal of Ethnobiology and Ethnomedicine.* **2007,** *3*, 24–10. Doi: 1186/1746-4269-3-24.

CHAPTER 10

Medicinal Plants of India and Their Antimicrobial Property

IFRA ZOOMI,[1] HARBANS KAUR KEHRI,[1] OVAID AKHTAR,[2] PRAGYA SRIVASTAVA,[1] DHEERAJ PANDEY,[1] and RAGHVENDRA PRATAP NARAYAN[3]

[1]*Sadasivan Mycopathology Laboratory, Department of Botany, University of Allahabad–211002, UP, India*

[2]*Department of Botany, Kamla Nehru Institute of Physical and Social Sciences, Sultanpur–228118, U.P., India*

[3]*Netaji Subhash Chandra Bose Government Girls P.G. College Lucknow–201010, UP, India, E-mail: narayan.raghvendra@gmail.com*

ABSTRACT

India has a great diversity of medicinal plants with pharmacological activity. Hundreds of the Indian plants have been used by the local inhabitants to treat and cure the various infectious diseases since ancient time. Medicinal value of the plants is due to the active principles they harbor. Phenols, alkaloids, essential oils, flavonoids, tannins, etc. are the major groups of bioactive compounds found in plants. Hundreds of Indian plants extract show antimicrobial, antifungal, antiviral properties in-vitro as well as in-vivo. Besides this, several nanoparticles (NPs) of gold, silver, and zinc have been made from medicinal plants extract (green synthesis). These NPs have been found very effective against the various human pathogens. Plants extract and/ or NPs synthesized from plant extract target the membranes, DNA or RNA, complexes of oxidative phosphorylation of the pathogen and destructed them. This literature documents about the medicinal plants of India and their antibiotic activities. Also, the types and nature of bioactive compounds and their target site in the pathogen is also discussed. In the last, plants derived NPs, their source and the activity is also discussed.

10.1 INTRODUCTION

Microorganisms including bacteria, fungi, protozoa and virus which have been considered as an important source of various infectious diseases. Though, synthetic antibiotics have been developed over the years to control these microorganisms but, this leads to the development of resistant microbes (Cohen et al., 2013). In addition to this, synthetic antibiotics are quite expensive, inadequate for the treatment and may cause side effects to the host, including immune-suppression and allergic reactions (Saxon et al., 1987). Therefore, there is a need to develop new antimicrobial agents derived from medicinal plants. Medicinal plants are those plants, which has pharmacological activity and are being used to cure the disease (Balunas and Kinghorn, 2005). According to the World Health Organization (WHO), about 80% of the population in developing countries rely on medicinal plants to cure many infectious diseases.

Among developing countries, India has great diversity of medicinal plants and possess antimicrobial property, including antibacterial, antimycobacterial, antiviral and antiprotozoal (Ahmad and Beg, 2001; Chander et al., 2016). The antimicrobial properties of medicinal plants are due to the presence of several bioactive compounds. The important bioactive compounds are alkaloids, flavonoids, tannins, essential oil and phenolic compounds which are synthesized and store in specific parts of the plant (Edeoga et al., 2005). Antimicrobial properties of medicinal plant are further enhanced with the development of new technology, for example, nanotechnology. The biological molecules derived from plant extract used for the synthesis of nanoparticles (Green synthesis), exhibiting superiority over chemical and physical methods (Ahmed et al., 2016). Therefore, in the present chapter, we describe antimicrobial properties of Indian medicinal plants along with their bioactive compounds. The antimicrobial property of nanoparticles synthesized from medicinal plants and their mode of action are also discussed.

10.2 ANTIMICROBIAL PROPERTIES OF INDIAN MEDICINAL PLANTS

India is rich in natural biodiversity and 7th rank among the 16 mega-diversity countries where 70% of the world species occur collectively. According to Sati et al. (2010), it comprises 11% of total known world flora with medicinal property. Traditionally, in India great diversity of medicinal plants with pharmacological activity have been used by the local inhabitants to treat and cure the various infectious diseases (Acharya and Rai, 2011; Parthiban et al., 2016). Medicinal plants are known to produce a variety of bioactive compounds with pharmacological/therapeutic properties. Bioactive compounds of medicinal plants possess various properties, such as antioxidant, stimulant and anticancerous (Nawab al., 2011; Kaur and Mondal, 2014; Batchu et al., 2017). One of the dynamic properties possessed by bioactive compounds of medicinal plants is antimicrobial property. Some of the Indian medicinal plants that possess antimicrobial were listed in table 10.1. The bioactive compounds of medicinal plants either inhibit the growth of pathogens or kill them and have no or least side effect to host cells. There are several mechanisms of bioactive compounds synthesized by medicinal plants that trigger the antimicrobial activity. These are (1) disruption of the microbial membranes or impairing cellular metabolism, (2) control biofilm formation, (3) inhibit bacterial capsule production, (4) can attenuate bacterial virulence by controlling quorum-sensing, (5) denaturation of proteins, (6) reduces the production of microbial toxin, (7) denaturation of DNA, (8) can also act as resistance-modifying agents and (9) inhibiting synthesis of DNA, RNA and protein (Ahmed et al., 2016; Ginovyan et al., 2017). Antimicrobial properties of medicinal plants are being progressively reported from different parts of the India (Ahmed et al., 2016) and may varied from plant to plant (Srinivasan et al., 2001). Thus, medicinal plants serve good substitutes over synthetic antimicrobial medications (Vaseeharan and Thaya, 2014).

10.2.1 ANTIBACTERIAL PROPERTY

Bacterial disease becoming a pivotal concern for public health that has accelerated the demand for plant-based antibacterial agent. Plant-derived antibacterial agents are non-narcotic natural products, economical and have no side effects. Antibacterial properties of plant species have been investigated since ancient time (UNESCO, culture and health, 1996; Kumari and Pandey, 2017). Parekh et al. (2005) screened 12 medicinal plants against *Bacillus subtilis* ATCC6633, *Staphylococcus epidermidis* ATCC12228, *Pseudomonas pseudoalcaligenes* ATCC17440, *Proteus vulgaris* NCTC8313 and *Salmonella typhimurium* ATCC23564 and reported that *Caesalpinia pulcherrima* Swartz showed best antibacterial activity among tested plants. Joshi et al. (2011) evaluated the antibacterial activity of 4 medicinal plants (*Eugenia caryophyllata, Achyranthes bidentata, Azadirachta indica* and *Ocimum sanctum*) against *Escherichia coli, Salmonella typhi, Salmonella paratyphi, Staphylococcus aureus, Klebsiella pneumonia* and *Pseudomonas*

TABLE 10.1 Medicinal Plants of India that Possess Antimicrobial Property

S. No.	Botanical Name	Common Name	Family	Reference
1.	*Abrus Precatorius* (L.)	Gunja	Fabaceae	Janakiraman et al. (2014)
2.	*Abutilon indicum* G. Don	Indian Mallow	Malvaceae	Mata et al. (2015)
3.	*Abutilon indicum* L. (Swart).	Indian mallow	Malvaceae	Srvidya et al. (2009)
4.	*Acacia nilotica* (L.) Willd. ex Del	Babul	Fabaceae	Patel et al. (2015)
5.	*Achillea millefolium* L.	Yarrow	Asteraceae	Verma et al. (2017)
6.	*Achyranthes aspera* L.	Chaff-flower	Amaranthaceae	Srivastav et al. (2011)
7.	*Adhatoda vasica* L.	Vasa, or Vasaka	Acanthaceae	Singh and Sharma (2013)
8.	*Aegle marmelos* (L.)	Bel	Rutaceae	Mujeeb et al. (2014)
9.	*Ailanthus excels* Roxb.	Mahaneem	Simaroubaceae	Manikandan et al. (2015)
10.	*Albezzia lebbeck* (L.) Benth	Lebbeck	Leguminosae	Jaiswal and Kumar (2016)
11.	*Aristolochia indica* L.	Birthwort	Aristolochiaceae	Abhishiktha et al. (2015)
12.	*Azadirachta indica* (Lin.)	Neem or Indian liliac	Meliaceae	Banerjee et al. (2014)
13.	*Caesalpinia bonducella* (L.) Fleming	Grey Nicker	Leguminosae	Lakshmidevi (2015)
14.	*Carica papaya* L.	Papaya	Caricaceae	Gupta et al. (2017)
15.	*Cassia fistula* L.	Indian laburnum	Leguminosae	Duraipandiyan et al. (2007)
16.	*Eclipta alba* (L.) Hassk	Bhringaraj	Asteraceae	Thakur and Pathak (2015)
17.	*Euphrbia hirta* L.	Garden spurge	Euphorbiaceae	Kumari and Pandey (2017)
18.	*Ficus benghalensis* L.	Banyan or Indian fig	Moraceae	Murti and kumar (2011)
19.	*Ficus racemosa* L.	Cluster fig tree	Moraceae	Murti and kumar (2011)
20.	*Hemidesmus indicus* (L.) R.Br.	Indian Sarsaparilla	Asclepiadaceae	Darbari et al. (2016)
21.	*Lantana camara* L.	Lantana	Verbanaceae	Dubey et al. (2013)

TABLE 10.1 *(Continued)*

S. No.	Botanical Name	Common Name	Family	Reference
22.	*Luffa echinata* Roxb.	Bindal	Cucurbitaceae	Modi and Kumar (2014)
23.	*Mallotus philippinensis* (Lam.) Muell. Arg.	Kamala	Euphorbiaceae	Kumar et al. (2006)
24.	*Mentha piperita* L.	Mentha	Lamiaceae	Singh et al. (2015)
25.	*Murraya exotica* L.	Orange Jessamine	Rutaceae	Krishnamoorthy et al. (2013)
26.	*Ocimum sanctum* L.	Tulsi	Lamiaceae	Kumar et al. (2011)
27.	*Psidium guajava* L.	Guava	Myrtaceae	Ranjan et al. (2017)
28.	*Quisqualis indica* L.	Rangoon creeper	Combretaceae	Kumar et al. (2014)
29.	*Raphanus sativus* L.	Radish	Cruciferae	Ahmad and Beg (2001)
30.	*Rosa indica* L.	Rose	Rosaceae	Manikandan et al. (2015)
31.	*Sesbania aegyptiaca* (Poir.) Pers.	Egyptian riverhemp	Fabaceae	Kumar et al. (2006)
32.	*Sesbania sesban* (L) Merr.	Sesban	Fabaceae	Mani et al. (2011)
33.	*Solanum nigrum* L.	Black nightshade	Solanaceae	Ramesh et al. (2015)
34.	*Sphaeranthus indicus* L.	Gorkha mundi	Asteraceae	Kumar et al. (2006)
35.	*Thespesia populnea* (L.) Correa	Indian tulip	Malvaceae	Prakash et al. (2016)
36.	*Vitex negundo* L.	Horseshoe vitex	Verbenaceae	Balasubramani et al. (2017)
37.	*Withania somnifera* (L.) Dunal	Ashawagandha	Solanaceae	Singh et al. (2011)

aeruginosa. Among all the tested pathogenic bacteria, *E. coli* and *K. pneumonia* were found to be resistant against all plants. However, *E. caryophyllata* was found to be potentially effective against *S. typhi* and *Achyranthes bidentata* was found to be ineffective against all the tested organisms. Maximum inhibition was obtained with *E. caryophyllata* against *S. typhi* and minimum with *Azadirachta indica* against *S. typhi.* Seeds extract of *Cuminum cyminum* possesses antibacterial activity against various strains of Gram-positive and Gram-negative bacteria with variable minimum inhibitory concentrations (MIC) (Arora and Kaur, 1999). Likewise, Pandey and Singh (2011) reported the *Syzygium aromaticum* (clove) extract to be potentially effective against *Staphylococcus aureus*, *Pseudomonas aeruginosa* and *Escherichia coli* with MIC ranged from 0.1 to 2.31 mg/ml. Recently, infections caused by multi-drug resistant (MDR) bacteria reported more frequently and are responsible for causing mortality and morbidity worldwide (Worthington and Melander, 2013). MDR bacteria develop due to the involvement of various transporter gene and efflux pump that confers resistance toward many antibiotics (Lubelski et al., 2007; Nikaido and Pagès, 2012; Blair et al., 2014). Singh et al. (2014) reported the antibacterial activity of *Tinospora cordifolia* against MDR strains of *Pseudomonas aeruginosa*. Similarly, *Aegle marmelos* was documented for its antibacterial property against *Escherichia coli* (MDR) (Ganapathy and Karpagam, 2016).

Tuberculosis (TB), a serious infectious disease and is the leading cause of death worldwide. According to the World Health Organization (WHO, 2004), TB infects about 9 million people and nearly 2 million people mortality is seen per annum. *Mycobacterium tuberculosis*, *Mycobacterium tuberculosis complex* including *Mycobacterium bovis* and *Mycobacterium africanum* are the causative microorganisms responsible for the disease (TB) (Sensi and Grassi, 2003). In developed countries (between 1950 and 1970s), decline cases of TB were reported due to the discovery of active antimycobacterial agent viz., isoniazid, ethambutol, rifampicin, pyrazinamide, and streptomycin. Nevertheless, since 1980s, as a result of emergence of multi-drug resistant *Mycobacterium tuberculosis* the number of TB cases all over the world has been increasing rapidly (Chan and Iseman, 2002). Traditionally, plant-based drugs have been used all over the world for the cure of various diseases. In '*Ayurvedic Formulary of India*' 60 medicinal plant species of India mentioned different formulations for the treatment of TB (Anon, 2003). Goel et al. (2002) evaluated six medicinal plants, viz., *Desmodium umbellatum* (L.) Benth., *Piper longum* L., *Plectronia travancorica* Bedd., *Scaevola taccada* (Gaertn.) Roxb., *Syncarpia laurifolia* Tenore and *Vaccinium vacciniaceum* (Roxb.) Sleumer for antimycobacterial activity

(*Mycobacterium tuberculosis* H37Rv). According to Gautam et al. (2007) 25 most active plant species, viz., *Acorus calamus, Adhatoda vasica, Allium sativum, Alpinia galanga, Artocarpus lakoocha, Caesalpinia pulcherrima, Calotropis gigantea, Canscora decussata, Cinnamomum camphora, Cissampelos pareira, Citrullus colocynthis, Erythrina variegata, Glycyrrhiza glabra, Inula racemosa, Juniperus excelsa, Morinda citrifolia, Ocimum sanctum, Piper cubeba, Plantago major, Portulaca oleracea, Psoralea corylifoila, Sassurea lappa, Solanum dulcamara, Tinospora cordifolia* and *Zingiber officinale* are widely distributed in India from tropics to alpine Himalayas and exhibited significant in vitro antimycobacterial activity.

10.2.2 ANTIVIRAL PROPERTY

A virus is harmful, transmissible and causative agent of various life-threatening diseases and cause mortality globally. Various biologically active synthetic agents have been identified for their diverse antiviral activity. The synthetically active antiviral compounds are virazole, ribavirin, ganciclovir, foscarnet and cidofovir. Beside this, biological products from medicinal plants have also been an important source of effective compounds against viral infection (Ballarin et al., 2008). The mechanistic action is due to the inhibition of DNA, RNA and protein synthesis, inhibition of the viral entry and viral reproduction, etc. Various Indian medicinal plants have been studied for their activity against viruses (Subba Rao et al., 1974; Premanathan et al., 2000; Vedapathy, 2003; Vimalanathan et al., 2009). Further, Parida et al. (2002), reported in vitro and in vivo inhibitory potential of leaves extract of *Azadiracta indica* on Dengue virus type-2 replication. Similarly, Tiwari et al. (2010), reported aqueous extract of barks of *Azardirachta indica* as potential entry inhibitor of HSV-1 infection in vivo.

10.2.3 ANTIFUNGAL PROPERTY

Fungi are ubiquitous in distribution and diseases due to fungal pathogens reported more frequently all over the world. Notably, it has been reported by many workers that several fungi have the capability to undergo genetic recombination (Zheng et al., 2011; Zhang et al., 2013), hybridization (Stukenbrock, 2016) and horizontal gene transfer (Cheeseman et al., 2014), which result in acquisition of novel traits. It is therefore, challenging to control the fungal pathogen due to their ability to use various substances as a carbon, nitrogen and energy source. There are various synthetic antifungal agents such as azoles (itraconazole (ITC), voriconazole (VRC), polyenes (amphotericin B) and echinocandins (caspofungin, micafungin and anidulafungin) (Groll et al., 1998; Kathiravan et al., 2012). These antifungal antibiotics no doubt, play major role in health care but also lead to the emergence of acquired drug resistance in fungi (Cowen, 2008; Meneau et al., 2016).

Medicinal plants play a greater role in health care and are considered to provide resistance against fungal diseases (Bansod and Rai, 2008; Murtaza et al., 2015). Vonshak et al. (2003) screened twenty-eight medicinal plants for their antifungal activity against *Trichophyton mentagrophytes*, *T. rubrum*, *T. soudanense*, *Candida albicans*, *C. krusei* and *Torulopsis glabrata.* Several studies have shown that medicinal plants of India have significant antifungal activity. These plants are *Azadirachta indica* (Biswas et al., 2002; Natarajan et al., 2003), *Aegle marmelos* (Balakumar et al., 2011) and *Euphrbia hirta* (Ahmad et al., 2017).

Antifungal activity of medicinal plants also varies with concentration. Very good antifungal activity of various concentration of methanolic extract of *Saraca indica* was reported against *Alternaria cajani*, *Helminthosporium* sp., *Bipolaris* sp., *Curvularia lunata* and *Fusarium* sp by Dabur et al. (2007). Furthermore, extracts of medicinal plants are also found to inhibit the noxious toxin production by fungi. Essential oil of *Curcuma longa* were evaluated at different concentrations (0.01, 0.05, 0.1, 0.5, 0.75, 1.0 and 1.5% (v/v) against *Aspergillus flavus* and aflatoxin production and reported that oil at 1.0% and 1.5% exhibited excellent inhibitory effect on toxin production by fungi (Sindhu et al., 2011). Contrary to this, Parekh and Chanda (2008) evaluated in vitro antifungal activity of nine Indian medicinal plants against pathogenic yeasts and molds and found that activity was not concentration dependent it varies from plant to plant. Kumar et al. (2007) reported that *Chenopodium ambrosioides* oil exhibits anti-aflatoxigenic property. Likewise, *Ocimum sanctum* L. has also been documented for their antifungal activity and anti-aflatoxigenic activity (Kumar et al., 2010).

10.3 BIOACTIVE COMPOUNDS IN MEDICINAL PLANTS

Plant-derived chemical is categorized according to their metabolic functions in plant into two groups; (1) primary metabolites and (2) secondary metabolites or bioactive compounds. Primary metabolites include common sugars, amino acids, proteins and photosynthetic pigments (chlorophyll) that are used by plant for its own metabolism; whereas bioactive compounds consist of phenols, alkaloids, essential oils, flavonoids, tannins and so on. Bioactive compounds possess antimicrobial properties (Figure 10.1) that are used to cure disease in various ways and mechanism of action some of the bioactive compounds (Figure 10.2) were discussed below.

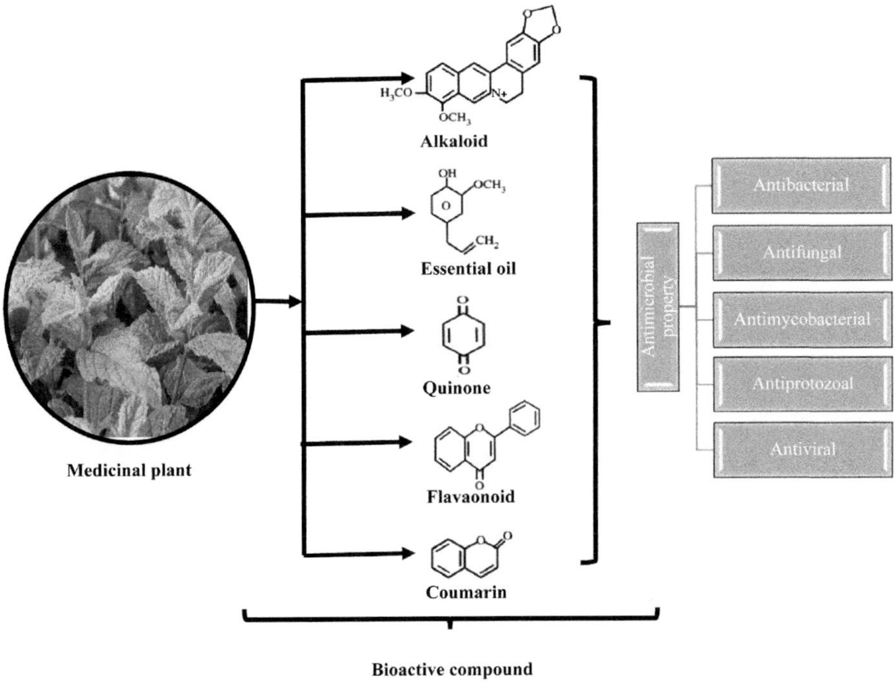

FIGURE 10.1 Diagrammatic representation of antimicrobial property of medicinal plant and their bioactive compounds.

10.3.1 FLAVONOIDS

Chemically, flavonoids are the compounds made of fifteen-carbon skeleton (C6-C3-C6) which consists of two phenyl rings connected by a three-carbon bridge. Flavonoids are synthesized by plants and can be categorized according to their biosynthetic origin, (1) as an intermediate, for examples, flavan-3-ols, flavan-3,4-diols, flavanones and chalcones and (2) as end products, for example, anthocyanidins, proanthocyanidins, flavones, and flavonols. In addition to flavonoids, there are isoflavones, isoflavonoids, and neoflavonoid, which are known to be synthesized by the plants also.

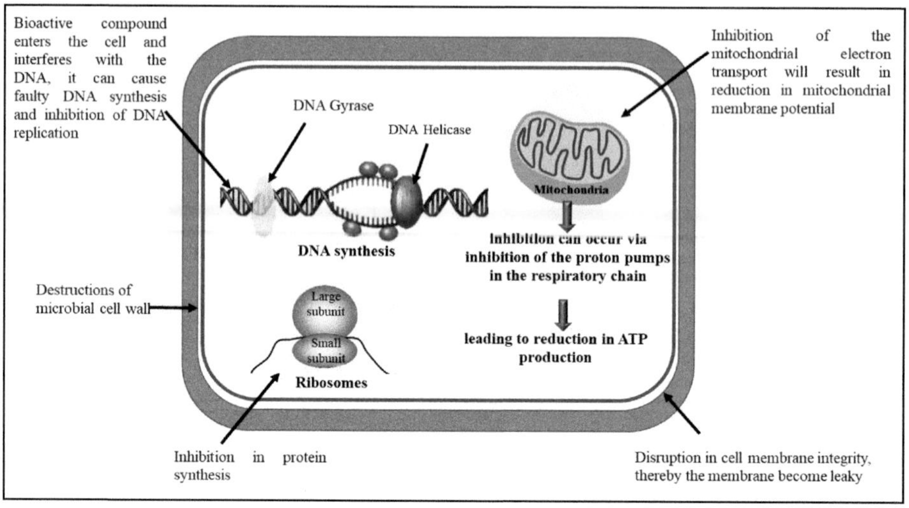

FIGURE 10.2 Mechanisms of action and target sites of bioactive compounds (coumarins, flavonoids, quinones, alkaloids and essential oils) on microbial cells.

Vijaya et al. (1995) and Sathiamoorthy et al. (2007) reported flavonoid glycoside, apigenin, pongaflavonol, galangin, flavanones, isoflavones, and chalcones isolated from the medicinal plants very effective against a number of microorganisms. Lipophilic flavonoids may also disrupt microbial membranes (Tsuchiya et al., 1996) by reducing its fluidity (Tsuchiya and Iinuma, 2000).

Most reduced form of C_3 unit in flavonoid compounds is the catechin. Catechins inactivated the bacterial cholera toxin in *Vibrio cholera* (Borris, 1992). Nakahara et al. (1993) reported catechins as inhibitory to bacterial enzyme glucosyltransferases in *Streptococcus mutans*. Flavonoids are also reported to inhibit electron transport chain between ubiquinone (CoQ) and cytochrome *c* in microorganisms (Haraguchi et al., 1998).

Mori et al. (1987) suggested that the flavonoids containing B ring may inhibit DNA and RNA synthesis in bacterial cell by forming H-bond or by intercalating with the stacking of nitrogenous bases. Inhibitory action of some flavones against *Escherichia coli* DNA gyrase were reported by Ohemeng et al. (1993).

10.3.2 COUMARINS

Coumarins consist of a large class of phenolic substances made of fused benzene and alpha pyrone ring. Based on their chemical structure coumarins are mainly categorized into six types: (1) simple coumarins, (2) furano coumarins, (3) dihydrofurano coumarins, (4) pyrano coumarins (are of two types; linear type and angular type), (5) phenyl coumarins, and (6) bicoumarin. Coumarins are reported most commonly in families such as Rutaceae, Umbelliferae, Clusiaceae, Guttiferae, Caprifoliaceae, Oleaceae, Nyctaginaceae and Apiaceae (Venugopala et al., 2013). The ecological conditions and seasonal variations could also affect its occurrence in various portions of the plant. Coumarins are distributed in whole plant but varied with different plant portion such as fruits > seeds > roots> leaves> latex. Various coumarins compound such as anthogenol, imperatorin, grandivittin, agasyllin, aegelinol benzoate, and osthole. The roles of plant derive coumarin compounds against microbial growth has been reviewed by Venugopala et al. (2013). The mechanistic action of coumarins on microbial activity is due to their interaction with eukaryotic DNA (Silva and Fernandes, 2010).

10.3.3 ALKALOIDS

Alkaloids are a group of small organic compounds derived from amino acids containing one or more nitrogen atom in a heterocyclic ring. These compounds are derived from amino acids such as ornithine, lysine, phenylalanine, tyrosine, tryptophan, histidine, aspartic acid, and anthranilic acid. They are nonvolatile, colorless, water-soluble crystalline solids and tend to have a bitter taste. Various types of alkaloids (more than 3000) have been identified. These are commonly found in families like Ranunculaceae, Solanaceae, Papaveraceae, and Amaryllidaceae and are found in different parts such as leaf, stem, bark and seed. Some examples of alkaloids are morphine, cocaine, atropine, quinine, nicotine, caffeine, atropine, ephedrine and strychnine. Alkaloids in medicinal plants reported to possess the antimicrobial property (Marimuthu et al., 2012; Singh et al., 2015). The mechanistic action of alkaloids is due to their ability to intercalate in between purine and pyrimidine thereby disrupting hydrogen bonds of DNA (Phillipson and O'Neill, 1987).

10.3.4 QUINONES

Quinones are a class of organic compounds containing an aromatic unsaturated ring with two ketone substitutions. They are widespread in occurrence.

They are highly reactive and by a redox reaction, they switch between hydroquinone (diphenol) and quinone (diketone). Saxena et al. (1996) isolated compounds 8-chloro- 2,7-dimethyl-1,4 naphthoquinone (8- chlorochimaphilin), together with chimaphilin and 3-hydroxychimaphilin from *Moneses uniflora* and found that these compounds possess antimicrobial activity. Similarly, Shanker et al. (2007) reported antifungal potential of a compound, 11-hydroxy-16-hentriacontanone, isolated from *Annona squamosa.* The mechanistic action of quinones is due to their ability to form a complex with surface-exposed adhesins, cell wall polypeptides and membrane-bound enzyme that often leads to denaturation of protein (Arif et al., 2008).

10.3.5 ESSENTIAL OILS

Essential oils are volatile, natural, complex compounds characterized by a strong odor and are formed by aromatic plants as secondary metabolites. Essential oils are synthesized by all plant organs, i.e., roots, stems, bark, buds, twigs, leaves, flowers, fruits or seeds and stored in different secretory cells, including epidermal cells or glandular trichomes. The major components of essential oils are panicutine, cineole, geranial, linalool, carvacrol, thymol, p-cymene, and 1,8-cineole and are reported to have antimicrobial activity (Pattnaik et al., 1997; Shankaracharya et al., 2000; Ghosh et al., 2008; Murthy et al., 2009). Chemical compounds from essential oils may act on proteins involved cytoplasmic membrane (Knobloch et al., 1989). Sikkema et al. (1995) reported that cyclic hydrocarbons act on enzyme ATPases, known to synthesize ATP (energy currency of a cell) located at the cytoplasmic membrane and surrounded by lipid molecules. Possibly, hydrocarbons may alter the interaction between lipid and protein molecules and the direct interaction of lipophilic compounds with hydrophobic parts of the protein. There are certain essential oils that stimulate the development of pseudo-mycelia, supporting that essential oils possibly have an effect on enzymes involved in the synthesis of bacterium structural components (Conner and Beuchat, 1984). Gutierrez et al. (2008) reported that most of the essential oils have a more dominant effect on Gram-positive bacteria than a Gram-negative species because of dissimilarities in cell membrane composition. Contrary to this, it has also been found that antimicrobial activity considerably reduced by increasing the hydrophilicity of kaurene diterpenoids (Ultee et al., 2002; Ahmad and Aqil, 2007).

10.4 GREEN SYNTHESIS OF NANOPARTICLES FROM MEDICINAL PLANTS AND THEIR ANTIMICROBIAL PROPERTY

Nanotechnology (Feynman, 1991) is an emerging field that deals with the synthesis of multifunctional nanoparticles (NPs) of different sizes (10–100 nm), shapes and chemical compositions (Ahmed at al., 2016). Multifunctional NPs play an imperative role in various research fields nowadays because of their unique particle size and shape dependent physical, chemical and biological properties. Catalytic reactivity and other related properties depend on the specific surface area of NPs, as specific surface area increases the biological efficiency of synthesized NPs (Mahdieh et al., 2012). According to Parashar et al. (2009), the biological molecule is suitable for the synthesis of NPs as they are reliable and environmentally benign processes. The number of microorganisms is known for the synthesis of NPs but microbes mediated synthesis at large scale is not feasible as it is costly and required aseptic conditions too (Sathishkumar et al., 2009). Raveendran et al. (2003) have initiated the NPs synthesis from a plant sample called green synthesis. Green synthesis of NP from the plant extract has gained much attention during the past 10 years. Various medicinal plants are reported to facilitate the green syntheses of nanoparticles that possess antimicrobial property (Table. 2). Jagtap and Bapat (2012) reported the green synthesis of silver nanoparticles (AgNP) from seed powder extract of *Artocarpus heterophyllus* Lam. The average size of synthesized AgNP was 10.78 nm and potentially effective toward *Bacillus cereus*, *Bacillus subtilis*, *Staphylococcus aureus* and *Pseudomonas aeruginosa*. *Ficus benghalensis* was also used for the synthesis of AgNPs which are effective against *E. coli* MTCC1302 (Saxena et al., 2012). Likewise, Sre et al. (2015) synthesized AgNP from root extract of *Erythrina indica*. The synthesized AgNP exhibited excellent antibacterial activity against Gram-positive and Gram-negative bacteria. Moreover, the antimicrobial property of synthesized NPs also depends on the susceptibility of microorganism. Savithramma et al. (2011) synthesized AgNP from the bark extract (*Boswellia ovalifoliolata* and *Shorea tumbuggaia*) and leaf extract (*Svensonia hyderobadensis*). By using the disc diffusion method, they tested the synthesized NP against bacteria (*Proteus, Pseudomonas, Klebsiella, Bacillus* and *E. coli*) and fungi (*Aspergillus, Fusarium, Curvularia* and *Rhizopus* sp.). The synthesized AgNP from bark extracts of *B. ovalifoliolata* and *S. tumbuggaia* were affective against *Klebsiella, Pseudomonas, Aspergillus* and *Fusarium* sp. respectively. However, silver NPs synthesized from leaf extract of *S. hyderobadensis* was found to be effective against

TABLE 10.2 Nanoparticles Synthesized by Medicinal Plants and their Antimicrobial Activity

Medicinal Plants	Nanoparticle	Activity	Plant Parts	References
Acorous calamus	Silver nanoparticles	Anti-bacterial activity	Rhizome extract	Nakkala et al. (2014)
Boerhaavia diffusa	Silver nanoparticles	Anti-bacterial activity	Leaf extract	Kumar et al. (2014)
Cocos nucifera	Silver nanoparticles	Antibacterial activity	Inflorescence	Mariselvam et al. (2014)
Trianthema decandra	Silver nanoparticles	Anti-microbial activities	Root extract	Geethalakshmi et al. (2010)
Citrus sinensis	Silver nanoparticles	Antibacterial activity	Peel extract	Kaviya et al. (2011)
Aclypha indicum	Silver nanoparticles	Antibacterial activity	Leaf extract	Ashok Kumar et al. (2015)
Cymbopogon citratus	Silver nanoparticles	Antibacterial activity	Leaf extract	Geetha et al. (2014)
Ficus benghalensis and *Azadirachta indica*	Silver nanoparticle	Antibacterial activity	Bark extract	Nayak et al. (2016)
Ocimum sanctum	Silver nanoparticles	Antibacterial activity	Leaf extract	Singhal et al. (2011)
Ocimum basilicum	Zinc oxide Nanoparticles	Antibacterial activity	Leaf extract	Parthasarathy et al. (2017)
Tribulus terrestris	Silver nanoparticles	Antibacterial activity	Fruit extract	Gopinath et al. (2012)
Bougainvillea spectabilis	Silver nanoparticles	Antibacterial activity	Flower extract	Bharati et al. (2016)
Azadirachta indica	Zinc oxide nanoparticles	Antibacterial activity	Leaf extract	Mankad et al. (2016)
Acorus calamus	Gold nanoparticles	Antibacterial activity	Rhizome extract	Ganesan et al. (2015)

Pseudomonas and *Rhizopus* sp. More recently, Jeyabharathi et al. (2017) synthesized zinc oxide nanoparticles (ZnONP) from the extract *Amaranthus caudatus*. The synthesized ZnONP were found to exhibit anti-bacterial activity more towards Gram-positive than Gram-negative bacteria. The antimicrobial activity of NP is due to the presence of positively charged ions that are able to form complexes with nucleic acids. Positively charge NP ions intercalates between the purine and pyrimidine base pairs disturbing the hydrogen bonding between the two antiparallel strands of DNA. Moreover, NPs cause damage to the cell wall or cell membranes and also found to inhibit the growth of microorganisms by modifying phosphotyrosine profile that in turn affects signal transduction. Therefore, the use of medicinal plant extracts for the syntheses of NPs offers numerous benefits over chemical and physical method as it is economical and eco-friendly (Kumar et al., 2014; Ahmed et al., 2016).

10.5 CONCLUSION

In all, it can be concluded that the crude extracts of Indian medicinal plants exhibit significant antimicrobial properties. In many scientific reports, it has been well documented that medicinal plants are potentially effective against pathogenic microbes (bacteria, fungi, and virus) both in vivo and in vitro. The antimicrobial properties of the medicinal plants are due to the presence of various bioactive compounds viz., alkaloids, quinones, flavonoids, coumarins, and essential oils. Antimicrobial properties of medicinal plants extracts are further boosted with the discovery of new technology (green synthesis), which is eco-friendly, economical and can be easily synthesized on a large scale. But, the major disadvantages for green synthesis is variation in chemical compositions of similar plant extract when collected from different localities and there is a need to resolve this problem. To overcome this problem identification, isolation and characterization of bioactive compounds from medicinal plants with specific antimicrobial property can give a new facelift to nanoparticles synthesis that can be used for the welfare of human beings.

ACKNOWLEDGMENTS

The authors gratefully acknowledge the Head, Department of Botany, University of Allahabad, Allahabad, for providing the necessary facilities and also grateful to acknowledge the University Grant Commission and

CSIR for providing financial assistance to Ifra Zoomi, Pragya Srivastava, Dheeraj Pandey, and Ovaid Akhtar.

KEYWORDS

- **bioactive compounds**
- **medicinal plants**
- **microorganisms**

REFERENCES

Abhishiktha, S. N.; Saba, S.; Shrunga, M. N.; Sunitha, K. L.; Prashith, K. T. R.; Raghavendra, H. L. Antimicrobial and Radical Scavenging Efficacy of Leaf and Flower of *Aristolochia indica* Linn. *Sci. Tech. Arts Res. J.* **2015,** *4*(1), 103–108.

Acharya, D.; Rai, M. Traditional knowledge about Indian antimicrobial herbs: retrospects and prospects. *Ethnomed. Plants: Revital. Trad. Know. Herbs,* **2011,** 212–237.

Ahmad, I.; Aqil, F. In vitro efficacy of bioactive extracts of 15 medicinal plants against ESβL-producing multidrug-resistant enteric bacteria. *Microbiol. Res.* **2007,** *162*(3), 264–275.

Ahmad, W.; Singh, S.; Kumar, S. Phytochemical Screening and antimicrobial study of *Euphorbia hirta* extracts. *J. Med. Plants* **2017,** *5*(2), 183–186.

Ahmed, S.; Ahmad, M.; Swami, B. L.; Ikram, S. A review on plants extract mediated synthesis of silver nanoparticles for antimicrobial applications: a green expertise. *J. Adv. Res.* **2016,** *7*(1), 17–28.

Anon. Ayurvedic Formulary of India. Part I. Department of ISM & H, Ministry of Health and Family Welfare, Govt. of India. 2003.

Arif, T.; Bhosale, J. D.; Kumar, N.; Mandal, T. K.; Bendre, R. S.; Lavekar, G. S.; Dabur, R. Natural products–antifungal agents derived from plants. *J. Asian Nat. Prod. Res.* **2009,** *11*(7), 621–638.

Arora, D; Kaur, J. Antimicrobial activity of spices. *Int. J. Antimicrob. Agents.* **1999,** *12,* 257–262.

Ashokkumar, S.; Ravi, S.; Kathiravan, V.; Velmurugan, S. Retracted: Synthesis of silver nanoparticles using *A. indicum* leaf extract and their antibacterial activity. *Spectrochim. Acta A Mol. Biomol. Spectrosc.* **2015,** 34–39.

Balakumar, S.; Rajan, S.; Thirunalasundari, T.; Jeeva S. Antifungal activity of *Aegle marmelos* (L.) Correa (Rutaceae) leaf extract on dermatophytes. *J. Trop. Med.* **2011,** *1,* 309–312.

Balasubramani, S.; Rajendhiran, T.; Moola, A. K.; Diana, R. K. B. Development of nanoemulsion from *Vitex negundo* L. essential oil and their efficacy of antioxidant, antimicrobial and larvicidal activities (*Aedes aegypti* L.). *Environ. Sci. Poll. Res.* **2017,** 1–9.

Ballarin, R.; Di Benedetto, F.; Masetti, M.; Spaggiari, M.; De Ruvo, N.; Montalti, R.; Gerunda, G. E. Combined liver-kidney transplantation in an HIV–HCV-coinfected patient with hemophilia. *Aids* **2008,** *22*(15), 2047–2049.

Balunas, M. J.; Kinghorn, A. D. Drug discovery from medicinal plants. *Life Sci.* **2005,** *78*(5), 431–441.

Banerjee, S.; Banerjee, R. P.; Pradhan, N. K. A Comparative Study on Antimicrobial Activity of Leaf Extract of Five Medicinal Plants and Commonly Used Antibiotics. *Am. J. Phytomed. Clinical Thera-peut.* **2014,** *2*(6), 788–795.

Bansod, S.; Rai, M. Antifungal Activity of Essential Oils from Indian Medicinal Plants against Human Pathogenic *Aspergillus fumigatus* and *A. niger*. *World J. Med. Sci.* **2008,** *3*(2), 81–88.

Batchu, U. R.; Mandava, K.; Bhargav, P. N. V.; Maddi, K. K.; Syed, M.; Rasamalla, S. P.; Madhira, S. Evaluation of Antibacterial and Antioxidant activities of Essential oil from *Michelia champaka*. *J. Appl. Pharm. Sci.* **2017,** *7*(03), 113–116.

Bharathi, D.; Kalaichelvan, P. T.; Atmaram, V.; Anbu, S. Biogenic synthesis of silver nanoparticles from aqueous flower extract of *Bougainvillea spectabilis* and their antibacterial activity. *J. Med. Plants*. **2016,** *4*(5), 248–252.

Biswas, K.; Chattopadhyay, I.; Banerjee, R. K.; Bandyopadhyay, U. Biological activities and medicinal properties of neem (*Azadirachta indica*). *Curr. Sci.* **2002,** *82,* 1336–1345.

Blair, J. M.; Richmond, G. E.; Piddock, L. J. Multidrug efflux pumps in Gram-negative bacteria and their role in antibiotic resistance. *Future Microbiol.* **2014,** *9*(10), 1165–1177.

Borris, R. P. Natural products research: perspectives from a major pharmaceutical company. *J. Ethnopharmacol.* **1996,** *51*(1-3), 29–38.

Chan, E. D.; Iseman, M. D. Current medical treatment for tuberculosis. *BMJ: British Medical J.* **2002,** *325*(7375), 1282.

Chander, M. P.; Pillai, C. R.; Sunish, I. P.; Vijayachari, P. Antimicrobial and antimalarial properties of medicinal plants used by the indigenous tribes of Andaman and Nicobar Islands, India. *Microb. Pathog.* **2016,** *96*, 85–88.

Cheeseman, K.; Ropars, J.; Renault, P.; Dupont, J.; Gouzy, J.; Branca, A.; Malagnac, F. Multiple recent horizontal transfers of a large genomic region in cheese making fungi. *Nat. Commun.* **2014,** 1–*5*.

Cohen, N. R.; Lobritz, M. A.; Collins, J. J. Microbial persistence and the road to drug resistance. *Cell Host Microbe.* **2013,** *13*(6), 632–642.

Conner, D.; Beuchat, L. R. Effects of essential oils from plants on growth of food spoilage yeasts. *J. Food Sci.* **1984,** *49*(2), 429–434.

Cowen, L. E. The evolution of fungal drug resistance: Modulating the trajectory from genotype to phenotype. *Nat. Rev. Microbiol.* **2008,** *6,*187–198.

Dabur, R.; Gupta, A.; Mandal, T. K.; Singh, D. D.; Bajpai, V. Antimicrobial activity of some Indian medicinal plants. *Afr. J. Trad. CAM.* **2007,** *4*(3), 313–318.

Darbari, S.; Agrawal, A.; Verma, P.; Rai, T. P.; Garg, R.; Chaudhary, S. B. antimicrobial activity of root extracts of the medicinal plant *Hemidesmus indicus* (l.) r. br. var. pubescens (w. & a.) hkf. *World J. Pharm. Pharmac. Sci.* **2016,** *5*(4), 1556–1562.

Dubey, D.; Padhy, R. N. Antibacterial activity of *Lantana camara* L. against multidrug-resistant pathogens from ICU patients of a teaching hospital. *J. Herb. Med.* **2013,** *3*(2), 65–75.

Duraipandiyan, V.; Ignacimuthu, S. Antibacterial and antifungal activity of *Cassia fistula* L.: An ethnomedicinal plant. *J. Ethnopharmacol.* **2007,** *112*(3), 590–594.

Edeoga, H. O.; Okwu, D. E.; Mbaebie, B. O. Phytochemical constituents of some Nigerian medicinal plants. *Afr. J. Biotechnol.* **2005,** *4,* 685–688.

Feynman, R. P. There's plenty of room at the bottom. *Science* **1991,** *254,* 1300–1301.

Fischer, G.; Dott, W. Relevance of airborne fungi and their secondary metabolites for environmental, occupational and indoor hygiene. *Arch. Microbiol.* **2003,** *179*(2), 75–82.

Ganapathy, S.; Karpagam, S. In vitro antibacterial and phytochemical potential of *Aegle marmelos* against multiple drug resistant (MDR) *Escherichia coli. J. Pharmacogn. Phytochem.* **2016,** *5*(1), 253.

Gautam, R.; Saklani, A.; Jachak, S. M. Indian medicinal plants as a source of antimycobacterial agents. *J. Ethnopharmacol,* **2007,** *110*(2), 200–234.

Geetha, N.; Geetha, T.S.; Manonmani, P.; Thiyagarajan, M. Green synthesis of silver nanoparticles *using Cymbopogan Citratus* (Dc) Stapf. Extract and its antibacterial activity. *Aus. J. Basic Appl. Sci.* **2014,** *8*(3), 324–31.

Geethalakshmi, R.; Sarada, D. V. L. Synthesis of plant-mediated silver nanoparticles using *Trianthema decandra* extract and evaluation of their antimicrobial activities. *Int. J. Eng. Sci. Technol.* **2010,** *2*(5), 970–975.

Ghosh, A.; Das, B. K.; Chatterjee, S. K.; Chandra, G. Antibacterial potentiality and phytochemical analysis of mature leaves of *Polyalthia longifolia* (Magnoliales: Annonaceae). *South Pac. J. Nat. Appl. Sci.* **2008,** *26*(1), 68–72.

Ginovyan, M.; Petrosyan, M.; Trchounian, A. Antimicrobial activity of some plant materials used in Armenian traditional medicine. *BMC Complement. Altern. Med.* **2017,** *17*(1), 50.

Goel, A. K.; Kulshreshtha, D.K.; Dubey, M. P.; Rajendran, S. M. Screening of Indian plants for biological activity: part XVI. *Ind. J. Exp. Biol.* **2002,** 40, 812–827.

Gopinath, V.; MubarakAli, D.; Priyadarshini, S.; Priyadharsshini, N. M.; Thajuddin, N.; Velusamy, P. Biosynthesis of silver nanoparticles from *Tribulus terrestris* and its antimicrobial activity: a novel biological approach. *Colloids Surf. B Biointerfaces.* **2012,** *96*, 69–74.

Groll, A. H.; Piscitelli, S. C.; Walsh, T. J. Clinical pharmacology of systemic antifungal agents: A comprehensive review of agents in clinical use, current investigational compounds, and putative targets for antifungal drug development. *Adv. Pharmacol.* **1998,** *44,* 343–500.

Gupta, A.; Patil, S.; Pendharkar, N. Antimicrobial and anti-inflammatory activity of aqueous extract of *Carica papaya. J. HerbMed. Pharmacol.* **2017,** *6*(4).

Gutierrez, J.; Barry-Ryan, C.; Bourke, P. The antimicrobial efficacy of plant essential oil combinations and interactions with food ingredients. *Int. J. Food Microbiol.* **2008,** *124*(1), 91–97.

Haraguchi, H.; Tanimoto, K.; Tamura, Y.; Mizutani, K.; Kinoshita, T. Mode of antibacterial action of retrochalcones from Glycyrrhiza inflata. *Phytochem.* **1998,** *48*(1), 125–129.

Jagtap, U. B.; Bapat, V. A. Biosynthesis, characterization and antibacterial activity of silver nanoparticles by aqueous *Annona squamosa* L. leaf extract at room temperature. *J. Plant Biochem. Biotech.* **2013,** *22*(4), 434–440.

Jaiswal, P.; Kumar, P. Evaluation of antimicrobial efficacy of extracts from the bark of a semiarid plant *Albizia lebbeck* (L.) Benth. *Indian J. Nat. Prod. Resour.* **2016,** *7,* 287–292

Janakiraman, N.; Jasmin Jansi, J.; Johnson, M.; Zahir Hussain, M. I.; Jeeva, S. Antibacterial Efficacy of *Abrus precatorius* L. and *Asystasia gangetica* (L.) T. Anderson. *Antiinfect. Agents* **2014,** *12*(2), 165–170.

Jeyabharathi, S.; Kalishwaralal, K.; Sundar, K.; Muthukumaran, A. Synthesis of zinc oxide nanoparticles (ZnONPs) by aqueous extract of *Amaranthus caudatus* and evaluation of their toxicity and antimicrobial activity. *Mater Lett.* **2017,** *209*, 295–298.

Joshi, B.; Sah, G. P.; Basnet, B. B.; Bhatt, M. R.; Sharma, D.; Subedi, K.; Malla, R. Phytochemical extraction and antimicrobial properties of different medicinal plants: *Ocimum sanctum* (Tulsi), *Eugenia caryophyllata* (Clove), *Achyranthes bidentata* (Datiwan) and *Azadirachta indica* (Neem). *J. Microbiol. Antimicrob.* **2011,** *3*(1), 1–7.

Kathiravan, M. K.; Salake, A. B.; Chothe, A. S.; Dudhe, P. B.; Watode, R. P.; Mukta, M. S.; Gadhwe, S. The biology and chemistry of antifungal agents: A review. *Bioorg. Med. Chem.* **2012,** *20,* 5678–5698.

Kaur, S.; Mondal, P. Study of total phenolic and flavonoid content, antioxidant activity and antimicrobial properties of medicinal plants. *J. Microbiol. Exp.* **2014,** *1*(1), 1–6.

Kaviya, S.; Santhanalakshmi, J.; Viswanathan, B. Green synthesis of silver nanoparticles using *Polyalthia longifolia* leaf extract along with D-sorbitol: study of antibacterial activity. *J. Nanotechnol.* **2011,** 5.

Krishnamoorthy, S.; Chandrasekaran, M.; Raj, G. A.; Jayaraman, M.; Venkatesalu, V. Identification of chemical constituents and larvicidal activity of essential oil from *Murraya exotica* L. (Rutaceae) against *Aedes aegypti*, *Anopheles stephensi* and *Culex quinquefasciatus* (Diptera: Culicidae). *Parasitol. Res.* **2015,** *114*(5), 1839–1845.

Kumar, P. V.; Pammi, S. V. N.; Kollu, P.; Satyanarayana, K. V. V.; Shameem, U. Green synthesis, and characterization of silver nanoparticles using *Boerhaavia diffusa* plant extract and their antibacterial activity. *Ind. Crops Prod.* **2014,** *52*, 562–566.

Kumar, R.; Mishra, A. K.; Dubey, N.; Tripathi, Y. Evaluation of *Chenopodium ambrosioides* oil as a potential source of antifungal, antiaflatoxigenic and antioxidant activity. *Int. J. Food Microbiol.* **2007,** *115,*159–164.

Kumar, R.; Shukla, P.; Singh, N. K.; Dubey Chemical composition, antifungal and antiaflatoxigenic activities of *Ocimum sanctum* L. essential oil and its safety assessment as plant-based antimicrobial. *Food Chem. Toxicol.* **2010,** *48,* 539–543.

Kumar, V. P.; Chauhan, N. S.; Padh, H.; Rajani, M. Search for antibacterial and antifungal agents from selected Indian medicinal plants. *J. Ethnopharmac.* **2006,** *107*(2), 182–188.

Kumari, I.; Pandey, R. K. Antibacterial Activity of Euphorbia hirta L. In: Applications of Biotechnology for Sustainable Development. Springer, Singapore. **2017,** pp. 1–5.

Lakshmidevi, N. Evaluation of in vitro antimicrobial activity of *Caesalpinia bonducella* and *Cyclea peltata* extracts against opportunistic microbes. *Asian J. Phytomed. Clinical Res.* **2015,** *3*(2), 55–63.

Lubelski, J.; Konings, W. N.; Driessen, A. J. Distribution and physiology of ABC-type transporters contributing to multidrug resistance in bacteria. *Microbiol. Mol. Biol. Rev.* **2007,** *71*(3), 463–476.

Mahdieh, M.; Zolanvari, A.; Azimee, A. S. Green biosynthesis of silver nanoparticles by *Spirulina platensis*. *Scientia Iranica* **2012,** *19*(3), 926–929.

Mani, R. P.; Awanish, P.; Shambaditya, G.; Poonam, T.; Kumudhavalli, V.; Pratap, S. A. Phytochemical Screening and In-vitro Evaluation of Antioxidant Activity and Antimicrobial Activity of the Leaves of *Sesbania sesban* (L) Merr. *Free Radic. Antiox.* **2011,** *1*(3), 66–69.

Manikandan, A.; Rajendran, R.; Balachandar, S.; Sanumol, M. S.; Sweety, M. M. Antimicrobial activity of *Ailanthus excelsa* Roxb. collected from Coimbatore District, Tamil Nadu, India. *World J. Pharm. Pharmac. Sci.* **2015,** *4*(3), 697–704.

Manikandan, R.; Manikandan, B.; Raman, T.; Arunagirinathan, K.; Prabhu, N. M.; Basu, M. J.; Munusamy, A. Biosynthesis of silver nanoparticles using ethanolic petals extract of *Rosa indica* and characterization of its antibacterial, anticancer and anti-inflammatory activities. *Spectrochim. Acta A Mol. Biomol. Spectrosc.* **2015,** *138*, 120–129.

Mankad, M.; Patil, G.; Patel, S.; Patel, D.; Patel, A. Green synthesis of zinc oxide nanoparticles using *Azadirachta indica* A. Juss. leaves extract and its antibacterial activity against *Xanthomonas orzyae* pv. Oryzae. *Ann. Phytomed.* **2016,** *5*(2), 76–86.

Marimuthu, J.; Aparna, J. S.; Jeeva, S.; Sukumaran, S.; Anantham, B. Preliminary phytochemical studies on the methanolic flower extracts of some selected medicinal plants from India. *Asian Pac. J. Trop. Biomed.* **2012,** *2*(1), 79–82.

Mariselvam, R.; Ranjitsingh, A. J. A.; Nanthini, A. U. R.; Kalirajan, K.; Padmalatha, C.; Selvakumar, P. M. (2014). Green synthesis of silver nanoparticles from the extract of the inflorescence of *Cocos nucifera* (Family: Arecaceae) for enhanced antibacterial activity. *Spectrochim. Acta A Mol. Biomol. Spectrosc.* **2014,** *129*, 537–541.

Mata, R.; Nakkala, J. R.; Sadras, S. R. Biogenic silver nanoparticles from *Abutilon indicum*: Their antioxidant, antibacterial and cytotoxic effects in vitro. *Colloids Surf. B Biointerfaces* **2015,** *128,* 276–286.

Meneau, I.; Coste, A. T.; Sanglard, D. Identification of *Aspergillus fumigatus* multidrug transporter genes and their potential involvement in antifungal resistance. *Med. Mycol.* **2016,** *54*(6), 616–627.

Modi, A.; Kumar, V. *Luffa echinata* Roxb.-A review on its ethanomedicinal, phytochemical and pharmacological perspective. *Asian Pac. J. Trop. Dis.* **2014,** *4*, 7–12.

Mori, A.; Nishino, C.; Enoki, N.; Tawata, S. Antibacterial activity, and mode of action of plant flavonoids against *Proteus vulgaris* and *Staphylococcus aureus*. *Phytochemistry* **1987,** *26* (8), 2231–2234.

Mujeeb, F.; Bajpai, P.; Pathak, N. Phytochemical evaluation, antimicrobial activity, and determination of bioactive components from leaves of *Aegle marmelos*. *BioMed. Res. Int.* **2014,** *2014*, 1–5.

Murtaza, G.; Mukhtar, M.; Sarfraz, A. A review: Antifungal potentials of medicinal plants. *J. Bioresour. Manag.* **2015,** *2*(2), 4.

Murthy, P. S.; Ramalakshmi, K.; Srinivas, P. Fungitoxic activity of Indian borage (*Plectranthus amboinicus*) volatiles. *Food Chem.* **2009,** *114*(3), 1014–1018.

Murti, K.; Kumar, U. Antimicrobial activity of *Ficus benghalensis* and *Ficus racemosa* roots L. *Am. J. Microbiol.* **2011,** *2*(1), 21–24.

Nakahara, K.; Kawabata, S.; Ono, H.; Ogura, K.; Tanaka, T.; Ooshima, T.; Hamada, S. Inhibitory effect of oolong tea polyphenols on glycosyltransferases of *mutans Streptococci*. *App. Environ. Microbiol.* **1993,** *59*(4), 968–973.

Nakkala, J. R.; Mata, R.; Gupta, A. K.; Sadras, S. R. Biological activities of green silver nanoparticles synthesized with *Acorous calamus* rhizome extract. *Eur. J. Med. Chem.* **2014,** *85*, 784–794.

Natarajan, V.; Venugopal, P. V.; Menon, T. Effect of *Azadirachta indica* (neem) on the growth pattern of dermatophytes. *Indian J. Med. Microbiol.* **2003,** *21,* 98–10.

Nawab, A.; Yunus, M.; Mahdi, A. A.; Gupta, S. Evaluation of anticancer properties of medicinal plants from the Indian sub-continent. *Mol. Cell. Pharmacol.* **2011,** *3*(1), 21–29.

Nayak, D.; Ashe, S.; Rauta, P. R.; Kumari, M.; Nayak, B. Bark extract mediated green synthesis of silver nanoparticles: evaluation of antimicrobial activity and antiproliferative response against osteosarcoma. *Mat. Sci. Eng. C.* **2016,** *58*, 44–52.

Nikaido, H.; Pagès, J. M. Broad-specificity efflux pumps and their role in multidrug resistance of Gram-negative bacteria. *FEMS Microbiol. Rev.* **2012,** *36*(2), 340–363.

Ohemeng, K. A., Schwender, C. F., Fu, K. P., & Barrett, J. F. DNA gyrase inhibitory and antibacterial activity of some flavones (1). *Bioorga. Med. Chem. Lett.* **1993,** *3*(2), 225–230.

Pandey, A.; Singh, P. Antibacterial activity of *Syzygium aromaticum* (clove) with metal ion effect against food borne pathogens. *Asian J. Plant Sci. Res.* **2011,** *1*(2), 69–80.

Parashar, V.; Parashar, R.; Sharma, B.; Pandey, A. C. *Parthenium* leaf extract mediated synthesis of silver nanoparticles: a novel approach towards weed utilization. *Digest J. Nanomat. Biostru.* **2009,** *4*(1).

Parekh, J.; Chanda, S. Phytochemicals screening of some plants from western region of India. *Plant Arch.* **2008,** *8,* 657–662.

Parekh, J.; Jadeja, D.; Chanda, S. Efficacy of Aqueous and Methanol Extracts of Some Medicinal Plants for Potential Antibacterial Activity. *Turk. J. Biol.* **2005,** *29*, 203–210.

Parida, M. M.; Upadhyay, C.; Pandya, G.; Jana, A. M. Inhibitory potential of neem (*Azadirachta indica* Juss) leaves on dengue virus type-2 replication. *J. Ethnopharmacol.* **2002,** *79*(2), 273–278.

Parthasarathy, G.; Saroja, M.; Venkatachalam, M.; Evanjelene, V. K. Characterization and Antibacterial activity of Green Synthesized ZnO Nanoparticles from *Ocimum basilicum* Leaf Extract **2017,** *8* (3), 29–35.

Parthiban, R.; Vijayakumar, S.; Prabhu, S.; Yabesh, J. G. E. M. Quantitative traditional knowledge of medicinal plants used to treat livestock diseases from Kudavasal taluk of Thiruvarur district, Tamil Nadu, India. *Rev. Bras. Farmacogn.* **2016,** *26*(1), 109–121.

Patel, J. D.; Sahay, N. S.; Kumar, V. Fingerprinting of the *Acacia nilotica* (L.) Bark Extract Having Antibacterial Property. *J. Health Sci.* **2015,** *3,* 123–127.

Pattnaik, S.; Subramanyam, V.R.; Bapaji, M.; Kole. C.R. Antibacterial and antifungal activity of aromatic constituents of essential oils. *Microbios.* **1997,** *89,* 39–46

Phillipson, J. D.; O'Neill, M. J. Antimalarial and amoebicidal natural products In: Biologically Active Natural Products; K. Hostettmann, P.J. Lea. Eds.; *Proceeding of the Phytochemical Society of Europe,* Clarendon Press, Oxford, 1987, pp. 49–64.

Prakash, S.; Ramasubburayan, R.; Ramkumar, V. S.; Kannapiran, E.; Palavesam, A.; Immanuel, G. In vitro-Scientific evaluation on antimicrobial, antioxidant, cytotoxic properties and phytochemical constituents of traditional coastal medicinal plants. *Biomed. Pharmacother.* **2016,** *83*, 648–657.

Premanathan, M.; Rajendran, S.; Ramanathan, T.; Kathiresan, K. A survey of some Indian medicinal plants for anti-human immunodeficiency virus (HIV) activity. *Indian J. Med. Res.* **2000,** *112*, 73.

Ramesh, M.; Anbuvannan, M.; Viruthagiri, G. Green synthesis of ZnO nanoparticles using *Solanum nigrum* leaf extract and their antibacterial activity. *Spectrochim. Acta A: Mol. Biomol. Spectrosc.* **2015,** *136*, 864–870.

Ranjan, D.; Deogam, A.; Polu, P. R.; Khan, S.; Rao, J. V. Preliminary Studies on Chemical Constituents and Antimicrobial Activities of *Psidium guajava* Linn. *Adv. Sci. Lett. 23*(3), **2017,** 1753–1757.

Rao, G. S.; Sinsheimer, J. E.; Cochran, K. W. Antiviral activity of triterpenoid saponins containing acylated β-amyrin aglycones. *J. Pharm. Sci.* **1974,** *63*(3), 471–473.

Raveendran, P.; Fu, J.; Wallen, S. L. Completely "green" synthesis and stabilization of metal nanoparticles. *J. Am. Chem. Soc.* **2003,** *125*(46), 13940–13941.

Sathiamoorthy, B.; Gupta, P.; Kumar, M.; Chaturvedi, A. K.; Shukla, P. K.; Maurya, R. New antifungal flavonoid glycoside from *Vitex negundo*. *Bioorg. Med. Chem. Lett.* **2007,** *17*(1), 239–242.

Sathishkumar, M.; Sneha, K.; Won, S. W.; Cho, C. W.; Kim, S.; Yun, Y. S. *Cinnamon zeylanicum* bark extract and powder mediated green synthesis of nano-crystalline silver particles and its bactericidal activity. *Colloids and Surf. B: Biointerfaces* **2009,** *73*(2), 332–338.

Sati, S. C.; Joshi, S. Antibacterial potential of leaf extracts of *Juniperus communis* L. from Kumaun Himalaya. *Afr. J. Microbiol. Res.* **2010,** *4*(12), 1291–1294.

Savithramma, N.; Linga Rao, M.; Rukmini, K.; Suvarnalatha Devi P. Antimicrobial activity of silver nanoparticles synthesized by using medicinal plants. *Int. J. Chem. Tech. Res.* **2011,** *3*(3), 1394–1402.

Saxena, A., Tripathi, R. M., Zafar, F., & Singh, P. (2012). Green synthesis of silver nanoparticles using aqueous solution of *Ficus benghalensis* leaf extract and characterization of their antibacterial activity. *Mater. Lett.* **2012,** *67*(1), 91–94.

Saxena, G.; Farmer, S. W.; Hancock, R. E. W.; Towers, G. H. N. Chlorochimaphilin: a new antibiotic from *Moneses uniflora. J. Nat. Prod.* **1996,** *59*(1), 62–65.

Saxon, A.; Beall, G. N.; Rohr, A. S.; Adelman, D. C. Immediate hypersensitivity reactions to beta-lactam antibiotics. *Ann. Intern. Med.* **1987,** *107*(2), 204–215.

Sensi, P.; Grassi, G. G. Chemotherapeutic agents. *Burger's Med. Chem. Drug Discov.* **2003,** *5*, 821–824.

Shankaracharya, N. B.; Nagalakshmi, S.; Naik, J. P.; Rao, L. M. Studies on chemical and technological aspects of ajowan (*Trachyspermum ammi* (L.) Syn. *Carum copticum* Hiern) seeds. *J. Food Sci. Technol.* **2000,** *37*(3), 277–281.

Shanker, K. S.; Kanjilal, S.; Rao, B. V. S. K.; Kishore, K. H.; Misra, S.; Prasad, R. B. N. Isolation and antimicrobial evaluation of isomeric hydroxy ketones in leaf cuticular waxes of *Annona squamosa. Phytochem. Anal.* **2007,** *18*(1), 7–12.

Sikkema, J.; De Bont, J. A.; Poolman, B. Mechanisms of membrane toxicity of hydrocarbons. *Microbiol. Reviews.* **1995,** *59*(2), 201–222.

Silva, N. C. C.; Fernandes Júnior, A. Biological properties of medicinal plants: a review of their antimicrobial activity. *J. Venom. Anim. Toxins Incl. Trop. Dis.* **2010,** *16*(3), 402–413.

Sindhu, S.; Chempakam, B.; Leela, N. K.; Bhai, R. S. Chemoprevention by essential oil of turmeric leaves (*Curcuma longa* L.) on the growth of *Aspergillus flavus* and aflatoxin production. *Food Chem. Toxicol.* **2011,** *49*(5), 1188–1192.

Singh, K.; Panghal, M.; Kadyan, S.; Chaudhary, U.; Yadav, J. P. Antibacterial activity of synthesized silver nanoparticles from Tinospora cordifolia against multi-drug resistant strains of *Pseudomonas aeruginosa* isolated from burn patients. *J. Nanomed. Nanotechnol.* **2014,** *5*(2), 1.

Singh, R.; Shushni, M. A.; Belkheir, A. Antibacterial and antioxidant activities of *Mentha piperita* L. *Arabian J. Chem.* **2015,** *8*(3), 322–328.

Singhal, G.; Bhavesh, R.; Kasariya, K.; Sharma, A. R.; Singh, R. P. Biosynthesis of silver nanoparticles using *Ocimum sanctum* (Tulsi) leaf extract and screening its antimicrobial activity. *J. Nanopart. Res.* **2011,** *13*(7), 2981–2988.

Sre, P. R.; Reka, M.; Poovazhagi, R.; Kumar, M. A.; Murugesan, K. Antibacterial and cytotoxic effect of biologically synthesized silver nanoparticles using aqueous root extract of *Erythrina indica* lam. *Spectrochimi. Acta A Mol. Biomol. Spectrosc.* **2015,** *135*, 1137–1144.

Srinivasan, D.; Nathan, S.; Suresh, T.; Perumalsamy, P. L. Antimicrobial activity of certain Indian medicinal plants used in folkloric medicine. *J. Ethnopharmacol.* **2001,** *74*(3), 217–220.

Srivastav, S.; Singh, P.; Mishra, G.; Jha, K. K.; Khosa, R. L. *Achyranthes aspera*-An important medicinal plant: A review. *J. Nat. Prod. Plant Res.* **2011,** *1*(1), 1–14.

Srvidya, A. R.; Yadev, A. K.; Dhanbal, S. P. Antioxidant and antimicrobial activity of rhizome of Curcuma aromatica and *Curcuma zeodaria*, Leaves of *Abutilon indicum. Arch. Pharm. Sci. Res.* **2009,** *1*(1), 14–19.

Stukenbrock, E. H. Hybridization speeds up the emergence and evolution of a new pathogen species. *Nat. Genet.* **2016,** *48*(2), 113–115.

Thakur, M.; Pathak, S. Phytochemical and Anti-Bacterial Activity of *Eclipta Alba. Asian Reson,* **2015,** 108–112.

Tiwari, V.; Darmani, N. A.; Yue, B. Y.; Shukla, D. In vitro antiviral activity of neem (*Azardirachta indica* L.) bark extract against herpes simplex virus type-1 infection. *Phytother. Res.* **2010,** *24*(8), 1132–1140.

Tsuchiya, H.; Iinuma, M. Reduction of membrane fluidity by antibacterial sophoraflavanone G isolated from *Sophora exigua*. *Phytomedicine*, **2000,** *7*(2), 161–165.

Tsuchiya, H.; Sato, M.; Miyazaki, T.; Fujiwara, S.; Tanigaki, S.; Ohyama, M.; Iinuma, M. Comparative study on the antibacterial activity of phytochemical flavanones against methicillin-resistant *Staphylococcus aureus*. *J. Ethnopharmacol.* **1996,** *50*(1), 27–34.

Ultee, A.; Bennik, M. H. J.; Moezelaar, R. The phenolic hydroxyl group of carvacrol is essential for action against the food-borne pathogen *Bacillus cereus*. *Appl. Environ. Microbiol.* **2002,** *68*(4), 1561–1568.

UNESCO. Culture and Health, Orientation texts, World Decade for Cultural Development 1988-1997. Document CLT/DEC/PRO-1996, UNESCO, Paris, 1996, pp. 129.

Vaseeharan, B.; Thaya, R. Medicinal plant derivatives as immunostimulants: an alternative to chemotherapeutics and antibiotics in aquaculture. *Aquac. Int.* **2014,** *22*(3), 1079–1091.

Vedapathy, S. Scope, and importance of Traditional medicine. *Indian J. Trad. Know*. **2003,** *2,* 236–239.

Venugopala, K. N.; Rashmi, V.; Odhav, B. Review on natural coumarin lead compounds for their pharmacological activity. *Biomed. Res. Int.* **2013,** 1–14.

Verma, R. S.; Joshi, N.; Padalia, R. C.; Goswami, P.; Singh, V. R.; Chauhan, A.; Sundaresan, V. Chemical composition and allelopathic, antibacterial, antifungal and in vitro acetylcholinesterase inhibitory activities of yarrow (*Achillea millefolium* L.) native to India. *Ind. Crops Prod.* **2017,** *104,* 144–155.

Vijaya, K.; Ananthan, S.; Nalini, R. Antibacterial effect of theaflavin, polyphenon 60 (*Camellia sinensis*) and *Euphorbia hirta* on *Shigella* spp.—a cell culture study. *J. Ethnopharmacol.* **1995,** *49*(2), 115–118.

Vimalanathan, S.; Ignacimuthu, S.; Hudson, J. B. Medicinal plants of Tamil Nadu (Southern India) are a rich source of antiviral activities. *Pharm. Biol.* **2009,** *47*(5), 422–429.

Vonshak, A.; Barazani, O.; Sathiyamoorthy, P.; Shalev, R.; Vardy, D.; Golan-Goldhirsh, A. Screening of South Indian medicinal plants for antifungal activity against cutaneous pathogens. *Phytother. Res.* **2003,** *17,* 1123–1125.

Worthington, R. J.; Melander, C. Combination approaches to combat multidrug-resistant bacteria. *Trends Biotechnol.* **2013,** *31*(3), 177–184.

Zhang, Y.; Qiao, M.; Xu, J.; Cao, Y.; Zhang, K. Q.; Yu, Z. F. Genetic diversity and recombination in natural populations of the nematode-trapping fungus *Arthrobotrys oligospora* from China. *Ecol. Evol.* **2013,** *3*(2), 312–325.

Zheng, W.; Huang, L.; Huang, J.; Wang, X.; Chen, X.; Zhao, J.; Liu, H. High genome heterozygosity and endemic genetic recombination in the wheat stripe rust fungus. *Nat. Commun.* **2013,** *4,* 2673.

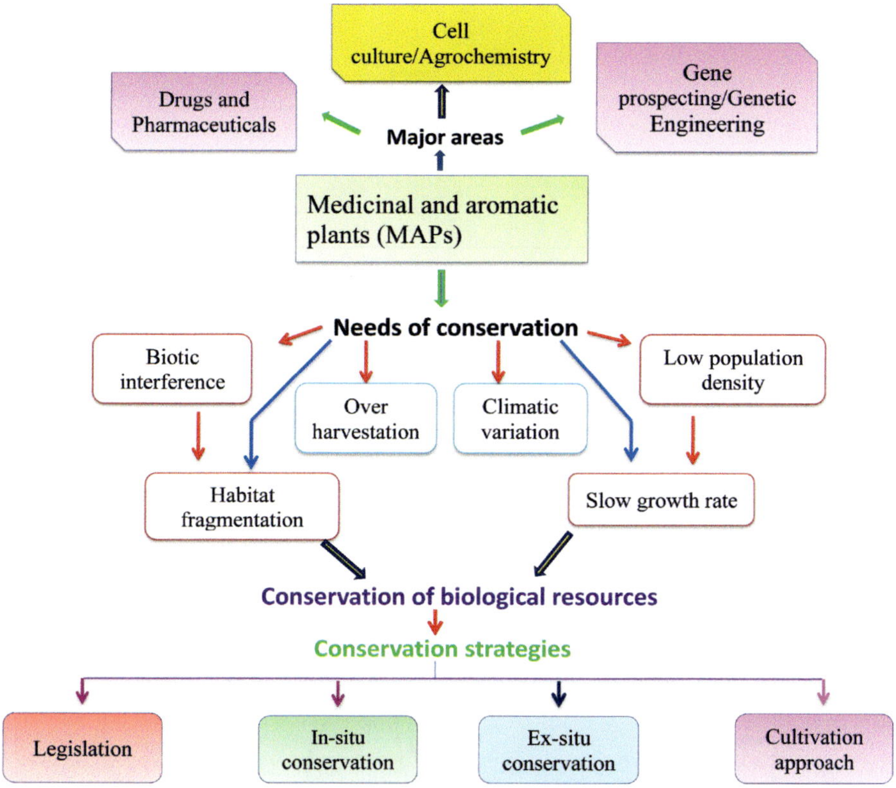

FIGURE 1.2 An overview of the need for different conservation strategies of MAPs.

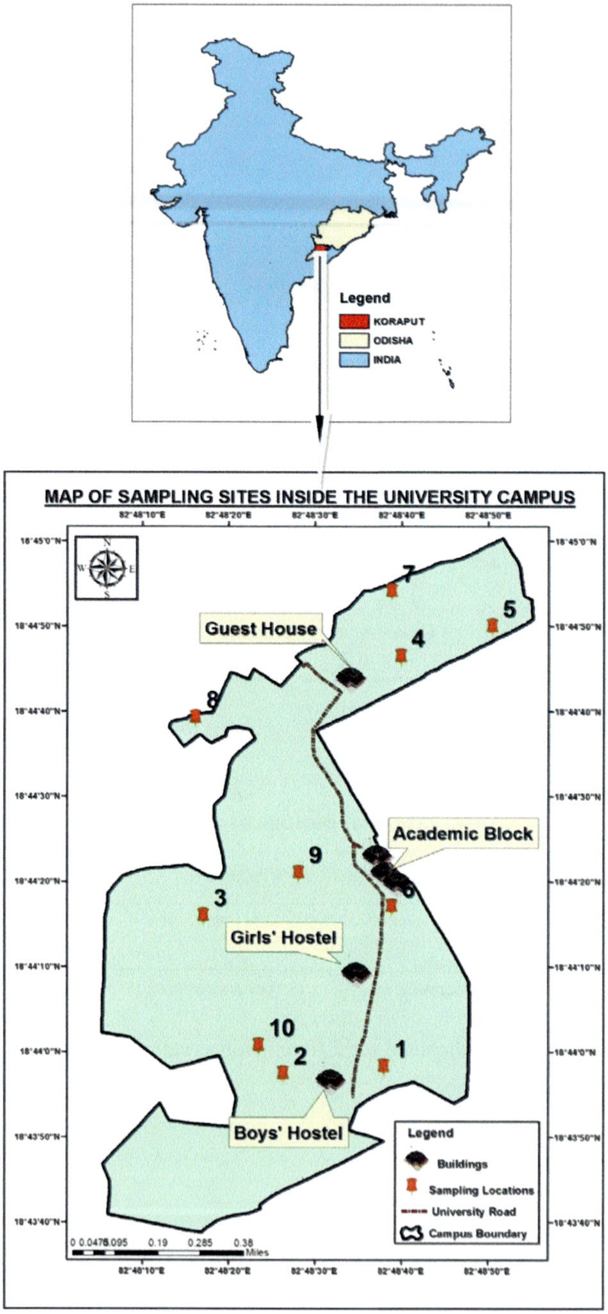

FIGURE 3.1 Map of the sampling locations.

Clerodendron bungii Steud.

Lantana camara L.

Anogeissus latifolia (Roxb. ex DC.) Wall.

Nerium indicum Mill.

Anisomeles indica (L) R. Br.

Chromolaena odorata (L.) R.M.King & H.Rob.

Diospyros peregrina (Gaertn.) Gürke

Wrightia arborea (Dennst.) Mabb.

Murraya koenigii (L.) Spreng.

FIGURE 3.10 Dominant shrub species found in the campus.

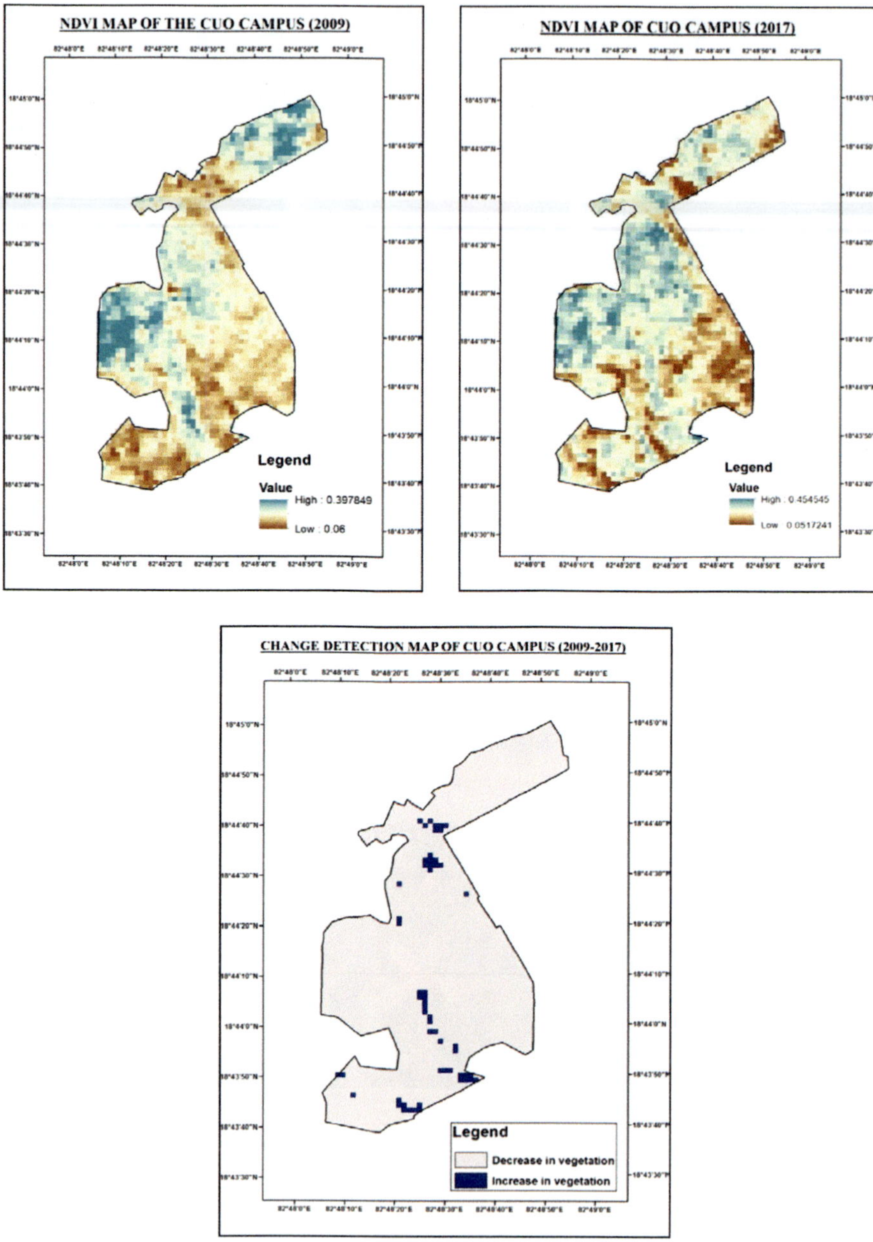

FIGURE 3.12 Vegetational Change Detection of CUO Campus using NDVI.

PLATE 4.1 Enumerated *Solanum* species of Tripura, India, 1: *S. torvum*, 2: *S. sisymbriifolium*, 3: *S. tuberosum*, 4: *S. erianthum*, 5: *S. violaceum*, 6: *S. viarum*, 7: *S. nigrum*, 8: *S. aethiopicum*, 9: *S. ovigerum*, 10: *S.violaceum*, 11: *S. ovigerum*, 12: *S. melongena*, 13: *S. aethiopicum*, 14: *S. torvum*.

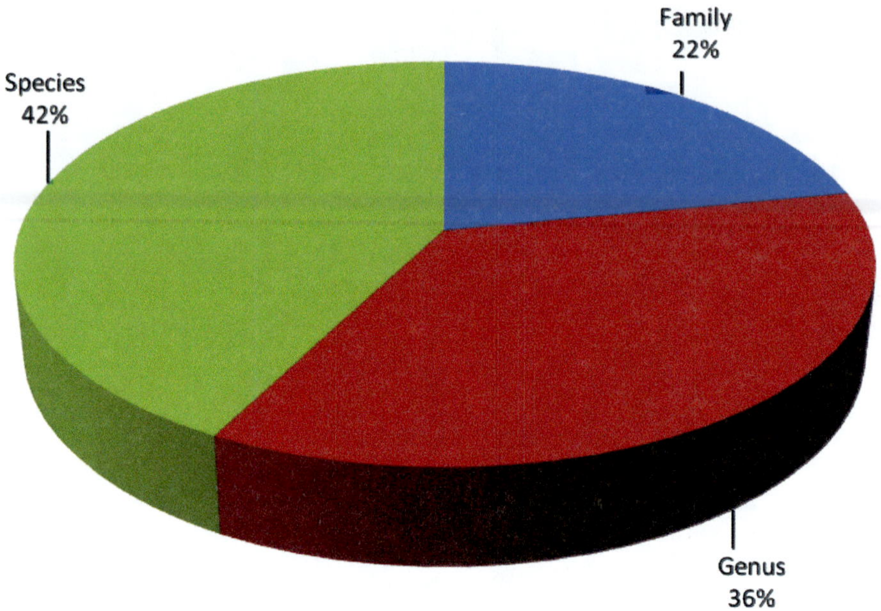

FIGURE 5.1 The diversity of enumerated plant species.

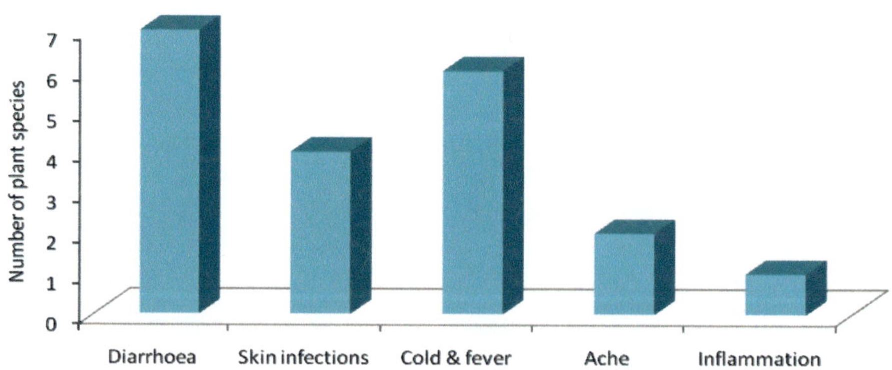

FIGURE 5.2 Plants used against some most common diseases/disorders.

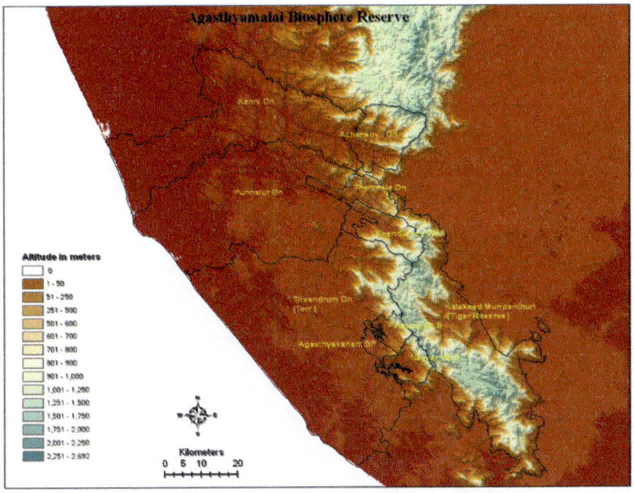

FIGURE 7.1 Map of Agasthyamala Biosphere Reserve.

FIGURE 7.2 (a–c) *Alocasia longiloba* Miq.; (d) *Amorphophallus bulbifer* (Roxb.) Blume; (e) *Amorphophallus bonaccordensis* Sivad. & N. Mohanan; (f–g) *Amorphophallus nicolsonianus* Sivad.; and (h) *Amorphophallus paeoniifolius* (Dennst.) Nicolson.

FIGURE 7.3 (a–c) *Anaphyllum wightii* Schott; (d–f) *Anaphyllum beddomei* Engl.; (g–h) *Ariopsis peltata* Nimmo; (i–j) *Ariopsis peltata* var. *brevifolia* J. Mathew & Kad. V. George.; and (k–m) *Arisaema agasthyanum* Sivad. & Sathish.

FIGURE 7.4 (a–c) *Arisaema barnesii* C.E.C. Fisch.

FIGURE 7.5 (a–e)*Arisaema leschenaultii* Blume; (f–h) *Arisaema sarracenioides* Barnes & C.E.C. Fisch.; and (i–l) *Arisaema madhuanum* Nampy & Manudev.

FIGURE 7.6 (a–d) *Arisaema tortuosum* var. *tortuosum* Hook. f.; and (e–g) *Arisaema tylophorum* C.E.C. Fisch.

FIGURE 7.7 (a) *Lagenandra meeboldii* (Engl.) C.E.C. Fisch.; (b) *Lagenandra ovata* (L.) Thw.; (c–d) *Pothos crassipedunculatus* Sivad. & N. Mohanan; and (e– f) *Remusatia vivipara* (Roxb.) Schott.

PLATE 8.2 **G.** *Galearis roborovskyi*; **H.** *Pegaeophytonscapiflorum*; **I.** *Rhodiola sedoides*; **J.** *Cirsium eriophoroides*; **K.** *Corydalis polygalina*; **L.** *Codonopsis foetens*; **M.** *Aconitum fletcherianum*; **N.** *Geranium polyanthes*; **O.***Impatiens chungtienensis.*

PLATE 9.2 (A–F): An overview of traditional formulae: (A&B) decoction, (C) infusion, (D) juice, (E) paste, and (F) powder.

FIGURE 11.2 Most threatened plant species of Kendujhar district, (1) *Gloriosa superba*; (2) *Scindapsus officinalis*; (3) *Rauvolfia serpentine*; (4) *Oroxylum indicum*.

A – *Azadirachta indica* L., B – *Ocimum sanctum* L., C – *Nyctanthes arbor-tristis* L. D – *Cynodondactylon* (L.) Pers., E – *Calotropis procera* W.T. Aiton F – *Tagetes patula* L., G – *Tridax procumbens* L., H – *Achyranthes aspera* L. I – *Aegle marmelos* L., J – *Tamarindus indica* L., K – *Ricinus communis* L., L – *Lawsonia inermis* L., M – *Cymbopogon flexuosus* W.Watson N – *Rauvolfia serpentina* L., O – *Phyllanthus emblica* L., P – *Punica granatum* L., Q – *Justicia adhatoda* L., R – *Syzygium cumini* L. S – *Aloe vera* (L.) Burm.f, T – *Curcuma longa* L.Most of the people of Bisoi block of Mayurbhanj district depend upon herbal medicines traditionally due to lesser side effects and cost-effective. This is an easy process to record the information on general uses of plants because a large number of people either consuming or sold in the nearby markets. But collecting indigenous information from tribal healers is a tough job for us. Tribal people are generally very shy in nature and don't want to spread their valuable knowledge to others. By creating a good rapport with these people, we can enable us to document this valuable information. The fast disappearance of indigenous knowledge and biodiversity due to population explosion, urbanization and industrialization result in the loss of valuable indigenous knowledge on medicinal plants forever. Hence, it is on a priority basis to document the valuable information and also conserve this knowledge for the next generation.

FIGURE 16.3 Some important medicinal plants of Bisoi block, Mayurbhanj district.

FIGURE 17.1 Ethnobotanical data and plant parts collections, (a) Panoramic view of Similipal Biosphere Reserve, (b) Leaves and fruits of experimental plant, (c) GPS data collection, (d) Fruits, (e–f) Ethnobotanical data collection.

PLATE 18.1 Studied plant species, (1) *Curcuma longa*, (2) *Costus speciosus*

CHAPTER 11

Ethnopharmacological Properties of Some Threatened Medicinal Plants

SANJEET KUMAR,[1] PADMA MAHANTI,[2] GITISHREE DAS,[3] and JAYANTA KUMAR PATRA[3]

[1]School of Life Sciences, Ravenshaw University, Cuttack, Odisha, India
E-mail: sanjeet.biotech@gmail.com

[2]Department of Forest and Wildlife, Thiruvananthapuram, Kerala, India

[3]Research Institute of Biotechnology and Medical Converged Science, Dongguk University-Seoul, Gyeonggi-do 10326, Republic of Korea

ABSTRACT

Nearly half of the area of Odisha state of India is covered with forest. The forest patches of Kendujhar district located towards the northern part of Odisha, India, harbors wide varieties of medicinal plants. They play an important role in the lives of rural and tribal people for various purposes. They are used for their food values as well as medicinal properties and other household activities. Hence, an effort was made for documenting the ethnomedicinal values of the most threatened plant's species available in some selected areas of Kendujhar district of Odisha state. Field survey revealed that among the threatened species, 9 wild species which are available in the forest patches of the district having potent food and medicinal values are more threatened. This chapter also highlights the sustainable harvesting and the chemical constituents of the plants and their relationship with the curing properties.

11.1 INTRODUCTION

Ethno-medicine has been a multidisciplinary aspect comprising plant science, history, culture and literature and comparison of the traditional medicine practiced by various ethnical groups and aboriginal people. Many of the earliest remedies survived only verbally and are being practiced in many parts of the country particularly in remote or rural and tribal societies (Doley et al., 2014). Today, more than 25% of the modern day medicines have been derived from wild medicinal plants (Ji et al., 2009). Ever since ancient times, the tribal people looked for drugs in the forest in order to search for a rescue of their diseases. The beginning of the use of medicinal plants was instinctive, as is the case with animals. The oldest written evidence of medicinal plants used for the preparation of drugs has been found on a Sumerian clay slab from Nagpur, approximately 5000 years old. The Chinese book on roots and grasses "Pen T Sao" written by Emperor Shen Nung circa 2500 BC, treats 365 natural drugs (dried parts of medicinal plants), many of which are used even in the modern era, such as ginseng, cinnamon bark, etc. Theophrastus (371–287 BC) founded botanical science in his book "*De Caussi Planetarium,*" generated a classification of more than 500 medicinal plants known at that time. Ancient and contemporary Indian literature also showed the richness of medicinal plants in the country.

India is gifted with rich varieties of plant species among which there are enormous numbers of valuable wild medicinal plants. Its history of using plant-based remedies for treating various forms of health disorders and ailments provides an ideal platform for a study on wild medicinal plants and its uses. Different medicinal plants of great diversity are found to be distributed in different geographical and environmental conditions throughout the country (Kala, 2007). Mostly the wild plants are used by the rural and tribal communities throughout the country for various purposes. As medicines, they are used as single or multiple formulations such as juice, paste, decoction, smoked and other modes. Ethno-medicinal studies offer immense scope and opportunities for the development of new drugs and search of active substances such as alkaloids, tannin, flavonoids, phenolic compounds, etc. from different plant sources (Kumar & Satpathy, 2011).

Among all the states of India, Odisha is prominent for wild medicinal plants. It is situated in the eastern part of the country and is known for its rich floral diversity (Mallik et al., 2012). Many parts of the state are facing serious anthropogenic activities on floral wealth, and due to this, some plants come under the threatened group of flora. Among such areas, Keonjhar district is highly affected due to mining activities, but, it has also rich forest and inhabitants of various tribal communities. The forest houses a highly diverse species of flora. The area is also extensively exploited for its richness in the availability of minerals and coal. Such an act is hampering the biodiversity of the place. Since agriculture is the main mode of sustenance; the tribal people are highly benefited by these plants, as they are solely dependent on the forests for their livelihood. Anthropogenic activities in these areas threaten some of the threatened plant species that are found in the region and care should be taken to conserve them for the maintenance of ecology as well as to meet the needs of the tribal people. Keeping this in view an attempt was made to document the ethnobotanical values of some threatened plant species of the district. This paper also lists some of the highly threatened medicinal plant species found in this region and their medicinal importance so as to point out the need for the conservation of such plants.

11.2 MATERIALS AND METHODS

11.2.1 STUDY AREA

The study was carried out in different forest areas (Sanghaghara, Gonasika, Sarkunda, Bada Ghaghara) of Kendujhar district, Odisha, India (Figure 11.1). This region consists of a number of hamlets and low-lying valley. The study area is inhabited mainly by tribals and ethnic groups such as Bathudi, Bhuyan, Gond, Ho, Juang Kolha, and Munda. The plants were selected as per anthropogenic activities, less population and usage rate among the rural and tribal communities of the study area. The threatened plant species of this region were identified by Sanjeet Kumar, following Flora's book (Saxena, Brahmam, 1996). The results presented here, are based on the fieldwork conducted with the rural and tribal communities of study area and information collected from the literature.

11.2.2 ETHNOBOTANICAL DATA COLLECTION

The methodological frameworks for the ethnobotanical study were as per the standard techniques of exploration and germplasm collection (Christian, Brigitte, 2004), qualitative and quantitative ethnobiological approaches in the field, interviews, elicitation methods, data collection and further authentication (Cotton, 1996). Intensive and extensive field surveys were done in different landscapes and micro-ecological niches across forest types, adjoining valleys, homesteads, kitchen gardens, farmlands, fallow lands, etc. The field surveys including weekly markets (Haat) of the study area were undertaken by the authors during different seasons of the study periods. The standard participatory rural appraisal method (Gerique, 2006) was adopted for sampling and data collection to incorporate indigenous knowledge. Opinions of tribal people were taken regarding the uses of experimental plant species through questionnaires. Primary data collection were made mainly through semi-structured questionnaires (Passport Data Form) focusing on local name(s), collecting season(s), present knowledge on use(s), mode of use(s), part(s) of the plant used, degree of wildness, economic values, change in collecting pattern over time, possible threats and potential of the experimental plants. A clear expression of consent was obtained from the informants before the interview elaborating the aim of the study as per the guidelines and code of ethics of the International Society of Ethnobiology (ISE, 2006).

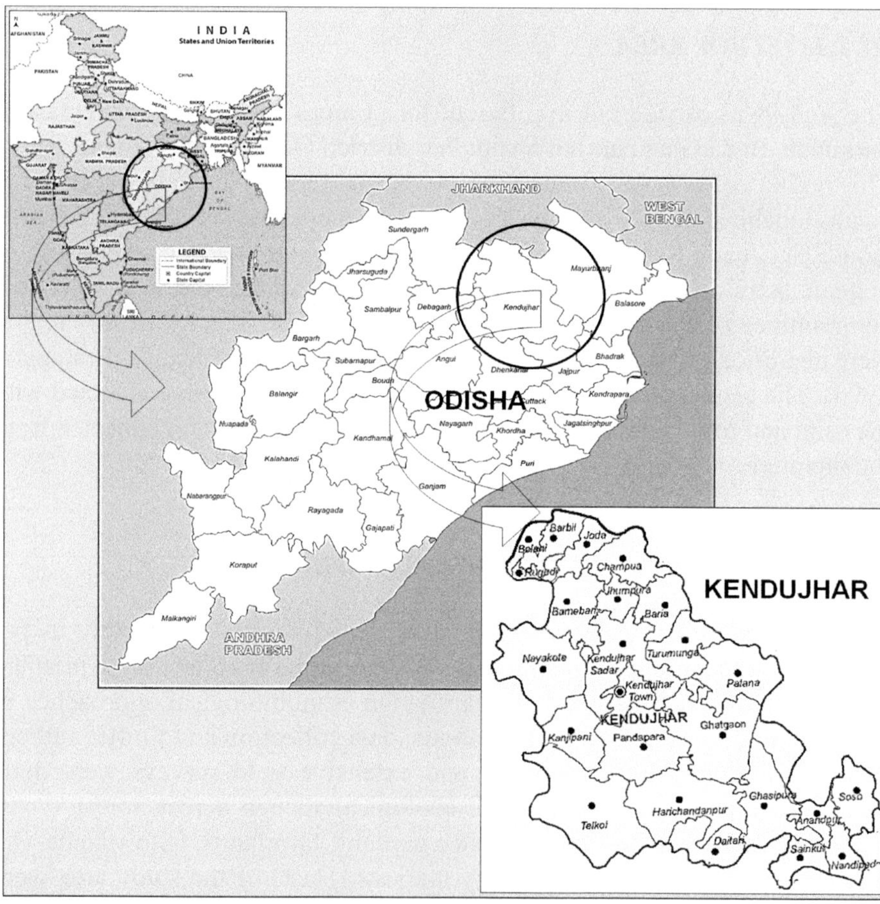

FIGURE 11.1 Map showing the Kendujhar district of Odisha state of India from where the ethnobotanical data of the plant species were collected.

11.3 RESULTS AND DISCUSSION

Field survey revealed that authors found 09 most threatened plant species belonging to eight families. Among them, 03 are trees, 05 are climbers, and 01 is herb (Table 11.1). The plants are *Celastrus paniculatus, Gardenia gummifera, Gloriosa superb* (Figure 11.2); *Scindapsus officinalis* (Figure 11.3); *Piper longum, Pterocarpus marsupium, Pueraria tuberosa, Rouvolfia serpentina* (Figure 11.4); and *Oroxylum indicum* (Figure 11.5). The present study also establishes the correlation between the tribal claims as collected and the presence of bioactive compounds reported (Table 11.1).

FIGURE 11.2 (See color insert.) Most threatened plant species of Kendujhar district, (1) *Gloriosa superba*; (2) *Scindapsus officinalis*; (3) *Rauvolfia serpentine*; (4) *Oroxylum indicum*.

The plant parts such as roots, leaves, stem, flowers, tubers, fruits, bark, gum or resinous extract are used against different diseases and disorders. When the seeds of Celastrus paniculatus were macerated, and oil was obtained. The oil is used to enhance the memory of abnormal children (Kulkarni et al., 2011). The juice of resin of Gardenia gummifera was used

TABLE 11.1 Ethno-Medicinal Uses, Correlation of Tribal Claims and Their Reported Bioactive Compounds of Some Selected Threatened Plants of Kendujhar District, Odisha, India

Botanical Name	Local Name	Family	Parts Used	Reported Uses	Tribal Claims	Identified Bioactive compounds	Justification
Celastrus paniculatus	Pengu	Celastraceae	Seed	Seed oil is used to enhance memory, tonic for nervous system and an anti-depressant.	Memory enhancer	Alkaloids	The presence of alkaloids indicate that the plants might be responsible to enhance the memory
Gardenia gummifera	Kurudu	Rubiaceae	Resin	Expels intestinal worm	Antihelmintic	Flavonoids	Presence of flavonoids indicate its antihelmintic properties
Gloriosa superba	Panchangulia	Liliaceae	Tuber	Induces abortion and relief from joint pain	Joint Pain	Flavonoids	Presence of flavonoids might be responsible
Oroxylum indicum	Fanfana	Bignoniaceae	Bark	Used to treat throat infections	Throat infections	Tannin	Tannin is responsible to cure throat infections
Piper longum	Pipli	Piperaceae	Fruits	Used to cure cough and asthma	Cough	Flavonoids	Presence of flavonoids might be responsible to cure cough
Pterocarpus marsupium	Piasala	Fabaceae	Leaves	Juice is used to cure diabetes	Diabetes	Tannin	Presence of tannin might be responsible to cure diabetes
Pueraria tuberosa	Bhui Kukharu	Fabaceae	Tuber	Raw tuber is used as emollient	Emollient	Saponin	Saponin is responsible
Rauvolfia serpentine	Sarpagandha	Apocynaceae	Root	Used to cure against snake and scorpion bites	Snake Bites	Alkaloids	Presence of alkaloids might be responsible to cure snake bites
Scindapsus officinalis	Gajapippali	Araceae	Leaves, fruits	Paste is used in inflammation	Inflammation	Flavonoids	Presence of flavonoids might be responsible to cure inflammation

to kill the intestinal worm (Nayak et al., 2011). It was also observed that tuber paste of Gloriosa superba was applied externally to reduce the joint pain among the old age people of tribal communities of Kendujhar (Akhtar, Haqqi, 2012). The tuber juice of this plant is taken for abortion. The bark decoction of Oroxylum indicum is used to treat throat infections (Payne et al., 2013). It was noted that fruits juice of Piper longum is used to cure cough and asthma (Okwu, Nnamdi, 2011). The leaf juice of Pterocarpus marsupium was taken to control diabetes (Liu et al., 2005) and tuber paste of Pueraria tuberosa was used as an emollient paste (Aburjai, Natsheh, 2003). The root paste of Rauvolfia serpentine is used against snake bites (Gomes et al., 2010) and leaves and fruits paste of Scindapsus officinalis is applied to reduce inflammation (Nijveldt et al., 2001).

It was found, that these plants were very useful for the cure of different diseases like dysentery, fever, diarrhea, respiratory distress, diabetes, insomnia, the antidote for snake venom and scorpion sting, etc. (Table 11.1). However, the investigation only focuses on the most threatened medicinal plants found in the district. The plants that have been listed in the Table-1 are found to be rich in bioactive compounds (Nayak et al., 2011; Nijveldt et al., 2001; Akhtar, Haqqi, 2012; Debnath et al., 2014). The bioactive compounds present in the plants have given them their characteristic medicinal values that have been adorned by the tribal people to a large extent.

Modern civilization has seen the improvement of medical science by leaps and bounds. In spite of such advancements, the remote areas remain deprived of such advantages. These people grossly depend on traditional medical therapies. Even to this date, modern medical sciences employ the help of traditional medicine when they are ineffective. Hence, care should be taken to retrieve as much data regarding traditional medicine as is possible, so that this knowledge does not go unidentified and underutilized. Therefore, it can be assumed that this investigation will be useful for the conservation of knowledge in the field of herbal treatment and traditional medicine.

ACKNOWLEDGMENTS

The authors are grateful to the DFO, Kendujhar and Chief Executive, Regional Plant Resource Centre, Bhubaneswar for their support during the field visits. We are grateful to the local communities for their help in the collection and identification of specimens.

KEYWORDS

- **biodiversity conservation**
- **ethnopharmacology**
- **sustainable harvesting**
- **threatened**

REFERENCES

Aburjai, T.; Natsheh, F. M. Plant used in cosmetics. *Phytotherapy Res.* **2003**, *17,* 987–1000.

Akhtar, N.; Haqqi, T. H. Current nutraceutical in the management of osteoarthritis: a review. *Therap. Adv. Mus. Dise.* **2012**, *4*(3), 181–207.

Christian, R. V.; Brigitte, V. L. Tools and methods for data collection in ethnobotanical studies of home gardens. *Field Method.* **2004**, *16*(3), 285–306.

Cotton, C. M. *Ethnobotany: Principles and Applications,* John Wiley and Sons Ltd., Chichester. **1996**, 1–80.

Debnath, M.; Biswas, M.; Shukla, V. J.; Nishteswar, K. Phytochemical and analytical evaluation of Jyotishmati (*Celastrus paniculatus* Willd.) leaf extracts. *Ayu.* **2014,** *35*(1), 54–57.

Doley, B.; Gajurel, P. R.; Rethy, P.; Buragohain, R. Uses of trees as medicine by the ethnic communities of Arunachal Pradesh, India. *J. Med. Plant. Res.* **2014**, *8*(24), 857–863.

Gerique, A. An introduction to ethnoecology and ethnobotany: theory and methods. In: Integrative assessment and planning methods for sustainable agroforestry in humid and semi-arid regions. *Adv. Sci. Training-Loja. Ecuador*. **2006**, 1–20.

Gomes, A.; Das, R.; Sarkhel, S.; Mishra, R.; Mukherjee, S.; Bhattacharya, S.; et al. Herbs and herbal constituents active against snake bite. *Ind. J. Exp. Biol.* **2010**, *48,* 865–878.

ISE (International Society of Ethnobiology). **2006**, International society of ethnobiology code of ethics (with 2008 additions). http://www.ethnobiology.net/ethics.php.

Ji, H. F. Li, X. J.; Zhang, H. Y. Natural products and drug discovery. Can thousands of years of ancient medical knowledge lead us to new powerful drug combinations in the fight against cancer and dementia? *EMBO. Rep.* **2009**, *10*(3), 194–200.

Kala, C. P. Local Preferences of Ethno-Botanical Species in the Indian Himalaya: Implications for Environmental Conservation. *Curr. Sci.* **2007**, *93,* 1828–1834.

Kulkarni, P. D.; Ghaisas, M. M.; Chivate, N. D.; Sankpal, P. S. Memory enhancing activity of *Cissampelos pariera* in mice. *Int. J. Pharmac. Pharmaceut. Sci.* **2011**, *3*(2), 206–211.

Kumar, S.; Satpathy, M. K. Medicinal plants in an urban environment; herbaceous medicinal flora from the campus of Regional Institute of Education, Bhubaneswar, Odisha. *Int. J. Pharm. Life. Sci.* **2011**, *2*(11), 1206–1210.

Liu, X.; Kim, J. K.; Li, Y.; Li, J.; Liu, F.; Chen, X. Tannic acid stimulates glucose transport and inhibits adipocyte differentiation in 3T3-L1 Cells. *J. Nut.* **2005**, *135*(2), 165–171.

Mallik, B. K.; Panda, T.; Padhy, R. N. Traditional Herbal Practices by the Ethnic People of Kalahandi District of Odisha, India. *Asian Pacific, J. Trop. Biomed.* **2012**, *2*(2), 988–994.

Nayak, S.; Manjari, S. A.; Kanti, C. C. Phytochemical screening and anthelmintic activity study of *Saraca indica* leaves extracts. *Int. Res. J. Pharmac.* **2011**, *2*(5), 194–207.

Nijveldt, R. J.; Nood, E. V.; Hoorn, D. E. V.; Boelens, P. G.; Norren, K. V.; Leeuwen, P. A. M. Flavonoids: a review of probable mechanisms of action and potential applications. *Amer. J. Clin. Nutri.* **2001**, *74*(4), 418–425.

Okwu, D. E.; Nnamdi, F. U. Two novel flavonoids from *Bryophyllum pinnatum* and their antimicrobial activity. *J. Chem. Pharm. Res.* **2011**, *3*(2), 1–10.

Payne, D. E.; Martin, N. R.; Parzych, K. R.; Rickard, A. H.; Underwood, A.; Boles, B. R. Tannic acid inhibits *Staphylococcus aureus* surface colonization in an IsaA-dependent manner. *Infect. Immu.* **2013**, *81*(2), 496–504.

Saxena, H. O.; Brahmam, M. The Flora of Orissa, Regional Research Laboratory, Orissa Forest Development Corporation. Bhubaneswar, Orissa. **1996**, 1372–1374.

CHAPTER 12

Medicinal Plants: A Potent Antimicrobial Source and An Alternative to Combat Antibiotic Resistance

SHRADDHA CHAUHAN, REECHA SAHU, and LATA S. B. UPADHYAY*

*Department of Biotechnology, National Institute of Technology, Raipur Chattisgarh–492010, India, *E-mail: lupadhyay.bt@nitrr.ac.in*

ABSTRACT

Microbial infections are one of the major issues in every country. Thousands of people are suffering from serious diseases like malaria, tuberculosis, diarrhea, typhoid, etc. Mostly antibiotics of microbial origin are used for the treatment of these diseases. However, some of these antimicrobial compounds depending on the individual patient, type and dose of antibiotic used have been reported to show harmful side effects such as fever, diarrhea, vomiting and allergic reactions (photodermatitis and anaphylaxis). Excessive use can also develop antibiotic resistance condition. Frequency of antibiotic-resistant bacteria strains is increasing day-by-day. To control the prevailing antibiotic resistance problems search for new antimicrobial substances with minimum side effects is needed. Plant and its products have been potentially used as medicine to treat infections for centuries. There are confirmations of herbs being utilized as a part of various disease treatments in Indian, Egypt, China, Greek, and Roman civic establishments. Medicinal plants are being used to treat diseases because of their antimicrobial properties. These phytochemicals are not only low price product than microbial antibiotics but also can serve to combat antibiotic resistance. The plant parts like root, leaf, stem and even a whole plant have been used for extraction of antimicrobial agents both in solid as well as in liquid form. Plants produce more than a thousand chemical compounds, each having different biological activity. These compound are active against pathogenic microorganisms exhibiting pharmacological property. The pharmacological activities of these medicinal plants are usually associated with its secondary metabolites. Some plants have shown the activity against many bacteria and fungi, for example, aloe vera, lemon, neem, garlic, tulsi, cranberry, and bearberry. It has been reported that plant extracts of *Punica granatum*, *Syzygium cordatum*, *Ozoroa insignis*, *Gymnosporia senegalensis*, *Indigofera daleoides*, *Elephantorrhiza elephantina*, *Elephantorrhiza burkei*, *Schotia brachypetala*, *Ximenia caffra*, and *Spirostachys africana* disclosed extraordinary antibacterial action against *Vibro cholera*, *Staphylococcus aureus*, *E. coli*, *Shigella* species, and *Salmonella typhi*. The antimicrobial activities of many plants have been studied still there is a room for possibilities with new species.

12.1 INTRODUCTION

Antibiotics also referred to as antibacterial or antimicrobial are biochemical compounds of microbial origin used for the treatment and prevention of microbial infection. They either kill the infection-causing bacteria or inhibit its growth. Some of the antibiotics also show antiprotozoal activity but are ineffective against the virus. During 20th century application/use of antibiotics emerged as a revolution in the field of medicine. Their usage in combination with vaccines eradiated disease like tuberculosis in many parts of the world. However, massive production, usage and rapid control of disease symptoms by antibiotics resulted in their over and misuse, this led to the development of antibiotic resistance by bacterial strains. According to the World Health Organization (WHO), antibiotic-resistant could be a great challenge for the future and thus classified as a serious threat to the medical field and practices. But in past years rather than a future prediction antibiotic resistance has emerged as a problem in almost every part of the globe and shown to adversely affect every individual irrespective of age and country. Thus, there is a need to identify new antibacterial agents so the problem associated with the development of resistant against existing antibacterial agents/antibiotics can be addressed. This has forced researchers to discover new drugs/antibacterial agents with less side effects at a comparable cost. Plants promise a source of natural antimicrobial agents. Antibacterial substances derived from plant sources have thus come up as an answer to the problem. These antimicrobial substances are natural and can provide an extensive range of application (Amenu, 2014).

Medicinal plants are usually referred as medicinal herbs as these plants find their application in herbology which is study of plants for medicinal purposes and are in traditional medicinal practices since prehistoric times.

Plants synthesize phytochemicals as molecules for defense against insects, fungi, diseases, and herbivorous mammals. Role of plants as medicine has been well documented and explained in the ancient manuscripts of Greco-Arab, Egypt, and China. In history of medicine since last 4000 years, Perso-Arabic medical practitioners and Indian Ayurveda doctors as well as European and Mediterranean cultures were using plant parts and extract for the wellbeing of society and to cure deadly diseases. According to WHO plant components are being used as traditional medicines from ancient time. According to WHO reports approximately 80 percent of world population use plant originated/extracted medicines to combat for one or the other aspect of their basic health care problems. According to WHO, around 21,000 plant species have the potential for being used as medicinal plants (Amenu, 2014). Herbal treatment methodologies show minimal side effects hence safe to use effectively.

Infectious diseases represent an important cause of illness and death among the general population, especially in developing countries. Therefore, the emphasis has to be focused on developing new antimicrobial drugs in recent years, especially due to the constant emergence of microorganisms resistant to conventional antimicrobials. Many plants species are known for their antimicrobial characteristics as a virtue of secondary metabolites they synthesis (Nascimento, Locatelli, Freitas and Silva, 2000). These bioactive compounds are mainly alkaloids, tannin, phenolic compound, and flavonoids. Their concentration in plant/plant part decides the level of activity and specificity of that particular plant against infection. For example, tea is active against many types of cancers (Mukhtar and Ahmad, 2000), leaves and bark of guava provide immunity against gastrointestinal disorders, toothaches. Neem leaves are well known for their activity against hepatitis, skin infection and diabetes (Sarmiento, Maramba and Gonzales, 2011). They have the potential to replace the existing antibiotics because of their specific activity, minimum or no side effect and cost-effectiveness. The need of discovering new antibiotics from medicinal plants has gained much interest because of the current issue of microbial resistance to antibiotics. Antimicrobial resistance is a huge problem worldwide; it interferes both the prevention as well as treatment. Therefore the discovery and identification of new antimicrobial substance are necessary so that new medicines can be produced (Farjana, Zerin and Kabir, 2014). Many new antibiotic drugs are synthesized by pharma companies to elucidate the microbial resistance.

For the past two decades, numerous institutions have instigated the selection of plants with antimicrobial activity to discover new compounds for drug synthesis. This has concluded with the finding of many phytochemicals.

These phytochemicals inhibit several microorganisms. Among 300,000 plant species present on earth, only 30 present of them is examined for their antibacterial activity. To find out new drugs is a need of time to cure and treat diseases. Phytotherapy is predicted to be one of the futuristic approaches of modern medicine to serve human health. This chapter deals with the pharmacological aspect of medicinal plant (Dash and Murthy, 2011).

12.2 MEDICINAL PLANTS DIVERSITY

Literature suggests that the globally all terrestrial plants account for around 500,000 species. Out of which 21,000 species are found to have medicinal properties (Seth and Sharma, 2004), but only about 0.5% of them are chemically investigated. Plants contain a wide range of biological active phytochemicals with numerous medicinal properties. For thousands of year's plant-derived medicinally active phytochemicals has been used across the globe. These sources of traditional medicines are very well practiced in present era too. In spite of the modernity, sophistication, and resources available for modern medicinal practice across the world, there is still room for these traditional practices to overcome the disadvantages modern medicines suffer from, like antibiotic resistance, lack of new antibiotics, etc. Medicinal properties and ethnopharmaceutical knowledge of around 5000 plant species have been reported in Traditional Chinese medicine supplies (Köberl et al., 2013). Miller and group investigated traditional Chinese endophyte herbs for its anticancer properties (Miller, Qing, Sze and Neilan, 2012). Such studies using the traditional medicine can be helpful to discover new pharmacological drugs, such as drugs against tumor and drug against multidrug resistance microbial strains.

12.2.1 WORLDWIDE DIVERSITY

Near about 500,000 species of plants, species have been estimated on earth. Reports suggest that species of plants that have a medicinal property outnumbers in total number of the plant species used as food by both humans and animals (Cowan, 1999). There have been significant mentions about plants with medicinal values. As early as in late 5th century BC, Hippocrates documented 300 to 400 medicinal plants and their application. A medicinal plant catalog named *De Materia Medica*, written by Dioscorides in the first century A.D. is available as a reference (Cowan, 1999). In a report by Moerman, Native American group uses 1,625 species of plants as food and 2,564 have medicinal usage (Moerman, 1996). A list of about 21,000 plants used for medicinal sake around the world has been reported by the WHO. Out of these 21,000, about 2,500 species of plants with medicinal value are found in India. And 150 species of them are used commercially for preparing drugs or some other medicinal purpose on a large scale. The prevailing favorable climatic conditions and biodiversity found in India has made it as the largest producer of medicinal herbs and therefore called as a botanical garden of the world (Seth and Sharma, 2004; Modak et al., 2007). In Asia-Pacific region, around 8,000 species of medicinal plants are found. Most of the medicinal plants found belong to the species of *Rauvolfia, Hyoscyamus, Cassia, Atropa, Podophyllum, Catharanthus, Papaver* and *Psoralea* (Batugal, Kanniah, Sy, and Oliver, 2004). Around 7,500 species are reported to be found in Indonesian archipelago, but only around 200 are used by the medicinal industries (Batugal, Kanniah, Sy, and Oliver, 2004; Hamid and Sitepu, 1990). In China, 4,000 species of medicinal plants are reported, India around 3,000 species and 2,000 species of medicinal plants are found in Malaysian region (Batugal, Kanniah, Sy, and Oliver, 2004). Table 12.1 sums up the total number of medicinal plants that have been used worldwide.

TABLE 12.1 Total Number of Medicinal Plants Present Worldwide and Percent Contribution of Major Countries Which Have Rich Medicinal Flora

Country	Total number of plant species	Plant species with medicinal value	% Contribution worldwide
China	26,092	4,941	18.9
Indonesia	22,500	1,000	4.4
Malaysia	15,500	1,200	7.7
India	15,000	3,000	20.0
USA	21,641	2,564	11.8
Vietnam	10,500	1,800	17.1
Thailand	11,625	1,800	15.5
Philippines	8,931	850	9.5
Nepal	6,973	700	10.0
Pakistan	4,950	300	6.1
Srilanka	3,314	550	16.6
Average	13,366		12.5
Worldwide	422,000	52,885	52,885

Taken from Schippmann, Leaman, and Cunningham (2002).

Source: Schippmann, Leaman and Cunningham, (2002); Jain and DeFilipps, (1991a); De Padua, Bunyapraphatsara and Lemmens, (1999b).

12.2.2 MEDICINAL PLANTS DIVERSITY IN INDIA

Indian forests are rich in medicinal biodiversity. India's mega centers of biodiversity, especially in plant species are Northeastern Himalayan region, Western Ghats, Andaman and Nicobar islands. About 3,000 species of plants have been known to have medicinal values which are used as ethnomedicine. Their information is passed over generations, with non-availability of written drafts about these plants. Nearly 700 species have been investigated pharmacologically and chemically, and these medicinal plants are used in modern medicine (Patil and Patil, 2010). Table 12.2 shows the various species of medicinal plants found in different geographical areas of India according to study conducted till 2000 (Samant et al., 2007).

TABLE 12.2 Estimated Number of Medicinal Plants Along with the Geographical Region of India in Which They Are Found

Geographical region of India	Number of medicinal plants (approx.)	Most common medicinal plants in that area
Trans-Himalaya or Gangdise–Nyenchen Tanglha range	Near about 800	*Arnebia euchroma* (Royle) John., *Ephedra geradiana.*
Himalayas	2500–3000	*Aconitum Ferox, Clematis Buchananiana, Nardostachys grandiflora, Nardostachys jatamansi, Rhus semilata, Drymaria cordata, Arctium Lappa.*
Semi-arid zones	900–1000	*Abutilon indicum, Acacia nilotica (l). delile, Calotropis procera, Commiphora wightii (arn.) bhandani,* *Balanites aegyptiaca* (L.)
Desert	500–600	*Tecomella undulata* (Sm.) Seem., *citrulus colocynthis* (L.), Schraderand *Cressa crertica* L.
Western Ghats	Near about 2000	*Mimosa pudica, Hibiscus angulosus, Rauvolfia serpentina, Saraca asoca, Leucas aspera, Phyllanthus neruri, Tridax procumbens, Parthenium hysterophorus, Calotropis gigantea, Ophiorhizza mungos..*
Northeastern region	2000–2100	*Oroxyllum indicum, Solanum kurzii Brace ex Prain, Poulzolzia hirta, Lysimachia racemosa Lam., Rhus javanica L., Cordyceps sinensis (Berk.) Sacc., Lemanea australis Alkins, Lycopodium pseudoclavatum.*

TABLE 12.2 *(Continued)*

Geographical region of India	Number of medicinal plants (approx.)	Most common medicinal plants in that area
Deccan Peninsular region	Near about 3000	*Andrographis paniculata, Withania somnifera, Cassia senna, Aloe vera, Ocimum sanctum, Abelmoschus moschatus, Ocimum basilicum, Eucalyptus, Piper longum, Adathoda vasica, Eclipta prostrata, Chlorophytum borivilianum, Curcuma longa,* and *Mucuna pruriens*
Gangetic Plain	1000–1100	*Holarrhenaq pubescens, Mallotus philippensis* (Lam.) Muell –Arg., *Pluchea lanceolata* C.B. Clarke
Coastal areas	Nearly 500	*Achyranthus aspera L., Calophyllum inophyllum L., Argemone mexicana L., Cissus quadraangularis L, Crotalaria verrucoasa L., Hibiscus tiliaceus L., Pongamia pinnata (L.)* Pierre, *Ziziphus mauritiana* Lam.
Islands (Andaman & Nicobar Islands, Lakshadweep Islands)	More than 500	*Adnanthera pavonina* L., *Barringtonia asiatica* (L.), *Alstonia macrophylla* Wall. ex G. Don, *Amomum fenzlii* Kurz., *Dioscorea vexans* Prain & Burk, *Costus speciosus* (Cone.), algal varieties like Ulva, Codium of Lakshadweep islands.

Taken from Katwal, Srivastava, Kumar, and Jeeva (2003).

12.3 PRINCIPLE COMPONENTS OF MEDICINAL PLANTS

Among thousands of plant species present on the earth, only a small percentage of these are used. Plants have endless capacity to synthesize aromatic compounds. These aromatic substances are mostly phenols (or their derivatives) and secondary metabolites (Geissman, 2004). These compounds are mainly responsible for defense mechanism against microbes or other pathogens (Cowan, 1999). Some are also responsible for imparting odor and pigment. Figure 12.1 represents the principle plant components with antimicrobial activities.

12.3.1 PHENOLIC COMPOUNDS

Phenolic compounds are chemical substances that contain a hydroxyl group bonded to an aromatic hydrocarbon. Some simple phytochemicals consist of single substituted phenolic ring. The two most common example of such phenolic compound includes cinnamic and caffeic acids. Some hydroxylated phenols such as catechol and pyrogallol are toxic against microorganisms.

The site and numbers of hydroxyl groups of phenols are responsible for their toxicity. It has been reported that highly oxidized phenols are more toxic to microbial cells (Urs and Dunleavy, 1975). These compounds work on enzyme inhibition mechanism and inhibiting the activity of the critical enzyme required for the survival of the microbes (Mason, 1987).

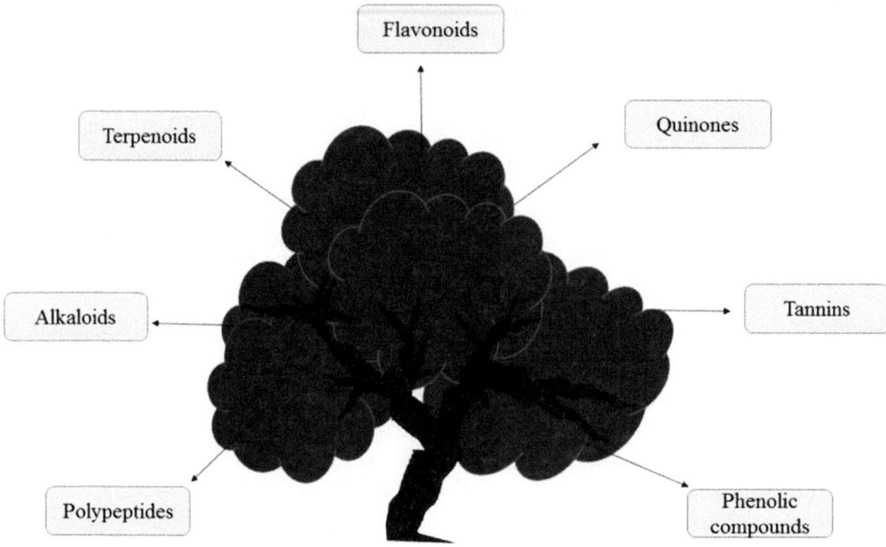

FIGURE 12.1 Principle components of medicinal plant.

12.3.2 FLAVONOIDS

A class of plant pigments with multiple functions is referred as flavonoids. Flavonoids are hydroxylated phenolic plant's secondary metabolites synthesized in plants also in response to infection (Dixon, Dey and Lamb, 1983). Catechines are most researched flavonoid compounds. They occur in green teas. Tea extracts are noticed to have antimicrobial activity, may be due to the high amount of catechin present in its extract (Toda, Okubo, Ohnishi, and Shimamura, 1989). They are reported to inhibit *Vibrio cholera* (Borris, 1996), *Shigella* (Vijaya, Ananthan, and Nalini, 1995) and many other toxic microorganisms. Antioxidant activity of plant flavonoids are more potential and effective than traditional antioxidant nutrients such as vitamin E, and vitamin C. Types of flavonoids with their examples, source, and property has been listed in Table 12.3.

TABLE: 12.3 Type of Flavonoids with Their Therapeutic Properties and Sources

Types of Flavonoid	Example	Clinical Significance	Source
Anthocyanidins	Pelargonidin, Cyanidin, Delphinidin, Peonidin, Petunidin, Malvidin, Apigeninidin, Aurantinidin, 6-hydroxy-cyanidin, 6-hydroxy-delphinidin, Rosinidin, Hirsutidin, 5-methyl-cyanidin, Luteolinidin, Tricetinidin,	Cellular Vitamin C levels are increased by action of anthocyanidins usage. Anthocyanidins also reduce the destruction of small blood vessels. It also prevents damage caused due to free-radical and thus protects collagen destruction ensuring healthy skin and connective tissue.	Blueberries, Blackberries, Plums, Cranberries, Raspberries, red onions, Red potatoes, Red radishes, and Strawberries
Proanthocyanidins	Monomers, Dimers, Trimers, 7–10 mers (heptamers, octamers, nonamers, and decamers), 4–6 mers (tetramers, pentamers, and hexamers), Polymers of flavan-3-ols	Due to potential antioxidant properties of proanthocyanidins, they reduce the risk of cardiovascular disease and also help to prevent cancer. Some proanthocyanidins are also consumed to control urinary tract infections. Growth of pathogenic bacteria like *Clostridium perfringens, C. difficile* and *Bacteroides* species are also profoundly inhibited by action of proanthocyanidins	Tea, Cocoa, Berries, Grape juice, Cranberries, and Red wine.
Flavones	Apigenin, Nobiletin, Luteolin, Tangeretin	Flavones are effective antitumor, anti-inflammatory, anti-allergic and anti-asthmatic agent. They also show anti-proliferative, anti-invasive, and anti-metastatic effect. They also control blood pressure and blood sugar levels.	Celery, Lettuce, Parsley, Beets, Bell peppers, Brussels sprouts, Cabbage, Cauliflower, Celery, Hot peppers, Lettuces, Spinach, Thyme, Citrus fruits
Flavonols	Myricetin, Quercetin, Isorhamnetin, Kaempferol, Isorhamnetin	Flavonols are anti-inflammatory and anti-tumor phytochemicals. They reduce the event of asthma and myocardial infarction. Flavonols are also beneficial against lung cancer.	Berries, grapes, parsley, spinach, onions, apples, broccoli, cranberries, grapes

TABLE: 12.3 *(Continued)*

Types of Flavonoid	Example	Clinical Significance	Source
Flavan-3-ols	Catechins and gallic acid esters of catechins, theaflavins and gallic acid esters of theaflavins, thearubigins, epicatechins and gallic acid esters of epicatechins	Flavan-3-ols lower the risk of coronary heart disease, effective against some types of cancers and promote lung health.	Tea, red wine, cocoa powder, dark chocolate, grapes, plums, fruits and legumes (beans)
Flavanones	Hesperetin, Eriodictyol, Naringenin	Flavanones have impact on cardiovascular system. They help in controlling atherothrombotic diseases. They also lower the levels of serum LDL cholesterol. Naringenin possesses antioxidant, anti-estrogen, and cholesterol-lowering properties. Naringenin is also active (with the lowest minimum inhibitory concentrations) against bacterial infection caused by *Lactobacillus rhamnosus*, *E. coli*, *Staphylococcus aureus*, *S. typhimurium*, and *H. pylori*.	Grapefruits and Oranges, Citrus fruits
Isoflavonoids	Genistein, Daidzein	Strong antioxidative properties	Soya, Soya products, and Red clover

12.3.3 QUINONES

Quinones are organic aromatic compounds with two ketone substitutions. They are highly reactive and abundant in nature. These compound are responsible for the brown color appeared on fruits and vegetables after being injured or damaged. They work as an intermediate compound in melanin synthesis pathway (Schmidt, 1988). They are often involved in the inactivation of protein, for that reason have a potential activity against microorganisms. Quinones restrict substrate availability to microorganisms. More research is needed to explore more probable anti-microbial effect of quinones (Cowan, 1999). Quinones compounds induce enzyme detoxification reactions, perform anti-inflammatory activities and can also modify the redox status of a cell and hence can induce cytoprotection.

12.3.4 TANNINS

Tannin is a polyphenolic substance work as an astringent and a common biomolecule used in tannery industries is plant product of its secondary metabolism. Tannin is one of the most abundant elements present in plant parts (Scalbert, 1991). The hydrophilic nature of tannins makes its extraction process simple and easy and thus its availability for pharmaceutical applications. They can form either by condensation of flavan derivatives or by polymerization of quinone units (Geissman, 1963). Tannin bearing plants exhibit astringent, hemostatic, antiseptic and toning properties. Tannins are usually responsible for many anti-infective actions in human (Haslam, 1996). It has been studied and reported that beverages that contain tannins have the capacity to prevent from many diseases (Serafini, Ghiselli, Ferro-Luzzi and Melville, 1994). Plant extracts containing tannins are used to block local hemorrhages/bleeding, prevent ulcer and swelling in the buccal cavity, bronchitis, burns, skin scares, wounds and to control diarrhea.

12.3.5 TERPENOIDS

Another name of terpenoid is isoprenoids. It is the most abundant and diverse plant product. The major portion of essential oils contains terpenoids (Mbaveng, Hamm and Kuete, 2014). Terpenoids are active against many microorganisms. They are reported to have an extensive range of pharmacological properties. Terpenoids perform a variety of functions in growth and function of plants, but majorly they provide protection to the plant in an adverse environment. Terpenoids obtained from plants are being used in food, medicine and in many chemical industries (Tholl, 2015). For humans, they play a major role for health and immune function. Few studies have reported that in human they prevent the formation and growth of ulcer (De Pasquale, Germano, Keita, Sanogo, and Iauk, 1995). Their antioxidant properties are responsible for providing health benefits. Formulation of terpenoids has been reported to be very effective against bacterial, fungal and viral infections. Table 12.4 summarizes information regarding different type of terpenoids in medicinal practice (Wang, Tang and Bidigare, 2005).

12.3.6 ALKALOIDS

Alkaloids diverse and abundant class secondary metabolites found at 10–15% concentration in almost all plants. Alkaloids are chemical compound containing basic nitrogen atom. They are colorless crystals. Their chemical structure is enormously dynamic. They are produced by wide variety of organisms such as plants, bacteria, fungus, etc. Many alkaloids have physiological effect that renders them valuable medicine against diseases like malaria, diabetics, cancer, and cardiac dysfunction. These are also practiced in local anesthesia and as analgesic compounds. The first example of alkaloid used in medicine was morphine isolated from *Papaver somniferum.* Morphine blunts the pain response secondary to the ischemic tissue damage along with providing anxiolysis. Codeine another alkaloid obtained from *Papaver somniferum* is a mild pain reliever and an effective cough suppressant. A number of alkaloids are being used as drugs for century. Among the oldest and best known of these is quinine, derived from the bark of the tropical cinchona tree which is used as an antimalarial drug. Berberine is another common example of alkaloid effective against *Plasmodia* and *Trypanosomes*. Mostly they control microbial regulation (Cowan, 1999). Harmine, a beta-carboline alkaloid, is widely distributed in the plants. Harmine has various types of pharmacological activities such as antimicrobial, antifungal, antitumor, cytotoxic, antiplasmodial, antioxidant, antimutagenic, antigenotoxic, and anti-HIV (Patel et al., 2012).

12.3.7 POLYPEPTIDES AND LECTINS

In 1942, the first peptides were reported having inhibitory effect on micro-organisms (Balls, Hale and Harris, 1942). They contain disulfide bond and

TABLE 12.4 Types of Terpenoid with Variety of Medicinal Properties and Potential Sources

Type of terpenoid	Example	Medicinal Property	Source
Monoterpenes	Limonene, Carvone, Carveol, Citronellol, Nerol, Pyrethrins, and Geraniol	Monoterpenes have antitumor activity, and they also exhibit ability to regress existing malignant tumors. Pyrethrins are used for treatment of skin parasites such as head lice. Acyclic monoterpene exhibit some activity against *Mycobacterium tuberculosis.*	Essential oils extracted from citrus fruits, cherry, mint, caraway seed oil, herbs, floral aromatic compounds, and plant resins
Sesquiterpenes	Artemisinin, Parthenolide, Chamazulene, Bisabolol, Bisabolol oxides A and B	Sesquiterpenes are effective against fever, headaches, menstrual difficulties, stomachaches, arthritis, migraine, asthma, and psoriasis. Also, show antimalarial activity and used as antibacterial and antifungal agents.	*Tanacetum parthenium, C. parthenium, Leucanthemum parthenium, Pyrethrum parthenium, Matricaria chamolilla* flowers
Diterpenes	Phytol, Taxines, Paclitaxel, Eleutherobin, Sarcodictyin A and B, Diterpene glycosidesm, Bromoditerpene Sphaerococcenol A, Diisocyanoadociane, Solenolide A and Spongiadiol	Phytol exhibits significant antituberculosis activity. Taxines show anticancer activity. Paclitaxel is currently used to treat ovarian, lung, and breast cancers, head and neck carcinoma, and melanoma. Sphaerococcenol A has antimalarial activity against the chloroquine-resistant Plasmodium falsciparum strain. Glycosidesm has analgestic properties and are used for promoting wound healing.	*Lucas volkensii, Mediterranean stoloniferan S. roseum, Gorgonian Pseudopterogorgia Sphaerococcus coronopifolius, Cymbastela hooperi, Indopacific gorgonian,*
Sesterterpenes	Scalaranes, Manoalide, Seco-manoalide, Luffariellolide, Luffariellins, Luffolide, Cacospongionolides, Petrosaspongiolides, Salmahyrtisol A, Halorosellinic acid, and Mangicols	Sesterterpenes are used for treatment of acute and/or chronic inflammation. Manoalide is a potent analgesic and anti-inflammatory agent. Acute and/or chronic inflammation. Salmahyrtisol A show significant cytotoxicity against murine leukemia, human lung carcinoma, and human colon carcinoma. Halorosellinic acid exhibits antimalarial activity and weak antimycobacterial activity.	Mostly originated from fungal and marine organism. *Cacospongia mollior, Luffariella variabilis,* Red Sea sponge *Hyrtios erecta, Halorosellinia oceanica, Fusarium heterosporum*

have positive charge (Zhang and Lewis, 1997). They inhibit the activity of microbes either by forming ion channels in microbial membrane or by inhibiting the action of enzymes and protein present inside the microbes. Thionins are active peptides used against yeast and bacteria (De Caleya et al., 1972). As per reports some of the thionin derivatives are potential agents against fungi but not for bacteria (Kragh et al., 1995). A newly identified peptide residue fabatin inhibit *E. coli*, *P. aeruginosa* as well as *Enterococcus hirae* but are not active against *Candida* or *Saccharomyces* (Zhang and Lewis, 1997). Lectins are proteins present in nature and in many plant species. They are generally active against viral infections (Balzarini et al., 1991). Few useful antimicrobial plant compounds are summarized in Table 12.5.

12.4 ANTIBIOTIC AND PROBLEM OF RESISTANCE

Antibiotics are antibacterial medicines that either kill the bacteria or stop the bacterial reproduction. They act as a powerful drug to treat number of diseases caused by bacteria. Our body's immune system work as a first defense system against microbial infection, when a microbe enters into the body white blood cells control their proliferation before they exhibit any pathological symptom. When the body's immune system fails to fight against infection, antibiotics are administered to control the growth and harmful effect of disease-causing microorganism on human health. Antibiotics have been found to be effective against microbial infection but cannot treat the infections caused by viruses. Some antibiotics are effective against a diverse group of organisms (bacterial species) and are referred as broad-spectrum antibiotics, while others are specific against a bacteria species. Some antibiotics are active against aerobic bacteria, and some kill anaerobic bacteria. Sometimes antibiotics such as cephalosporins, vancomycin are used for prevention of infection rather than for treatment; this is known as prophylactic way to use antibiotics. Such types of antibiotics are generally used before surgery to prevent the patient from infections. Antibiotics are usually administered orally; however, they can also be injected into the body or can be applied directly to the body. Excess use of these antibiotics causes few side effects such as diarrhea, fever, vomiting and vaginal itching. There are cases in which some patients have also developed allergic reactions (rashes, swelling, etc.) against antibiotics. Such allergic reactions can be extremely harmful and are known as an anaphylactic reaction (itchy rash, throat or tongue swelling, shortness of breath, vomiting, lightheadedness, and low blood pressure). Thus, antibiotics should be used in a proper dose-dependent way and carefully.

TABLE 12.5 Common Medicinal Herbs and Principle Metabolic Compounds Present in Them

Common name	Botanical name	Metabolic compound	Target microorganism	References
Aloe vera	*Aloe indica* Royle *A. vulgaris, A barbadensis*	Latex	*Corynebacterium, Salmonella, Streptococcus, S. aureus*	Martinez, Betancourt, Alonso-Gonzalez and Jauregui, 1996
Apple	*Malus sylvestris*	Phloretin a derivative of flavonoids	Microbes	Hunter and Hull, 1993
Ashwagandha	*Withania somniferum*	A class of Lactone-Withafarin A	Bacteria, fungi	Hunter and Hull, 1993
Basil	*Ocimum basilicum*	Terpenoids	*Salmonella*, bacteria	Wan, Wilcock and Coventry, 1998
Black pepper	*Piper nigrum*	Class of alkaloid Piperine	Fungi, *Lactobacillus, Micrococcus, E. coli, E. faecalis*	Ghoshal, Prasad and Lakshmi, 1996
Blueberry	*Vaccinium* spp.	Carbohydrate (Fructose)	*E. coli*	Ofek et al., 1991
Cashew	*Anacardium pulsatilla*	Salicylic acids a class of polyphenols	*P. acnes* Bacteria, fungi	Himejima and Kubo, 1991
Chili peppers, paprika	*Capsicum annuum*	Type of terpenoid Capsaicin	Bacteria	Cichewicz and Thorpe, 1996
Green tea	*Camellia sinensis*	Flavinoid class -Catechin	Bacteria and viruses	Toda et al., 1992
Henna	*Lawsonia inermis*	Phenolic compounds	*S. aureus*	Cowan, 1999
Lemon balm	*Melissa officinalis*	Tannins and polyphenols	Viruses	Wild, 1994
Olive oil	*Olea europaea*	Hexanal	Common	Kubo, Lunde and Kubo, 1995
Onion	*Allium cepa*	Allicin	Bacteria and *Candida*	Vohora, Rizwan and Khan, 1973
Orange peel	*Citrus sinensis*	Terpenoid	Fungi	Stange, Midland, Eckert and Sims, 1993
Papaya	*Carica papaya*	Latex	Common	Osato, Santiago, Remo, Cuadra and Mori, 1993
Peppermint	*Mentha piperita*	Menthol	Common	Cowan, 1999
Turmeric	*Curcuma longa*	Curcumin, a class of terpenoid	Bacteria and protozoa	Apisariyakul, Vanittanakom and Buddhasukh, 1995

Sir Alexander Fleming in 1928 started the era of antibiotics by discovering penicillin. Since then antibiotics are saving billions of lives. In the early 1940s, antibiotics were usually prescribed for hazardous infections. In a very short time, period resistance has been developed to almost all antibiotics (Ventola, 2015). The overuse of antibiotics can make the bacteria resistant towards that particular antibiotic making it ineffective against the bacteria. The European Centre for Disease Prevention and Control (ECDC) has stated antibiotic resistance to be a serious issue worldwide. In November 2012, the statement issued by ECDC stated that almost 25,000 people die every year in the European Union because of infections caused by antibiotic-resistant bacteria. The number of infections and the death rate is increasing day-by-day. In 1945 Sir Alexander Fleming the man who himself discovered the first antibiotic, gave the idea about the resistance of microbes towards the antibiotics.

The question is what is the root cause of increase in antibiotic resistance cases, why bacteria are becoming more resistant to antibiotics in use. Overuse of antibiotics forces bacteria to acquire DNA from other resistant bacterial strains by horizontal gene transfer phenomenon. The presence of antibiotics exposes bacterial cell to experience enormous stress, and this stress triggers random behavior (Figure 12.2). Thus, the non-resistant bacteria are developing into resistant one by accident or by ingesting DNA. The research is still trying to evaluate the actual cause/ parameter responsible for uptake of DNA from other resistant bacteria in order to evolve resistance towards antibiotics. Mostly bacteria avoid to take up a foreign DNA from other bacterium because they do not have DNA reader system to identify the utility and usefulness of the foreign molecule taken up. The random selection of DNA is lethal in most of the cases studied. If the rate of antibiotic resistance will persist at similar scale within few years, there will be no antibiotic left that can effectively deal with microbial infections and would be a dark era in the field of modern medicine.

12.4.1 SOURCES OF ANTIBIOTIC RESISTANCE

The excessive use of antibiotics is responsible for the evolution in bacterium (Control, 2015a; Read and Woods, 2014b). The genes in the bacteria can either be inherited or acquired from genetic elements such as plasmids (Read and Woods, 2014). These mechanisms of gene transfer are permitting the transfer of antibiotic resistance among different bacterial species and strains. This resistance can occur spontaneously too in certain cases (Read and Woods, 2014). Despite warnings about overuse, antibiotics are still overprescribed worldwide. A lot of research must be done to decrease the utilization of these drugs (Gross, 2013). According to the analysis of IMS Health Midas database, approximately 22.0 standard units (a unit equaling one dose, e.g., one pill, capsule, or ampoule) of antibiotics were prescribed per person in the U.S. in 2010 (Van Boeckel et al., 2014). In many places, antibiotics are available to people without any prescription. Such availability of antibiotics increases the rate of antibiotic resistance (Michael, Dominey-Howes and Labbate, 2014). Improper prescription also promotes resistance to bacteria from antibiotics (Control, 2015).

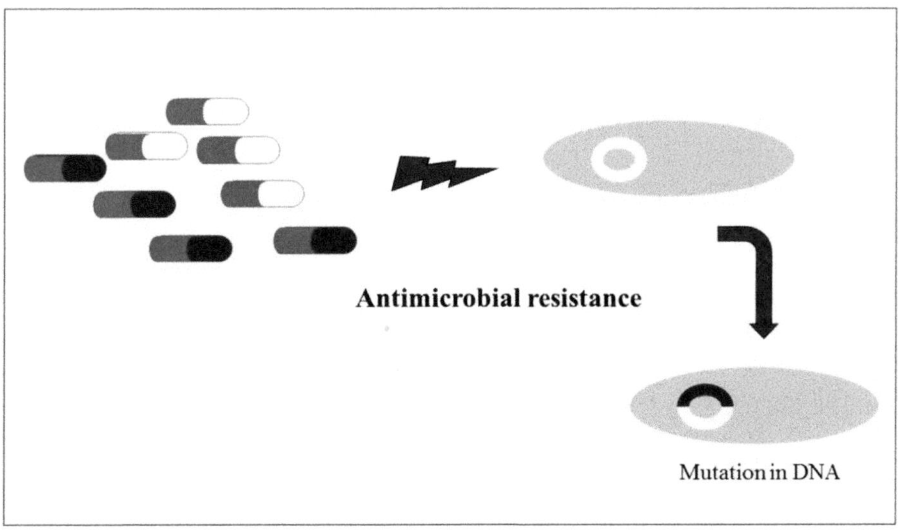

FIGURE 12.2 Antimicrobial resistance.

Nowadays a good amount of antibiotics are also being used in veterinary practices to protect and treat animals from microbial infection. Such practices are leading to contamination of animal products with antibiotics. These products on consumption are serving as source/means to develop antibiotics resistance in human microbial pathogens.

The extreme use of antibiotics is not the only reason of antibiotic resistance, but few more factors are also responsible (Figure 12.3). Lack of availabilities of new drugs, lack of proper sanitation practices, incomplete dosage, excessive use in agriculture practices, etc. are few additional reasons why antibiotic resistance is spreading so fast.

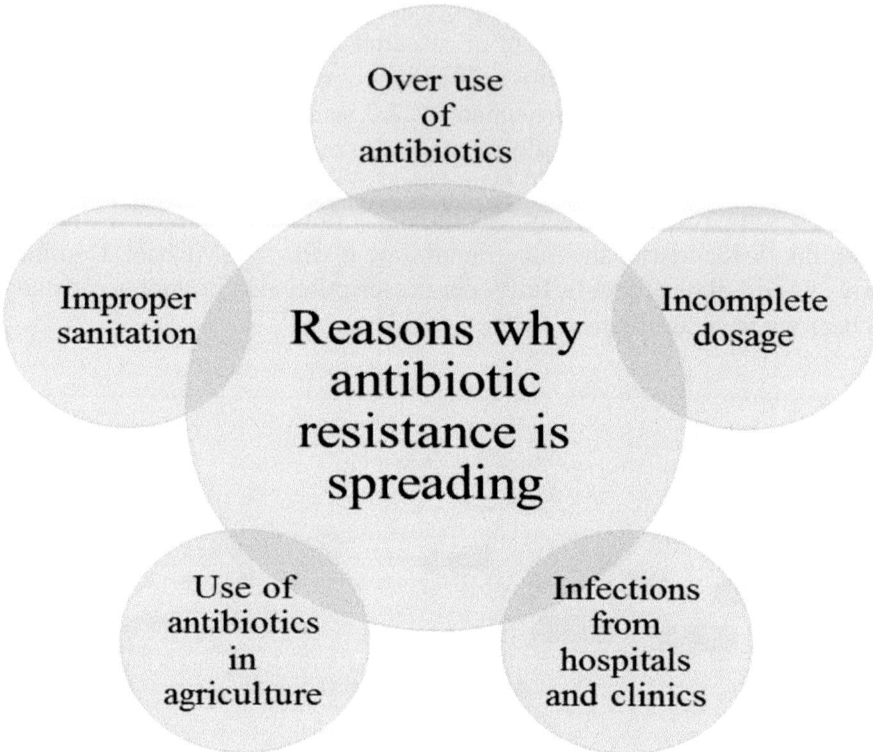

FIGURE 12.3 Causes of antibiotic resistance.

12.5 MECHANISM OF ANTIMICROBIAL PLANT METABOLITES

Microbial infectious diseases continue to be one of the leading causes of morbidity and mortality and pose a great threat to humankind. The genetic makeup, metabolic and physiological diversity of microorganisms are extraordinary, and it makes them a major cause of infections across the world. Development of antibiotic resistance, the emergence of new pathogens in addition to the resurgence of old ones, and the lack of new effective therapeutic agent has exaggerated the problems. To improve theranostics and to combat antibiotic resistance there is a need to discover and develop new antimicrobial agents. Plant secondary metabolites offer particular promise in this sense. Most of the plant metabolites are found to have antimicrobial activity as a mechanism for self-defense against harmful microorganisms and various infections. These active compounds in plant metabolites play role in disinfection; for example, clove contains eugenol and isoeugenol as active compounds. In few cases, soluble compound like terpenes has higher antimicrobial activity than other compounds (Knobloch et al., 1989). Antimicrobial activity and mechanism of action of different plant compounds depend majorly on the type of the targeted microorganism, e.g., bacterial or fungal and if bacterial then whether it is Gram-positive or Gram-negative. The second parameter which affects the antimicrobial activity is the environment condition of the antimicrobial action, like its hydrophilicity, temperature, pH, etc. The above-mentioned two parameters majorly decide the final effect of plant secondary metabolite as useful and effective an antimicrobial agent (Denyer and Stewart, 1998a; Radulovic, Blagojevic, Stojanovic-Radic and Stojanovic, 2013b).

Many phytochemicals have the ability to inhibit microbial growth; some interferes the metabolic system and gene expression of microbes. The mode of action of plant constituents are antimicrobial agent is different for different bacteria. For example, in case of enterotoxigenic bacteria, they act by damaging phospholipid cell membrane that causes loss of cell components, destruction of cellular enzymes and genetic material. Overall the mechanism of action is to cause the disruption in cytoplasmic membrane, interference in electron flow and transport mechanism (Sakagami and Kajimura, 2002). The antibacterial components of *M. oleifera* cause damage to the pathogens by inhibiting the bacterial enzymes like sortase, interference in DNA replication and cell lysis. Many studies proposed that pterygospermin inhibit the activity of transaminase enzyme that create cell membrane distress (Chartone-Souza, 1998). Plant antimicrobial peptides interrelate with cellular membrane probably by following two steps. In the first stage negatively charged phospholipoidal groups on the surface of microbial membrane attract cationic amino acids present in plant antimicrobial peptide. These peptides and hydrophobic acid together react with aliphatic fatty acid and anionic constituents of microbial cell in the second step. This persuades membrane deterioration in infectious bacteria. Thus, bacterial cell is killed by loss of cytoplasmic content and membrane potential.

Tannins block essential metabolic enzymes such as proteolytic enzymes thus destroy bacterial cell proliferation. Saponins are also supposed to exert toxic effects by creating alteration in permeability of cell walls thus damage cell wall organization. The antibacterial activity of saponins stimulate alteration in cell morphology too which lead to cell lysis (Cowan, 1999). Polyphenols present in medicinal plants such as gallic acids probably bind to dihydrofolate reductase (DHFR) bacterial enzymes. They inhibit the supercoiling activity of gyrase by binding to the ATP binding active site of gyrase B thus interfere with the effective DNA packaging (Alzoreky and Nakahara, 2003), The mechanism how plant secondary metabolites are affecting pathogen is illustrated in Figure 12.4 (Nazzaro, Fratianni, De Martino, Coppola, and De Feo, 2013).

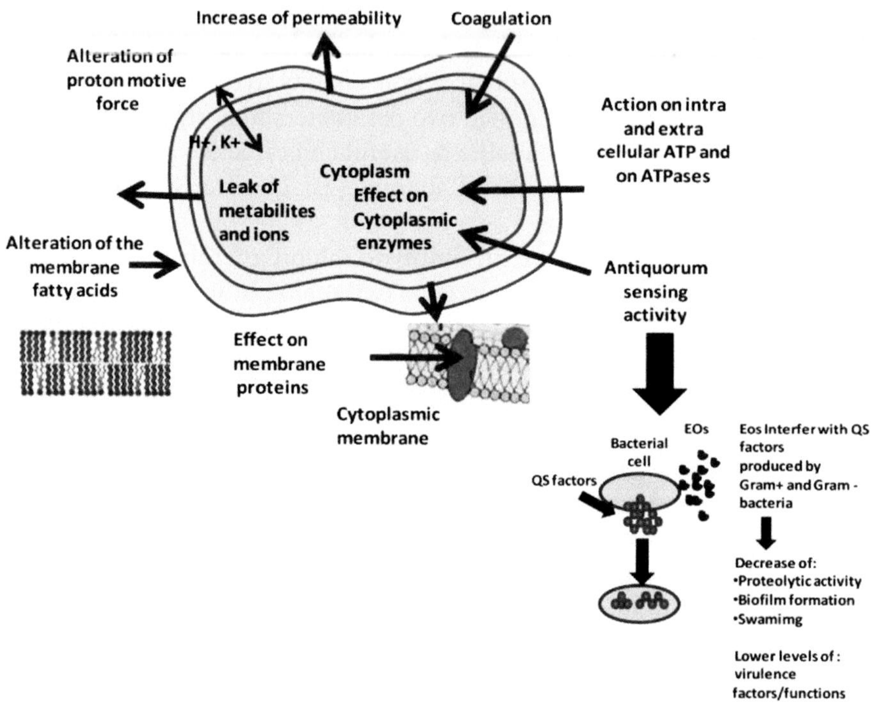

FIGURE 12.4 Mechanism of antimicrobial activity of plants secondary metabolites.

Plants have an extensive range of antioxidant molecules that destroy free radicals comprising phenols, flavonoids, terpenoids, and many vitamins, with different mechanism of antimicrobial activity (Rice-Evans, Miller and Paganga, 1997). A list of various plants secondary metabolites, which act as an antimicrobial agent against a wide range microorganism is shown in Figure 12.5 along with their site of action (Silva and Fernandes, 2010; Cowan, 1999).

Numerous plant species fruits and vegetables serve as a source of ascorbic acid, vitamin E and other antioxidant molecules that removes free radicals from human body. Phytochemicals present in plant are usually involved in providing protection to the plant from fungi, bacteria and other pathogens. These phytochemicals can be used as human medicines (Jakhetia et al., 2010). Phenolic acids reduce the adherence of organisms to the cell thus decrease the chances of urinary infection. Plant's essential oils can also inhibit the growth of microbes. Phytochemical obtained from green tea epigallocatechin-3-gallate inhibit the verotoxin released from *E. coli*. Ethanol extract from *Punica granatum* was also reported to inhibit verotoxin. Although the exact mechanisms of plant phytochemicals are still unclear but it is supposed that they obstruct the transcription and translation process of bacteria (Jakhetia et al., 2010). All plants have beneficial compounds, but usually, secondary metabolites such as alkaloids, tannins, phenol compounds, and steroids are involved in antibacterial activities (Omojate, Enwa, Jewo, and Eze, 2014). These secondary metabolites have medicinal effect, as they resemble many hormones, metabolites, signal transduction molecules, etc. So screening and selection of plants for their medicinal properties are necessary.

Class	Subclass	Examples	Mechanism
Phenolics	Simple phenols	Catechol	Substrate deprivation
		Epicatechin	Membrane disruption
	Phenolic acid	Cinnamic acid	?
	Quinones	Hypericin	Adhesin binding, complex with cell wall, enzyme inactivation
	Flavonoids	Chrysin	Adhesin binding
	Flavones	–	Complex with cell wall
		Abyssinone	Enzyme inactivation
			HIV reverse transcriptase inhibition
	Flavonols	Totarol	?
	Tannins	Ellagitannin	Protein binding
			Adhesin binding
			Enzyme inhibition
			Substrate deprivation
			Complex with cell wall
			Membrane disruption
			Metal-ion complexation
	Coumarins	Warfarin	Interaction with eucaryotic DNA (antiviral activity)
Terpenoids, essential oils	–	Capsaicin	Membrane disruption
Alkaloids	–	Berberine	Intercalation into cell wall and/or DNA
		Piperine	
Lectins and polypeptides	–	Mannose-specific agglutinin	Block of viral fusion or adsorption
		Falxatin	Disulfide bridge formation
Polyacetylenes	–	8s-heptadeca-2(Z),9(Z)-diene-4,6-diyne-1,8-diol	?

FIGURE 12.5 Major groups of phytochemicals having antimicrobial activity.

Plants secondary metabolites have emerged as a great alternative to conventional antibiotics that are being used because of many reasons already discussed above, but combating antibiotic resistance and being eco-friendly tops the list (Radulovic, Blagojevic, Stojanovic-Radic, and Stojanovic, 2013b; Gupta and Birdi, 2017). Literature suggests that instead of using a single compound, using a mixture of plant secondary metabolites helps the cause in a better way. Antimicrobial property of plants secondary metabolites do not depend on a single phytochemical, rather its effectiveness against group of pathogen can be achieved by formulating a combination of different plant metabolites together (Gertsch, 2011a; Van Vuuren and Viljoen, 2011b). It can be well concluded that still a long way to go before we can use a plant-derived antibiotic commercially in place of microbial antibiotic.

12.6 CONCLUSION

Over the centuries plants are being used to treat several diseases. Medicinal plants are the potential source for the antimicrobial agents that can be used for the production of new medicines. The search for new antimicrobial compounds is being necessary to overcome the problem of antimicrobial resistance caused by existing medicines. Medicinal plants can be used as an alternative towards this. Numerous studies have proved the antibacterial activity of medicinal plants either in the form of essential oil or extract. Sometimes plant's part can also be used as a source of medicine. This property of medicinal plants has the potential to serve as a drug compound, which is a need of the time. Numerous studies have reported to show the potential application of plant components, but still, there is a place to find new plant species for their medicinal properties.

KEYWORDS

- **antibiotics**
- **antimicrobial agents**
- **medicinal plants**
- **secondary metabolites**

REFERENCES

Alzoreky, N.; Nakahara, K. Antibacterial activity of extracts from some edible plants commonly consumed in Asia. *International journal of food microbiology* **2003,** *80* (3), 223–230.

Amenu, D. Antimicrobial activity of medicinal plant extracts and their synergistic effect on some selected pathogens. *American Journal of Ethnomedicine* **2014,** *1* (1), 18–29.

Apisariyakul, A.; Vanittanakom, N.; Buddhasukh, D. Antifungal activity of turmeric oil extracted from Curcuma longa (Zingiberaceae). *Journal of Ethnopharmacology* **1995,** *49* (3), 163–169.

Balls, A.; Hale, W.; Harris, T. A crystalline protein obtained from a lipoprotein of wheat flour. *Cereal Chem* **1942,** *19* (19), 279–288.

Balzarini, J.; Schols, D.; Neyts, J.; Van Damme, E.; Peumans, W.; De Clercq, E. Alpha-(1–3)-and alpha-(1–6)-D-mannose-specific plant lectins are markedly inhibitory to human immunodeficiency virus and cytomegalovirus infections in vitro. *Antimicrobial agents and chemotherapy* **1991,** *35* (3), 410–416.

Batugal, P. A.; Kanniah, J.; Sy, L.; Oliver, J. T. *Medicinal Plants Research in Asia-Volume I: The Framework and Project Workplans*. Biodiversity International: 2004.

Borris, R. P. Natural products research: perspectives from a major pharmaceutical company. *Journal of Ethnopharmacology* **1996,** *51* (1–3), 29–38.

Chartone-Souza, E. Bactérias ultra-resistentes: uma guerra quase perdida. *Cienc Hoje* **1998,** *23* (138), 27–35.

Cichewicz, R. H.; Thorpe, P. A. The antimicrobial properties of chile peppers (Capsicum species) and their uses in Mayan medicine. *Journal of Ethnopharmacology* **1996,** *52* (2), 61–70.

(a) Control, C. f. D.; Prevention, Office of Infectious Disease. Antibiotic resistance threats in the United States, 2013. April 2013. 2015; (b) Read, A. F.; Woods, R. J. Antibiotic resistance management. *Evolution, medicine, and public health* **2014,** *2014* (1), 147.

Cowan, M. M. Plant products as antimicrobial agents. *Clinical microbiology reviews* **1999,** *12* (4), 564–582.

Cowan, M. Plant products as antimicrobial agents. Clin. Microbial. Rev. 22: 564–582. 1999.

Dash, G.; Murthy, P. Antimicrobial activity of few selected medicinal plants. *International Research Journal of Pharmacy* **2011,** *2* (1), 146–152.

De Caleya, R. F.; Gonzalez-Pascual, B.; García-Olmedo, F.; Carbonero, P. Susceptibility of phytopathogenic bacteria to wheat purothionins in vitro. *Applied Microbiology* **1972,** *23* (5), 998–1000.

De Pasquale, R.; Germano, M.; Keita, A.; Sanogo, R.; Iauk, L. Antiulcer activity of Pteleopsis suberosa. *Journal of Ethnopharmacology* **1995,** *47* (1), 55–58.

(a) Denyer, S. P.; Stewart, G. Mechanisms of action of disinfectants. *International Biodeterioration & Biodegradation* **1998,** *41* (3–4), 261–268; (b) Radulovic, N.; Blagojevic, P.; Stojanovic-Radic, Z.; Stojanovic, N. Antimicrobial plant metabolites: structural diversity and mechanism of action. *Current Medicinal Chemistry* **2013,** *20* (7), 932–952.

Dixon, R.; Dey, P.; Lamb, C. Phytoalexins: enzymology and molecular biology. *Advances in Enzymology and Related Areas of Molecular Biology* **1983,** *55* (1).

Farjana, A.; Zerin, N.; Kabir, M. S. Antimicrobial activity of medicinal plant leaf extracts against pathogenic bacteria. *Asian Pacific Journal of Tropical Disease* **2014,** *4*, S920-S923.

Geissman, T. Chapter X-Flavonoid Compounds, Tannins, Lignins and, Related Compounds.

Geissman, T. Flavonoid Compounds, Tannins, Lignins and, Related Compounds. *Comprehensive Biochemistry*, Elsevier: **1963**, *9*, 213–250.

(a) Gertsch, J. Botanical drugs, synergy, and network pharmacology: forth and back to intelligent mixtures. *Planta medica* **2011,** *77* (11), 1086–1098; (b) van Vuuren, S.; Viljoen, A. Plant-based antimicrobial studies–methods and approaches to study the interaction between natural products. *Planta Medica* **2011,** *77* (11), 1168–1182.

Ghoshal, S.; Prasad, B. K.; Lakshmi, V. Antiamoebic activity of Piper longum fruits against *Entamoeba histolytica* in vitro and in vivo. *Journal of Ethnopharmacology* **1996,** *50* (3), 167–170.

Gross, M. *Antibiotics in Crisis*. Elsevier: **2013**.

Gupta, P. D.; Birdi, T. J. Development of botanicals to combat antibiotic resistance. *Journal of Ayurveda and integrative medicine* **2017**.

Hamid, A.; Sitepu, D. An understanding of native herbal medicine in Indonesia. *Industrial Crops Research Journal* **1990,** *3* (1), 11–17.

Haslam, E. Natural polyphenols (vegetable tannins) as drugs: possible modes of action. *Journal of Natural Products* **1996,** *59* (2), 205–215.

Himejima, M.; Kubo, I. Antibacterial agents from the cashew *Anacardium occidentale* (Anacardiaceae) nut shell oil. *Journal of Agricultural and Food Chemistry* **1991,** *39* (2), 418–421.

Hunter, M. D.; Hull, L. A. Variation in concentrations of phloridzin and phloretin in apple foliage. *Phytochemistry* **1993,** *34* (5), 1251–1254.

(a) Jain, S. K.; DeFilipps, R. A. *Medicinal Plants of India*. Reference Publications: 1991; (b) De Padua, L.; Bunyapraphatsara, N.; Lemmens, R. *Plant Resources of South-East Asia*. Backhuys Publ.: 1999; Vol. 12; (c) Govaerts, R. How many species of seed plants are there? *Taxon* **2001,** *50* (4), 1085–1090; (d) Groombridge, B.; Jenkins, M. *World Atlas of Biodiversity: Earth's Living Resources in the 21st Century*. University of California Press: **2002**.

Jakhetia, V.; Patel, R.; Khatri, P.; Pahuja, N.; Garg, S.; Pandey, A.; Sharma, S. Cinnamon: a pharmacological review. *Journal of Advanced Scientific Research* **2010,** *1* (2), 19–23.

Katwal, R.; Srivastava, R.; Kumar, S.; Jeeva, V. In *Status of Forest Genetic Resources Conservation and Management in India*, Forest Genetic Resources Conservation and Management: Proceedings of the Asia Pacific Forest Genetic Resources Programme (APFORGEN) Inception Workshop, Kepong, Kuala Lumpur, Malaysia, 15–18 July, 2003, Biodiversity International: **2004**, p. 49.

Knobloch, K.; Pauli, A.; Iberl, B.; Weigand, H.; Weis, N. Antibacterial and antifungal properties of essential oil components. *Journal of Essential Oil Research* **1989,** *1* (3), 119–128.

Köberl, M.; Schmidt, R.; Ramadan, E. M.; Bauer, R.; Berg, G. The microbiome of medicinal plants: diversity and importance for plant growth, quality and health. *Frontiers in Microbiology* **2013,** *4*, 400.

Kragh, K. M.; Nielsen, J. E.; Nielsen, K. K.; Dreboldt, S.; Mikkelsen, J. D. Characterization and localization of new antifungal cysteine-rich proteins from Beta vulgaris. *MPMI-Molecular Plant Microbe Interactions* **1995,** *8* (3), 424–434.

Kubo, A.; Lunde, C. S.; Kubo, I. Antimicrobial activity of the olive oil flavor compounds. *Journal of Agricultural and Food Chemistry* **1995,** *43* (6), 1629–1633.

Martinez, M.; Betancourt, J.; Alonso-Gonzalez, N.; Jauregui, A. Screening of some Cuban medicinal plants for antimicrobial activity. *Journal of Ethnopharmacology* **1996,** *52* (3), 171–174.

Mason, T. L. Inactivation of red beet β-glucan synthase by native and oxidized phenolic compounds. *Phytochemistry* **1987,** *26* (8), 2197–2202.

Mbaveng, A. T.; Hamm, R.; Kuete, V. 19-Harmful and protective effects of terpenoids from African medicinal plants. *Toxicological Survey of African Medicinal Plants. Oxford: Elsevier* **2014**, 557–576.

Michael, C. A.; Dominey-Howes, D.; Labbate, M. The antimicrobial resistance crisis: causes, consequences, and management. *Frontiers in Public Health* **2014,** *2*.

Miller, K. I.; Qing, C.; Sze, D. M. Y.; Neilan, B. A. Investigation of the biosynthetic potential of endophytes in traditional Chinese anticancer herbs. *PLoS One* **2012,** *7* (5), e35953.

Modak, M.; Dixit, P.; Londhe, J.; Ghaskadbi, S.; Devasagayam, T. P. A. Recent advances in Indian herbal drug research guest editor: Thomas Paul Asir Devasagayam Indian herbs and herbal drugs used for the treatment of diabetes. *Journal of Clinical Biochemistry and Nutrition* **2007,** *40* (3), 163–173.

Moerman, D. E. An analysis of the food plants and drug plants of native North America. *Journal of Ethnopharmacology* **1996,** *52* (1), 1–22.

Mukhtar, H.; Ahmad, N. Tea polyphenols: prevention of cancer and optimizing health. *The American Journal of Clinical Nutrition* **2000,** *71* (6), 1698s–1702s.

Nascimento, G. G.; Locatelli, J.; Freitas, P. C.; Silva, G. L. Antibacterial activity of plant extracts and phytochemicals on antibiotic-resistant bacteria. *Brazilian Journal of Microbiology* **2000,** *31* (4), 247–256.

Nazzaro, F.; Fratianni, F.; De Martino, L.; Coppola, R.; De Feo, V. Effect of essential oils on pathogenic bacteria. *Pharmaceuticals* **2013,** *6* (12), 1451–1474.

Ofek, I.; Goldhar, J.; Zafriri, D.; Lis, H.; Adar, R.; Sharon, N. Anti-Escherichia coli adhesin activity of cranberry and blueberry juices. *New England Journal of Medicine* **1991,** *324* (22), 1599.

Omojate Godstime, C.; Enwa Felix, O.; Jewo Augustina, O.; Eze Christopher, O. Mechanisms of antimicrobial actions of phytochemicals against enteric pathogens–a review. *J Pharm Chem Biol Sci* **2014,** *2* (2), 77–85.

Osato, J. A.; Santiago, L. A.; Remo, G. M.; Cuadra, M. S.; Mori, A. Antimicrobial and antioxidant activities of unripe papaya. *Life Sciences* **1993,** *53* (17), 1383–1389.

Patel, K.; Gadewar, M.; Tripathi, R.; Prasad, S.; Patel, D. K. A review on medicinal importance, pharmacological activity and bioanalytical aspects of beta-carboline alkaloid "Harmine". *Asian Pacific Journal of Tropical Biomedicine* **2012,** *2* (8), 660–664.

Patil, D.; Patil, M. Diversity and Concerns of Indian Medicinal Plants: A Scenario. *Journal of Ecobiotechnology* **2010,** *2* (8).

Rice-Evans, C.; Miller, N.; Paganga, G. Antioxidant properties of phenolic compounds. *Trends in Plant Science* **1997,** *2* (4), 152–159.

Sakagami, Y.; Kajimura, K. Bactericidal activities of disinfectants against vancomycin-resistant enterococci. *Journal of Hospital Infection* **2002,** *50* (2), 140–144.

Samant; Pant, S.; Singh, M.; Lal, M.; Singh, A.; Sharma, A.; Bhandari, S. Medicinal plants in Himachal Pradesh, north western Himalaya, India. *The International Journal of Biodiversity Science and Management* **2007,** *3* (4), 234–251.

Sarmiento, W. C.; Maramba, C. C.; Gonzales, M. L. M. An in-vitro study on the antibacterial effect of neem (*Azadirachta indica*) leaf extract on methicillin-sensitive and Methicillin-resistant *Staphylococcus aureus*. *PIDSP J* **2011,** *12* (1), 40–45.

Scalbert, A. Antimicrobial properties of tannins. *Phytochemistry* **1991,** *30* (12), 3875–3883.

Schippmann, U.; Leaman, D. J.; Cunningham, A. Impact of cultivation and gathering of medicinal plants on biodiversity: global trends and issues. *Biodiversity and the Ecosystem Approach in Agriculture, Forestry and Fisheries* **2002**.

Schmidt, H. Phenol oxidase (EC 1.14. 18.1) a marker enzyme for defense cells. *Progress in Histochemistry and Cytochemistry* **1988,** *17* (3), III1–VI186.

Serafini, M.; Ghiselli, A.; Ferro-Luzzi, A.; Melville, C. Red wine, tea, and antioxidants. *The Lancet* **1994,** *344* (8922), 626.

Seth, S.; Sharma, B. *Medicinal Plants in India.* **2004**.

Silva, N.; Fernandes Júnior, A. Biological properties of medicinal plants: a review of their antimicrobial activity. *Journal of Venomous Animals and Toxins Including Tropical Diseases* **2010,** *16* (3), 402–413.

Stange Jr, R. R.; Midland, S. L.; Eckert, J.; Sims, J. J. An antifungal compound produced by grapefruit and Valencia orange after wounding of the peel. *Journal of Natural Products* **1993,** *56* (9), 1627–1629.

Tholl, D. Biosynthesis and biological functions of terpenoids in plants. In *Biotechnology of Isoprenoids*, Springer: **2015**, pp. 63–106.

Toda, M.; Okubo, S.; Ikigai, H.; Suzuki, T.; Suzuki, Y.; Hara, Y.; Shimamura, T. The protective activity of tea catechins against experimental infection by Vibrio cholerae O1. *Microbiology and Immunology* **1992,** *36* (9), 999–1001.

Toda, M.; Okubo, S.; Ohnishi, R.; Shimamura, T. Antibacterial and bactericidal activities of Japanese green tea. *Nihon saikingaku zasshi. Japanese Journal of Bacteriology* **1989,** *44* (4), 669–672.

Urs, N.; Dunleavy, J. Enhancement of the bactericidal activity of a peroxidase system by phenolic compounds [*Xanthomonas phaseoli sojensis*, soybeans, bacterial diseases]. *Phytopathology* **1975**.

Van Boeckel, T. P.; Gandra, S.; Ashok, A.; Caudron, Q.; Grenfell, B. T.; Levin, S. A.; Laxminarayan, R. Global antibiotic consumption 2000 to 2010: an analysis of national pharmaceutical sales data. *The Lancet Infectious Diseases* **2014,** *14* (8), 742–750.

Ventola, C. L. The antibiotic resistance crisis: part 1: causes and threats. *Pharmacy and Therapeutics* **2015,** *40* (4), 277.

Vijaya, K.; Ananthan, S.; Nalini, R. Antibacterial effect of theaflavin, polyphenon 60 (*Camellia sinensis*) and Euphorbia hirta on *Shigella* spp.—a cell culture study. *Journal of Ethnopharmacology* **1995,** *49* (2), 115–118.

Vohora, S.; Rizwan, M.; Khan, J. Medicinal uses of common Indian vegetables. *Planta Medica* **1973,** *23* (04), 381–393.

Wan, J.; Wilcock, A.; Coventry, M. The effect of essential oils of basil on the growth of Aeromonas hydrophila and Pseudomonas fluorescens. *Journal of Applied Microbiology* **1998,** *84* (2), 152–158.

Wang, G.; Tang, W.; Bidigare, R. R. Terpenoids as therapeutic drugs and pharmaceutical agents. In *Natural Products*, Springer: **2005**, pp. 197–227.

Wild, R. The complete book of natural and medicinal cures. ISBN: **1994**, pp. 13–9780425152263.

Zhang, Y.; Lewis, K. Fabatins: new antimicrobial plant peptides. *FEMS Microbiology Letters* **1997,** *149* (1), 59–64.

CHAPTER 13

Lasia spinosa: Wild Nutraceutical for Formulation of Future Drugs

SABEELA BEEVI UMMALYMA[1] and RAJKUMARI SUPRIYA DEVI[2]

[1]*Institute of Bioresources and Sustainable Development (IBSD), A National Institute Under Department of Biotechnology Govt. of India, Takyelpat, Imphal–795001 Manipur, India,*
E-mail: sabeela.25@gmail.com

[2]*Ambika Prasad Research Foundation, Regional Centre, Imphal, Manipur*

ABSTRACT

Lasia spinosa is a potent wild medicinal plant having potential to fight against various diseases. Main active phytochemicals found in this plant such as alkaloids, saponins, glycosides, tannins, phenolic compounds and flavonoids are used for several activity studies. The phytochemical activities reported are anti-microbial, anti-inflammatory, anti-helminthic, antioxidant, anti-diabetic, anti-hyperlipidemic and anti-tumor. This chapter illustrates the importance of the enthano-medicinal value of *Lasia spinosa* plants along with some scientific data which is showing phytochemical compounds of this plant to prove pharmacological values of the plant.

13.1 INTRODUCTION

In the 20th century, continuous utilization of antibiotic for the past had substantially minimized the risk of hazardous diseases (Gould, 2008). Continuous use of antibiotics for the past several years, leads microbes to develop resistance for the available antimicrobials (Wise, 2011). Therefore antibiotic resistance is a big challenge in future health care and pharmaceutical industries. Hence lots of research is going on how to develop a drug to fight against multi antimicrobial resistance. This is an urgent need of researcher responsibility to produce antimicrobial compounds from natural origin (plant-based) to refill existing anti-infectives or drugs. The report says that on an average two to three antibiotics are released every year (Osbourn, 1996). After a decline in rate in recent decades, again hastening to scientists, they realized that the effective life span of any antibiotics is limited (Alper, 1998; Eisenberg et al., 1993; Moerman, 1996; Wise, 2011). Therefore it is necessary focus on antimicrobial resistance (AMR), the mechanisms of action and secondary metabolites screening from available bioresources (Osman et al., 2012). Plant materials are used as traditional ways of medicines and used different communities of the world in different methods in many countries (Fischbash and Walsh, 2009). Plants are the only resources available for the treatment for various bacterial infections in rural and tribal communities. Most reported plants species are having antimicrobial activities; but still, many plants are available in the wild which is not explored so far: having lots of ethnic values of medicinal properties (Dianella, 2012). There is an urgent necessity for the screening of bioactive compounds available in wild plants for antimicrobial activities from the plants secondary metabolites (Ginsburg and Deharo, 2011).

Among such unexplored plants wealth, tubers/root crops are very important. These are modified plant structures for the storage of nutrients and used by the plants to survive in harsh ecological conditions as energy and nutrients source for next favorable season for growth and multiplication. Tubers are two types of tubers; stem as well as root tubers. Major example of tuber crops are *Solanum tuberosum* (potato), *Manihot esculenta* (cassava), *Dioscorea* spp. (yam), *Ipomoea batatas* (sweet potato), edible aroids (*Colocasia esculenta*) and *Xanthosoma* spp.. These tubers crops are vital part of food supply, industrial products, and animal feed. According to worldwide scenario, approximately 45% crops produced are consumed as food, animal feed, and various industrial applications. Mode of consumption of tuber crops varied in different countries. In developing countries, moderately 20% is utilized for livestock feeding other than as food source. It is less attractive in urban cities primarily due cost associated with transportation; processing, storage as well as long time for cooking. Developed country it is mainly used as animal feed compared to developing countries. Tuber crops are good source of medicinal compounds and have ethnomedicinal values in tribal communities worldwide. Plants are basis of sophisticated Traditional Medicinal (TM) systems has been existing thousands of years and continuing to provide people with new remedies. Approximately 80% of global populations are primarily depends on traditional medicines for remedies for their health problems (Goud et al., 2005; Khan et al., 2009). Roughly 25% of modern medicinal drugs are developed from plants and synthetic analogs active compounds of plants. The World Health Organization (WHO) listed around 21,000 species of plants exploited for medicinal purposes and 25,000 species reported by International Union for Conservation of Nature and Natural Resources (IUCN) (Thatoi and Rout, 2011). Medicinal plants used in traditional medicines are organized and coded in written treatise book such as Siddha, Ayurveda, Amchi, Unani and Tibetan systems of medicines uses around 1200 to 2000 medicinal plant species (Dahanukar et al., 2000). It also forms tradition of local health practiced by villagers, folk healers, vaidyas and tribal people. Trends dependencies on complementary and traditional medicines are escalated throughout the world. As per reports from the USA, approximately 42.5 million are depends on herbalists in 1990 in contrast to 388 million depends to physicians of primary health (Gebre-Mariam and Asres, 1996). Since 1992, ~20 million peoples depend on homeopathy, acupuncture, chiropractic and herbal medicine which is the popular forms of complementary medicine in Germany. In Australia about 60% of the population depends on traditional medicine and 17,000 herbal products has registered along with the US $ 650 million is utilized for this purpose. The herbal medicine market has expanded immensely in the last 15 years, and the total sale of the product is still emerging every year." Europe's growth market" the sale of herbal products in Europe which earned up to USD 1.4 billion in 1992 (Gebre-Mariam and Asres, 1996). In Malaysia, it was reported that about US $500 million is utilized each year on TM compared with modern medicine (the US $300 million). In China dependency on traditional medicines relatively for 30–50% of their total medicinal usage and the total sales of their herbal medicinal products is around USD 2.5 billion. Furthermore to China has exported medicinal herbs with an estimated value of USD 40 million in 1993. This country is having more than 160 research institutions for traditional Chinese Materia Medica forming research system. China has 2,000 factories for herbal product manufacturing and producing 4,000 different types of ready-made Chinese herbal medicine every year (Xiang, 1990). As per the Indian scenario, 75% of the total population depends on herbal formulations, and 540 plant species are utilized for different formulations (Bhat, 1990). The world market for plant-derived drugs may account for about Rs. 2,00,000 crores. However plant-based drug constitutes around 25% of total drugs in the USA whereas China and India, the contribution 80%. Hence the economic values of medicinal plants are more in India than other countries.

Hence here we are discussing more about wild tuber crop *Lasia spinosa* (L) Thwaites as medicinal potential plant and its various potentials. *Lasia* is a flowering plant genus belongs to Araceae family. It contains only two species and native to Asia. *Lasia spinosa* (L) Thwaites and *Lasia concinna* (L). *Lasia* has been believed as monotypic genus in 1997, and then wild population of *Lasia concinna* (L) was again discovered in the paddy field in West Kalimantan in Indonesia. The farmers are growing it for edible young leaves. The species of *Lasia* had been known initially only from a single specimen from the Bogor Botanic Gardens. At that time species has been believed it is a hybrid between *Lasia spinosa* (L) Thwaites and *Cyrtosperm merkusii*. Then the subsequent discovery by Hambali and Sizemore in 1997, led to the realization that it was a new species. It is commonly found in India especially in the Eastern Ghats including Odisha. The rural and tribal communities of Odisha has been used it as medicine as well as food.

13.2 WILD TUBER CROPS

A plant is having storage root, or stem is called tuberous plant. Plants are having two types of tuber one stem modified tuber like *Solanum tuberosum*, and *Dioscorea hamiltonii* is an example for root modified tuber. Most of the tubers are edible and plays a vital as diet, and few numbers of tuberous plants are cultivated and known as tuber crops. Naturally, available tuber crops called a wild tuber crop which is unexploited and consumed by tribal communities and forest inhabitants. They are bitter in taste and rich with fiber, starch, and other nutrients. *Lasia spinosa* (L) Thwaites is a root modified tuber crops.

13.2.1 BOTANY AND DISTRIBUTION

Lasia spinosa (L.) Thwaites is a member of Araceae family. It is a stout, spinous herb with thick and branched rhizome. The leaves are petioled and long, and Spathe is very long, narrow which opens only at the base. Spadix is short and cylindrical. The flowers are bisexual. Perianth is obovate, incurved at the tip. The peduncle is spinous and long as the petioles. The petiole is erect and spinous which sheaths towards the base. The stamens are mainly 4–6 having short flat filament and shorter anthers. Ovary is ovoid with stout styles. Stigmas are depressed with solitary ovule and pendulous from top of the cell. Fruiting of *Lasia spinosa* (L.) Thwaites head oblong with hexagonal berries more or less 4-seeded which is minutely muricate at the apex. The seeds of this species are compressed, rugose and evanescent. The species of *Lasia* is native to tropical and temperate Asia. It is found mainly along muddy streams under shade. It is distributed in Bangladesh, Bhutan, Cambodia, China, India, Malaysia, Indonesia, Myanmar, Nepal, Sri Lanka, Thailand, and Taiwan. In India, it is distributed in Sikkim, Himalaya, Assam, Bengal, Bihar, and Odisha. In Odisha, mainly found in Similipahar, Deogarh, Bonai, Kalahandi, and Athmallick (Saxena and Brahmam, 1995).

13.3 POTENTIAL OF *LASIA SPINOSA* (L.) THWAITES

13.3.1 ETHNO-MEDICINAL AND PHARMACOLOGICAL VALUES

The medicinal plants are used as a remedy for health issues such as cholic, intestinal problems and rheumatism. The young leaves of the plants are utilized against worm infections, and throat infection. Corm of the plant is exploited in folk medicines of Naga Tribes of India. Leaves and corms are given for problem associated with for piles. The tubers are exploited for blood purification, constipation, and rheumatoid arthritis, in Nature and Rajshahi district of Bangladesh. The rhizome used for the study of activities like antioxidant, cytotoxic activities and antimicrobial properties. Shoots are used as traditional food in Southeast Asian communities. The stems are used as expectorant, antitussive, boiled water used for bath for relieving itching from measles, rubella and *Roseolar infantum*, other skin problem. The anticestodal effect of the plant was reported by Yadav and Temjenmongla (Yadav and Temjenmongla, 2006). The leaf of the plant is cooked and given for respiratory, intestinal and other gastric problems. Goshwami et al., (2012) proved the traditional uses of the crude and purified methanol extract of the plant leaf. The phytochemical screening qualitatively indicates the existence of flavonoids, alkaloids and glycosides tannins. From the above reports, it is clearly revealed that this plant has potent ethnomedicinal value with phytochemicals can be exploited as medicines (Gogoi and Zaman 2013). Das et al. (2014) studied the anti-diabetic activity of *Lasia spinosa* (L.) Thwaites stem with different dosages (200 and 400 mg/kg b.w.) in dexamethasone-induced diabetic albino rats and result indicated that it has potent antidiabetic activity against standard (Das et al., 2014). The *Lasia spinosa* (L.) Thwaites cultivated in muddy streams, marshy areas, and swampy grounds. The talk and leaves are used as anticestodal agents and rhizome is used against bleeding cough, lung inflammation and the whole plant used as anticancer agent. The local people give these plants for lactating mothers. This plant is also used as a source of multivitamins to boost the nutritional level (*Brahma* et al., 2014). Research showed the antinociceptive activities of methanol extract of *Lasia spinosa* (L.) Thwaites leaves (Goshwami et al., 2012). The whole plant is traditionally used in many Asian countries as a solution for wide range of diseases and ailments (Deb et al., 2010). It is used as antimalarial medicine in Vietnamese traditional medicine and showed potent antimalarial activity against *Plasmodium falciparum* (Tran et al., 2003). In the traditional medicine of India, Bangladesh, and Sri Lanka, the leaves and rhizomes of plant is continuously used as an antibacterial, antidiarrhoeal, antinociceptive and inflammatory medicine (Ngomdir et al., 2007; Deb et al., 2010), the rhizome extract has potent antimicrobial and cytotoxic activity against bacteria and fungi (Mahato et al., 2015; Alam et al., 2011; *Ngomdir* et al., 2007*)*. The traditional medicines of Naga tribes in India, the leaves of the plant used as anthelminthic remedy (Temjenmongla and Yadav, 2011) and have significant activity against *Hymenolepis diminuta* infections in rats (Temjenmongla and Yadav 2006). *Lasia spinosa* (L.) Thwaites has shown potent antidiabetic activity against Dexamethasone induced diabetic in rats (Das et al., 2014) (Figures 13.1–13.3).

FIGURE 13.1 Showing the morphology of *Lasia spinosa* (L) Thwaites (A: leaves and B: rhizomes).

13.3.2 ANTIOXIDANT AND ANTINOCICEPTIVE ACTIVITY

Shefana and Ekanayake (2009) showed that *Lasia spinosa* (L.) Thwaites rhizome has wide range of antioxidant capacity with total activity of 145–957 mol/g TEAC on a wet weight basis. Yong (2009) showed that plant extract with different solvents has total antioxidant activity by using 2,2-Diphenyl-1-picrylhydrazyl (DPPH) assay, ferric reducing/antioxidant power (FRAP), β-carotene bleaching methods along with total phenolic contents detection by using Folin-Ciocalteu method. The water extraction has good antioxidant activity compared to hexane extracts. The soluble fraction of methanol extract in acetone showed potent antioxidant activity with DPPH assay (Goshwami et al., 2012). Goshwami et al. (2013) evaluated the methanol extract of the plant for antinociceptive activity in animal model by using acetic acid writhing method. The crude extract enhances decrease in the number of writhes compared to standard and also observed inhibition of paw edema as compared to standard (diclofenac) in dose-dependent manner (Goshwami et al., 2012).

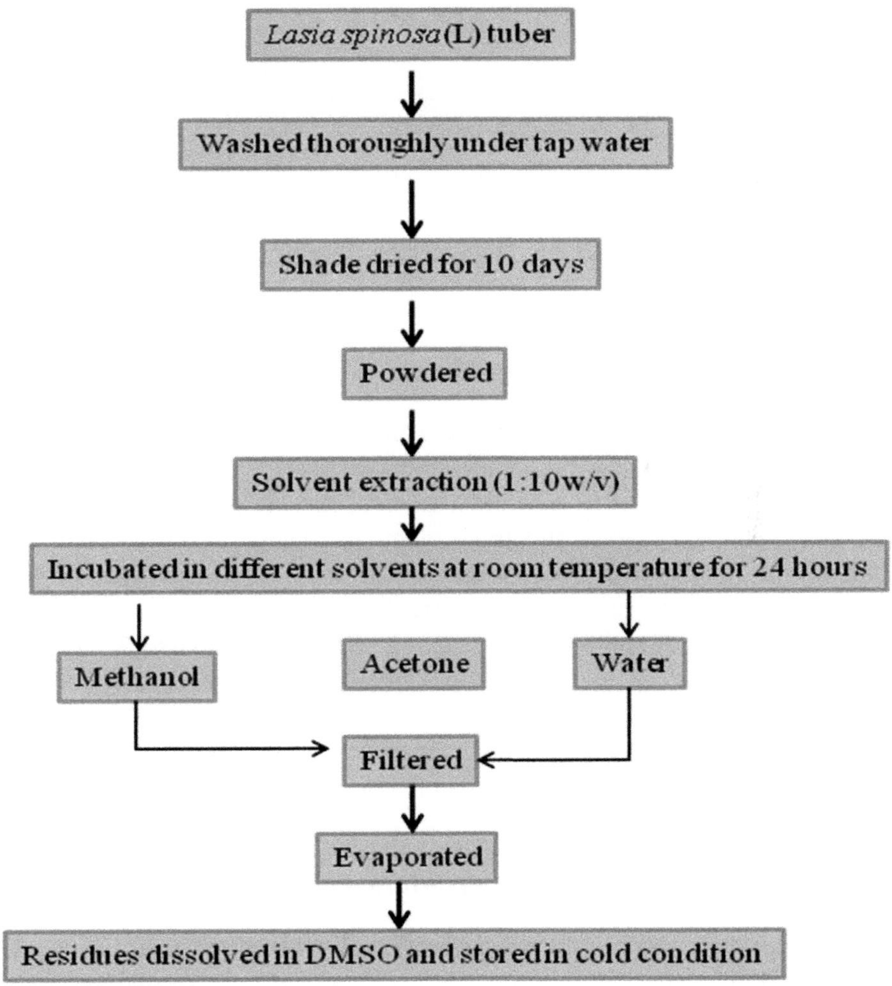

FIGURE 13.2 Schematic representation of extraction of phytochemicals from *Lasia spinosa* (L.) Thwaites.

FIGURE 13.3 Phytochemical screening of *Lasia spinosa* (L) tuber of methanol and acetone extract.

13.3.3 BIOACTIVE COMPOUNDS

Brahma et al. (2014) investigated phytochemical analysis qualitatively from *Lasia spinosa* (L.) Thwaites confirmed the presence of different phytochemicals like flavonoids, alkaloids, tannins, terpenoids, saponin, proteins, steroids, reducing sugars and trace elements. These bioactive chemicals play a vital role in traditional medicine. Other component such as protein, fats, ash, total solids, carbohydrate present in plant has some functional role. Nutritional analysis showed that protein 17.6 kcal/100 g, fats 1.16 kcal/100 g, ash 34 kcal/100 g, total solids 17 kcal/100 g, moisture 83 kcal/100 g, carbohydrate 35.7 kcal/100 g, and total nutritive value of 224 kcal/100 g was also reported. Other micronutrients of Zinc Magnesium, Molybdenum, Iron, Copper, and Manganese are also present (Brahma et al. 2014).

13.3.4 CYTOTOXICITY AND ANTIMICROBIAL ACTIVITY

The cytotoxicity studies of brine shrimp lethality bioassay like LC50 values obtained from the best fit line slope were 0.544, 3.181, and 4.096 μg/mL for standard vincristine sulfate, dichloromethane and ethyl acetate soluble fraction of methanol extract, respectively (Goshwami et al., 2012). The brine shrimp lethality bioassay is used for cytotoxic activity of the extracts as well as purified meridional which is showed cytotoxic activities. Alam et al. (2011) studied the LC 90 with different solvent extracts showed cytotoxicity values ranges from11.22 μg/ml, 12.3 μg/ml, 13.49 μg/ml, 11.57 μg/ml and 15.85 μg/ml, respectively.

The *Lasia spinosa* (L.) Thwaites rhizome extracted with different solvent showed antimicrobial activity against *Escherichia coli, Bacillus cereus, Staphylococcus aureus, Aspergillus niger, Candida albicans,* and *Vibrio parahemolyticus*. The crude extracts along with meridional was screened for antimicrobial activity against a wide range of bacteria and fungi (Alam et al., 2011).

13.3.5 ANTHELMINTIC ACTIVITY

The leaves of *Houttuynia cordata, Psidium guajava,* and stalk of *Lasia spinosa* (L.) Thwaites possess potential anticestodal efficiency as evidenced by means of mortality rate of *R. echinobothrida* which ranged from 1–4 hrs exposed with 5–40 mg/ml of extract. The leaves of *Lasia spinosa* (L.) Thwaites along with other plant such as *Clerodendrum colebrookianum, Centella asiatica* showed moderate activity while *Lasia spinosa* stem with other plants extract of *Cinnamomum cassia, Curcuma longa,* and *Aloe vera* showed negative effect on anticestodal activity (Temjenmongla et al., 2005). The extracts (methanol) of *Lasia spinosa* (L.) Thwaites leaf used against Pheretima posthuma various concentrations varied from 25, 50 and 100 mg/ml used for the determination of paralysis and time for death of worms. The results showed that extract has considerable effect on paralysis and death of worms with 100 mg/ml in comparison with standard albendazole (Goshwami et al., 2013). The efficacy of *Lasia spinosa* (L.) Thwaites leaf extract against all the three life cycle stages including adult, larval form and encysted muscle larvae of *Trichinella spiralis*. The potential of leaf extract as significant anthelmintic properties against the larva and adult worm *Trichinella spiralis* less sensitive in the case of encysted larva (Yadav et al., 2011).

13.3.6 NUTRIENT AND MINERAL CONTENT

Rhizome of *Lasia spinosa* (L.) Thwaites has plenty of fibers with total dietary fiber 40–75% including 35–60% insoluble and 4–18% of soluble fiber, respectively (*Shefana and Ekanayake 2009)***.** Fiber content of the leaves of was reported as 14.60 g which means that this plant has a role for reducing problem associated with various chronic degenerative diseases (Firdusi et al., 2013).

Changmora is the local name of *Lasia spinose* plant, which is a not a part of traditional plant Assam used in rural people as food vegetable and medicinal applications. The tender leaves of *Lasia spinosa*(L.) Thwaites. were evaluated for their nutrient contents which showed significantly good amount energy boosting nutrients such as carbohydrates–43.50 g, protein–14.50 g, and fat 8.52 g. The rhizome of *Lasia spinosa (L.) Thwaites*. are found to be rich source of macroelements like (Na, K, Ca, Mg and P) as well as trace minerals (Fe, Zn, Cu Mn, Cr, and Ni) content of fresh and cooked vegetables traditionally consumed in North-East India. Calcium is the most abundant macroelement, raw form it contains 543.2 mg/100 g and cooked 576.4 mg/100 g, other macroelement like Na, K, Mg and P of raw it is 6.9 mg, 170.4 mg/100 g, 85.7 mg/100 g and 43.8 mg/100 g, respectively. Similarly in cooked form Na, K, Mg and P content are 6.1 mg/100 g, 157.1 mg/100 g, 74.5 mg/100 g and 28.4 mg/100 g, respectively. The microelement like Zn, Fe, Cu, Mn, Cr and Ni present in raw is 0.82 mg/100 g, 12.16 mg/100 g, 0.12 mg/100 g, 0.23 mg/100 g, 0.171 mg/100 g and 0.141 mg/100 g, respectively whereas in cooked form contains 0.80 mg/100 g, 12.71 mg/100 g, 0.17 mg/100 g, 0.24 mg/100 g, 0.076 mg/100 g, and 0.143 mg/100 g, respectively (Saikia and Deka, 2013). The leaves of the plant contain plenty of minerals like potassium (109.41 mg), iron (19.45 mg), and calcium (416.00 mg) (Firdusi et al., 2013).

13.4 EXPERIMENTAL STUDIES

13.4.1 COLLECTION OF PLANT MATERIALS

Tubers of *Lasia spinosa* (L) Thwaites were collected from the members of Ambika Prasad Research Foundation (Dr. Sanjeet Kumar). Rhizomes were collected from mature plant of *Lasia spinosa* (L) Thwaites planted in the garden. They are washed, cut into small pieces and air-dried. The dried are powdered and were kept in airtight container for phytochemical screening and antibacterial activity.

13.4.2 PREPARATION OF PLANT EXTRACTS

The plant extract (50 g) was extracted from the powder with Soxhlet apparatus (Tiwari et al., 2011). The residues were collected, and air-dried crude sample of extract was stored in cold condition for further experimental work. Extraction was done with different solvent such as acetone, methanol, and water up to 5–6 siphons. The dried extract was weighed and stored under cold condition.

13.4.3 PHYTOCHEMICAL ASSAY

Phyto-chemical experiment was carried out on different extracts of *Lasia spinosa* (L). Thwaites rhizome using standard operating protocol to identify the bioactive compounds according to earlier researchers (Harborne, 1973; Sofowara, 1993).

13.5 FINDINGS

The experimental studies proved that is a medicinal food wild crop among tribal communities. Tubers showed morphological differences with respect to change in variations in ecological niche. Qualitative analysis of the rhizome showed this wild plant was rich in phytochemicals such as Tannin, phenolic compounds, flavonoids, terpenoids, saponin and reducing sugar (Table 13.1).

13.6 CONCLUSION

This wild plant can be used as food and medicinal purposes. The presence of various phytochemicals could be potential biological, pharmacological antimicrobial properties of the wild plant. The present study indicated some potential phytochemicals are present in this plant and can be exploited as new antimicrobial agents to fight against antimicrobial resistance. Furthermore, scientific studies are needed to evaluate the pharmacological properties of new compound from this wild plan which may lead as to new discovery of potent antimicrobial and pharmacological compounds.

ACKNOWLEDGMENTS

Authors are thankful to Dr. P. K. Jena, Department of Botany, Ravenshaw University, Cuttack; HOD, Department of Botany, Ravenshaw University, Cuttack, Dr. Sanjeet Kumar, Ambika Prasad Research Foundation for contributing part of the work and Institute of Bioresources and Sustainable

TABLE 13.1 Bioactive Compounds Present in Different Extracts of *Lasia spinosa* (L) Tuber

Extracts	Bioactive compounds detected						Color of Extract
	Tannin	Flavonoid	Phenolic Compound	Saponin	Terpenoid	Reducing Sugar	
Aqueous	+ve	+ve	–ve	+ve	+ve	+ve	Light brown
Methanol	+ve	+ve	+ve	+ve	–ve	+ve	Light brown
Acetone	+ve	+ve	+ve	+ve	+ve	+ve	Light brown
Hexane	–ve	–ve	–ve	–ve	–ve	–ve	Colorless
Chloroform	–ve	–ve	–ve	–ve	–ve	+ve	Colorless
Petroleum ether	–ve	–ve	–ve	+ve	–ve	+ve	Light brown
Ethanol	+ve	+ve	+ve	+ve	–ve	+ve	Light creamy
Methanol: Water	+ve	–ve	+ve	+ve	+ve	+ve	Light pink

Development, A National Institute under Department of Biotechnology Government of India for providing all help and support.

KEYWORDS

- **antimicrobials**
- **antioxidant**
- **ethnomedicine**
- ***Lasia spinosa***
- **phytochemicals**

REFERENCES

Aboaba, O. O.; Efuwape, B. M. Antibacterial properties of some Nigerian species. *Bio. Res. Comm.* **2001**, *13,* 183–188.

Aboaba, O. O.; Smith, S. I.; Olide, F. O. Antimicrobial effect of edible plant extract on *Escherichia coli. Pakistan J Nutrition.* **2006**, *5*(4), 325–327.

Aderotimi, B.; Samuel, A. Phytochemical screening and antimicrobial assessment of *Abutilon mauritianum, Bacopa monnifera*, and *Datura stramonium*. *Biokemistri.* **2006**, *18* (1), 39–44.

Alam, F.; Haque, M.; Sohrab, H.; Monsur, M. A.; Hasan, C. M.; Ahmed, N. Antimicrobial and cytotoxic activity from *Lasia spinosa* and isolated lignan. *Lat Am J Pharm.* **2011**, *30,* 550–553.

Allen, K. L.; Molan, P. C.; Reid, G. M. A survey of the antibacterial activity of some New Zealand honeys. *J. Pharm. Pharmac.* **1991**, *43,* 817–822.

Alper, J. Effort to combat microbial resistance lags. *Amer. Sci. Microb. News*. **1998**, *64,* 440–441.

Amanda, J. D.; Bhat, N.; Ruth, A. K.; Katherine, L. O.; David, R.M. Disc Diffusion Bioassays for the detection of antibiotic activity in body fluids: Applications for the Pneumonia Etiology Research for Child Health Project. *Clin. Inf. Dise*. **2012**, *54,* 159–164.

Aruoma, O. I. Free radicals, oxidative stress, and antioxidants in human health and disease. *J. American Oil Chem. Soc*. **1998**, *75,* 199–212.

Bhandari, M, Kawabata, J. Bitterness and Toxicity in wild yam tubers of Nepal. Plant Food Hum. *Nutri.* **2005**, *60,* 129–135.

Bhat, K. G. Preparation of herbal medicines. In: Proceedings of International Conference on Traditional Medicinal Plants. **1990**, *Arusha*. 218–220.

Bhattacharjee, I.; Chatterjee, S. K.; Chandra, G. Isolation and identification of antibacterial components in seed extracts of *Argemone mexicana* L. (Papaveraceae). *Asian Pacific J. Trop. Med.* **2010**, *3*(7), 547–551.

Blandrin, M. F.; Kjocke, A. J.; Wurtele, E. Natural Plant Chemicals: Source of industrial and mechanical materials. *Science*. **1985**, *228,* 1154–1160.

Bobbarala, V.; Vadlapudi, V. R.; Naidu, K. C. Antimicrobial potentialities of mangrove plant *Avicennia marina. J. Pharmacy Res*. **2009**, *2*(6), 1019–1021.

Brahma, J.; Chakravarty, S.; Rethy, P. Qualitative Estimation of the Presence of Bioactive and Nutritional Compound in *Lasia Spinosa*: An Important Vegetable Plant used by the Bodos of Kokrajhar District. *International Journal ChemTech Research*. **2014**, *6*(2), 1405–1412.

Brahma, J.; Singh, B.; Rethy, P.; Gajurel, P. Nutritional Analysis of Some Selected Wild Edible Species Consumed By The Bodos Tribes of Kokrajhar District, Btc, Assam. *Asian J Pharm Clin Res*. **2014**, *7*(3), 34–37.

Cheng, W. Y.; Kuoy, H.; Huango, C. J. Iso Latin, and Identification of Novel Estrogenic Compounds in Yam tuber- *Dioscorea alata. Journal Agricultural food Chemistry*. **2007**, *55*(18), 7350–7358.

Choudhury, R.; Datta, M.; Choudhury, B.; Paul, S. B. Importance of certain tribal edible plants of Tripura. *Indian J Traditional Knowledge*. **2010**, *9*(2), 300–302.

Dahanukar, S. A.; Kulkarni, R. A.; Rege, N. N. Pharmacology of medicinal plants and natural products. *Indian J Pharmacol*. **2000**, *32*, 81–118.

Das, S.; Baruah, M.; Shill, D. Evaluation of Antidiabetic Activity from the Stem of *Lasia spinosa* in Dexamethasone Induced Diabetic Albino Rats. *J Pharmaceutical Chemical Biological Sciences*. **2014**, *1*(1), 12–17.

Deb, D.; Dev, S.; Das, A.K.; Khanam, D.; Banu, H.; Shahriar, M. M. Antinociceptive, anti-inflammatory and antidiarrhoeal activities of the hydroalcoholic extract of *Lasia spinosa* Linn. (Araceae) Roots. *Lat. Am. J. Pharm*. **2010**, *29*(8), 1269–1276.

Dianella, S. Plant-derived antimicrobial compounds: alternatives to antibiotic. *Future Microbiol*. **2012**, *7*(8), 979–990.

Doble, B.; Dwivedi, S.; Dubey, K.; Joshi, H. Pharmacognostical and antimicrobial activity of leaf of *Curcuma angustifolia* Roxb. *Int. J. Drug Discovery Herbal Res*. **2011**, *1*(2), 46–49.

Duh, P. D. Antioxidant activity of burdock (*Arctium Lappa*): its scavenging effect on free radical and active oxygen. *J. American Oil Chem. Soc*. **1998**, *75*, 455–465.

Eisenberg, D. M.; Kessler, R. C.; Foster, C.; Norlock, F. E.; Calkins, D. R.; Delbanco, T. L. Unconventional medicine in the United States: prevalence, costs, and patterns of use. *New Eng. J. Med.* **1993**, *328*, 246–252.

Ellen, R. P.; Banning, D.W.; Fillery, E. D. Longitudinal microbiological investigation of a hospitalized population of older adults with a high root surface caries risk. *J. Dent. Res*. **1985**, *64*, 1377–1381.

Falkow, S.; Monack, D. M.; Mueller, A. Persistent bacterial infections: the interface of the pathogen and the host immune system. Nature Rev. *Microbiol*. **2004**, *2*, 747–765.

Finkel, T. Oxidants, oxidative stress and the biology of aging. *Nature*. **2000**, *408*, 239–248.

Fischbash, M. A.; Walsh, C. T. Antibiotics for emerging pathogens. Sci. **2009**, *325*, 1089–1093.

Franklin, W.; Martin. Tropical Yams and their potential, *Dioscorea alata. Agricultural handbook N*. **1976**, 495.

Gebre, M. T.; Asres, K. Applied Research of Medicinal plants. In: Proceedings of the National workshop of Biodiversity conservation and sustainable use of Medicinal plants in Ethiopia, Zewdu, M., Demissie, A. (Eds.). Institute of Biodiversity Conservation and Research. Addis Ababa. **1996**, 34–45.

Ginsburg, H.; Deharo, E. A cell for using natural compounds in the development of new antimalarial treatments- an introduction. *Malarial J*. **2011**, *10*, 1–11.

Goshwami, D.; Rahman, M.; Muhit, A.; *Islam, S.* Antinociceptive activity of leaves of *Lasia spinosa. Archives Applied Science Research*. **2012a**, *4* (6), 2431–2434.

Goshwami, D.; Rahman, M.; Muhit, A.; *Islam, S.* Antinociceptive, Anti-inflammatory and Antipyretic Activities of Methanolic Extract of *Lasia spinosa* Leaves. *International J Pharmaceutical Chemical Sciences*. **2013b,** *2*(1), 55–58.

Goshwami, D.; Rahman, M.; Muhit, A.; Islam, S. *In-vitro* evaluation of Anthelmintic Activity of *Lasia Spinosa* Leaves. *International J Current Pharmaceutical Research.* **2013a,** *5*(1), 64–67.

Goshwami, D.; Rahman, M.; Muhit, A.; *Islam, S.;* Ansari, M. Antioxidant Property, Cytotoxicity and Antimicrobial Activity of *Lasia spinosa* Leaves. *Nepal J Science Technology.* **2012b,** *13*(2), 215–218.

Goud, P. S. P.; Murthy, K.S.R.; Pillaiah, T.; Babu, G.V. A. K. Screening for Antibacterial and Antifungal activity of some medicinal plants of Nallamalais, Andhra Pradesh. *India. J. Econ. Taxon. Bot.* **2005***, 29*(3), 704–708.

Gould, I. M. The epidemiology of antibiotic resistance. *Int. J. Antimicrob. Agent.* **2008***, 32,* 52–59.

Haines, H. H. The Botany of Bihar and Orissa. Adlard & Son & West Newman Ltd., London. **1921–1925***, 5*(6), 1115–1124.

Hambali, G.G; Sizemore, M. The rediscovery of Lasia concinna Alderw. (Aracea: Lasioideae) in West Kalimantan, Borneo. 1997 Aroideana 20, 37–39.

Hasan, C. M.; Alam, F.; Haque, M.; Sohrab, M. H.; Monsur, M.A.; Ahmed, N. Antimicrobial and Cytotoxic Activity from *Lasia spinosa* and Isolated Lignan. *Lat. Am. J. Pharm.* **2011***, 30*(3), 550–553.

Hassan, M. R.; Alam, K. D.; Mahjabeen, S.; Rahman, M. F.; Akter, N.; Bushra, M.U. Free radical scavenging potential of methanol extract of *Smilax roxburghiana. Pharmacologyonline.* **2011***, 2,* 774–783.

Hay, D. I.; Ahern, J. M.; Schluckebier, S. K.; Schlesinger, D. H. Human salivary acidic proline-rich protein polymorphisms and biosynthesis studied by high-performance liquid chromatography. *J. Dent. Res.* **1994***, 73,* 1717–1726.

Hou, W. C.; Lee, M. H.; Chen, H. J.; Liang, W. L.; Han, C. H.; Liu, Y. W.; Lin, Y. H. Antioxidant activities of dioscorin, the storage protein of Yam (*Dioscorea batatus Dicre*) tuber. *Journal Agricultural Food Chemistry*. **2001***, 49*(10), 4950–4960.

Jansen, A. M.; Chefferm, J. J. C.; Svendsen, A. B. Antimicrobial activity of essential oils. Aspects of test methods. *Planta Med.* **1987***, 40,* 395–398.

Khan, M. Y.; Kumar, V.; Rajkumar, S. Recent advances in medicinal Plant Biotechnology. *Indian J. Biotechnol*. **2009***, 8,* 9–22.

Kumar, S. Qualitative studies of bioactive compounds in leaf of *Tylophora indica* (Burm.f.) Merr. *Int. J. Res. Pharmaceut. Biomed. Sci*. **2011a**, 2(3), 1188–1192.

Kumar, S.; Jena, P. K.; Kumari, M.; Patnaik. N.; Nayak, A. K.; Tripathy, P. K. Validation of tribal claims through pharmacological studies of *Helicteres isora* L. leaf extracts: an Empirical Research. *Int. J. Drug Dev. Res*. **2013***, 5*(1), 1–10.

Kumar, S.; Jena, P. K.; Sabnam, S.; Kumari, M.; Tripathy, P. K. Study of plants used against the skin diseases with special reference to *Cassia fistula* L. among the king (Dongaria Kandha) of Niyamgiri: A primitive tribe of Odisha, India. *Int. J. Drug Dev. Res.* **2012b**, *4*(2), 256–264.

Kumar, S.; Jena, P. K.; Tripathy, P. K. Study of wild edible plants among tribal groups of Simlipal Biosphere Reserve forest, Odisha, India; with special reference to *Dioscorea* species. *Int. J. Biol. Tech*. **2012a**, *3*(1), 11–19.

Kumar, S.; Satapthy, M. K. Medicinal plants in an urban environment; herbaceous medicinal flora from the campus of Regional Institute of Education, Bhubaneswar, Odisha. *Int. J. Pharm. Lif. Sci*. **2011b**, *2*(11), 1206–1210.

Lee, S. C.; Tsai, C. C.; Chen, J. C.; Lin, J. G.; Lin, C. C.; Hu, M. L.; Lus. Effects of "Chinese Yam" on hepato- nephrotoxicity of acetaminophen in rats. *Acta Pharamcologica Sinica.* **2002***, 23*(6), 503–506.

Lee, S. E.; Hyun, J. H.; Ha, J.S.; Jeong, H. S.; Kim, J. H. Screening of medicinal plant extracts for antioxidant activity. *Life Sci.* **2003**, *73,* 167–179.

Liu, Y. U.; Shang, H. F.; Wang, C. K.; Hsu, F.L.; Hou, W. C. Immunomodulatory activity of dioscorin, the storage protein of Yam *Dioscorea alata* tuber. *Food Chemical Toxicology.* **2007**, *45,* 2312–2318.

Magdolena, K. Chelation of Cu (II), Zn (II) and Fe (II) by tannin constituents of selected edible nuts. *Int. J. Mol. Sci.* **2009**, *10,* 5485–5497.

Mahato, K.; Kakoti, B. B.; Borah, S.; Kumar, M. In-vitro antioxidative potential of methanolic aerial extracts from three ethnomedicinal plants of Assam: A Comparative Study. *J Applied Pharmaceutical Science*. **2015**, *5*(12), 111–116.

Maithili, V.; Dhanabal, S. P.; Mahendran, S.; Vadivelen, R. Antidiabetic activity of ethanolic extract of tubers of *Dioscorea alata* in alloxan-induced diabetic rats. *Indian J Pharmacology.* **2011**, *43*(4), 455–459.

Majumdar, S. H.; Chakarborty, G. S.; Kulkarni, K. S. Medicinal potential of *Semecarpus anacardium* nut: a review. *J. Herb. Med. Toxicol.* **2008**, *2*(2), 9–13.

Misra, R. C.; Kumar, S.; Pani, D. R.; Bhandari, D. C. Empirical tribal claims and correlation with bioactive compounds: a study on *Celastrus paniculata* Wild, a vulnerable medicinal plant of Odisha. *Ind. J. Trad. Knowl.* **2012**, *11*(4), 615–622.

Misra, R. C.; Sahoo, H. K.; Mahapatra, A. K.; Reddy, R N. Addition to the flora of Simlipal Biosphere Reserve, Orissa, India. *J. Bom. Nat. Hist. Soc.* **2011**, *108*(1), 69–76.

Misra, R. C.; Sahoo, M. K.; Pani, D. R.; Bhandari, D. C. Genetic resources of wild tuberous food plants traditionally used in Simlipal Biosphere Reserve, Odisha, India. *Genet. Resour. Crop Evol.* **2013**, *13,* 9971–9976.

Moerman, D. E. An analysis of the food plants and drug plants of native North A. *J. Ethnopharmac.* **1996**, *52,* 1–22.

Mohan, V. R.; Kalidas, C. Nutritional and antinutritional evaluation of some unconventional wild edible plants. *Trop Subtrop. Agroecosys*. **2010**, *12,* 494–506.

Musa, A. M.; Aliyu, A. B.; Yaro, A. H.; Magaji, M. G.; Hassan, H. S.; Abdullahi, M. Preliminary phytochemical, analgesic and anti-inflammatory studies of the methanol extract of *Anisopus mannii* in rodents. *African J. Pharm. Pharmac*. **2009**, *3*(8), 374–378.

Nascimento, F. G. G.; Locatelli, J.; Freitas, P. C.; Silva, G. L. Antibacterial activity of plant extracts and phytochemicals on antibiotic-resistant bacteria. *Braz. J. Microbiol.* **2000**, *31,* 247–256.

Ngomdir, M.; Debbarma, B.; Debbarma, A.; Chanda, S.; Raha, S.; Saha, R.; Pal, S.; De, B. Antibacterial evaluation of the extracts of edible parts of few plants used by tribal people of Tripura, India. *J Pure Appl Microbiol.* **2007**, *1,* 65–68.

Niksusanti, Herlina, Masil K. I. Antibacterial and Antioxidant of Uwi (*Dioscorea alata* L.) Starch Edible film Incorporated with Ginger Essential oil. *International J Bioscience.* **2013**, *3*(4), 354–356.

Odebiyi, A.; Sofowora, A. E. Phytochemical Screening of Nigerian Medicinal Plants. Part III. Iloydia. **1978**, *41*(3), 234–246.

Okwe, D.E.; Okwe, M. E. Chemical composition of *Spondias mombin* Linn, plant parts. *J. Sustain. Agric. Env*. **2004**, *6,* 140–147.

Okwu, D. E. Phytochemicals and vitamin content of indigenous spices of southern *Nigeria. J. Sustain. Agric. Environ*. **2004**, *6*(1), 30–37.

Osbourn, A. E. Performed antimicrobial compounds and plant defense against fungal attack. *Plant Cell*. **1996**, *8,* 1821–1831.

Osman, K.; Evangelopoulos, D.; Basavannacharya, C. An antibacterial from Hypericum acmosepalum inhibis ATP-dependent MurE ligase from *Mycobacterium tuberculosis*. *Int. J. Antimicrob. Agents*. **2012**, *39,* 124–129.

Philippa, M. A.; Shanahan, J.; Mary, V.; Thomson, C. J.; Sebastia, G. B. Molecular analysis of and identification of antibiotic resistance genes in clinical isolates of *Salmonella typhi* from India. *A. Soc. Microbiol*. **1998**, *36,* 1595–1600.

Polycarp, D.; Afoakwa, E.O.; Budu, A. S. Characterization of Chemical Composition and Antinutritional factors in seven species within the Ghanaian Yam (*Dioscorea)* germplasm. *International Food Research Journal*. **2013**, *19*(3), 985–992.

Ravikumar, S.; Gnanadesigan, M.; Suganthi, P.; Ramalakshmi, A. Antibacterial potential of chosen mangrove plants against isolated urinary tract infectious bacterial pathogens. *International J. of Med. and Medical Sci*. **2010**, *2*(3), 94–99.

Sakthidevi, G.; Mohan, V. R. Total Phenolies, flavonoid contents and In vitro Antioxidant Activity of *Dioscorea alata* L. Tuber. *J Pharmaceutical Research*. **2013**, *5*(5), 115.

Samy, R. P.; Gopalakrishnakone, P. Therapeutic potential of plants as antimicrobials for drug discovery. *Evidence-Based Comp. and Alt. Med*. **2010**, *7*(3), 283–294.

Sandhu, A.; Bhardwaj, N.; Gupta, R.; Menon, V. Antimicrobil activity and phytochemical screening of *Tinospora cordifolia* and *Euphorbia hirta*. *Int J Agr Biol*. **2013**, *4,* 310–316.

Sansone, C.; Van, H.; Joshipura, K.; Kent, R.; Margolis, H. C. The association of *mutans streptococci* and non-mutans *streptococci* capable of acidogenesis at a low pH with dental caries on enamel and root surfaces. *Int. J. Dent. Res*. **1993**, *72,* 508–516.

Sartie. A.; Asiedu, R.; Franco, J. Genetic and phenotypic diversity in a germplasm working collection of cultivated tropical yams. Genet. *Resour. Crop Evol*. **2012**, *9*(3), 667–670.

Saxena, H. O.; Brahmam, M. The Flora of Orissa. Orissa Forest Development Corporation Ltd. and Regional Research Laboratory, Bhubaneswar. **1995**, *3,* 1940–1956.

Scalbert, A. Antimicrobial properties of tannins. *Phytochemistry*. **1991**, *30,* 3875–3883.

Seetharam, Y. N.; Jyothishwaran, G.; Sujeeth, H.; Barad, A.; Sharanabasappa, G.; Shivkumar, D. Antimicrobial activity of *Dioscorea Bulbifera* Bulbils. *Indian J Pharmaceutical Sciences*. **2003**, *65*(2), 195–196.

Senanayake, S. A.; Ranaweera, K. K. D. S.; Banunuarachchi, A.; Gunaratne, A. Proximate analysis and phytochemical and mineral constituents in four cultivars of yams and tuber crops in Srilanka. *Tropical Agricultural Research extension*. **2012**, *15*(1), 32–36.

Shajeela, P. S.; Mohan, V. R.; Jesudas, L. L.; Soris, P. T. Nutritional and antinutritional evaluation of wild yam. (*Dioscorea spp*). *Tropical and subtropical Agroecosystems*. **2011,** *14,* 723–730.

Shefana, A. G.; Ekanayake, S. Some nutritional aspects of *Lasia spinosa* (kohila) *Vidyodaya. J. Sci.* **2009**, *14, 59–64.*

Sinha, R.; Lakra, V. Wild tribal food plants of Orissa. *Ind. J. Trad. Knowl*. **2005**, *4*(3), 246 252.

Smith, M. A.; Perry, G.; Richey, P. L.; Sayre, L. M.; Anderson, V. E.; Beal, M. F.; Kowal, N. Oxidative damage in Alzheimer's. *Nature*. **1996,** *382,* 120–121.

Sofowora, A. Medicinal plants and traditional medicine in Africa. Spectrum Books limited. Ibadan. 1993.

Temjenmongla, Yadav, A. K. Anticestodal efficacy of *Lasia spinosa* extract against experimental *Hymenolepis diminuta* infections in rats. *Pharm Biol*. **2006**, *44,* 499–502.

Thatoi, H. N.; Rout, S. D. Medicinal Plants: Ethnomedicine and Bioethnological potential. Biotech Books, New Delhi, **2011**.

Tiwari, A. K. Imbalance in antioxidant defense and human diseases: Multiple approach of natural antioxidant therapy. *Curr Sci*. **2001**, *8,* 1179–1187.

Tiwari, P.; Kumar, B.; Kaur, M.; Kaur, G.; Kaur, H. Phytochemical screening and extraction: a review. *Int. Pharmaceut. Sci.* **2011**, *1*(1), 98–106.

Tran, Q. L.; Tezuka, Y.; Ueda, J. Y.; Nguyen, N. T.; Maruyama, Y.; Begum, K.; Kim, H. S.; Kadota, S. *In-vitro* antiplasmodial activity of antimalarial medicinal plants used in Vietnamese traditional medicine. *J Ethnopharmacol.* **2003**, *86,* 249–252.

Tyagi, N.; Bohra, A. Screening of phytochemicals of fruit plant and antibacterial potential against *Pseudomonas aeruginosa. Biochem Cellular Arch.* **2002,** *2,* 21–24.

Wanasundera, J.P.; Ranidran, G. Nutritional assessment of Yam (*Dioscorea alata*) tubers. *Plant foods Human Nutrition.* **1994**, *46,* 33–39.

Wang, T.; Lii, H. Y.; Chang, J.; Yang, F. Anticlastogenic effect of aqueous extract from water Yam (*Dioscore alata* L.). *J Medicinal Plants Research.* **2011**, *5*(26), 6192–6202.

Webster, J.; Beck, W.; Ternai, B. Toxicity, and bitterness in Australian *Dioscorea bulbifera* L. and *Dioscorea hispida* Dennst from Thailand. *J. Agric. F. Chem.* **1984,** *32,* 1087–1090.

Wise, R. Regenerating antibacterial drug discovery and development, the urgent need for antibacterial agents. *J Antimicrob Chemo.* **2011**, *66,* 1939–1940.

Xiang, Z. R. Utilization of traditional medicine in China. In: Proceedings of International Conference on Traditional Medicinal Plants. *Arusha.* **1990,** 229–232.

Yadav, A. K.; Temjenmongla. Anticestodal efficacy of *Lasia spinosa* extract against experimental Hymenolepis diminuta infections in rats. *Pharm. Biology.* **2006,** *44*(7), 499–502.

Yadav, A. K.; Temjenmongla. Efficacy of *Lasia spinosa* leaf extract in treating mice infected with *Trichinella spiralis*, Springer-Verlag. **2011.**

CHAPTER 14

Plant Diversity and Ethnobotanical Perspective of Odisha

SASWATI DASH, ICHHAMATI PRADHAN, SURAJA KUMAR NAYAK, and BIGHNESWAR BALIYARSINGH*

Department of Biotechnology, College of Engineering and Technology, Techno Campus, Ghatikia, Bhubaneswar, Odisha,
**E-mail: bighnesh.singh@gmail.com*

ABSTRACT

The variety and variation among species along with the number of living organisms and ecosystem confers diverse biotic system. In many ways, since immemorial ages, biodiversity have been benefited mankind by providing array of indirect and direct products and essential services and also modulating functionality of ecosystem. The large geographic area of India includes two out of world's eight biodiversity hotspots which can be grouped into 10 biogeographic regions. Moreover, the negative impact of chemosynthetic products in one hand and on the other, the potential of plants being the largest source of diverse biogenic products have renewed and emphasized the ethnobotanical studies. The studies on indigenous people's classification, use and management of indigenous plants lead to identification and evaluation of plant-human relations and their conservation. More than 70% of the population of India is reliant on traditional plant-based medicines and about 7,500 plant species have been reported to have medicinal utility in India. The forests of northern and western region of Odisha, India, more specifically the Mayurbhanj, Keonjhar and Baragarh districts are the rich sources of large number of diverse plants with medicinal and non-medicinal values. A diverse range of 2,727 indigenous plant species under 228 families was reported in Odisha. The challenges of plant-derived products being effective depend on appropriate approaches and problems in-hand that in-turn need help from indigenous knowledge, faith and traditional practices.

Therefore, analyzing and reporting the traditional knowledge of interaction between diverse plant species and human communities will improve socio-economic aspects of society.

14.1 INTRODUCTION

Since the beginning of mankind, the humans have fascinated about its environment, more specifically towards plants for their need and activity. By virtue of these, humans started examining beneficial qualities of plants and enriching the knowledge of useful and harmful plants. His survival and continuous interaction with nature and the experience ofhis practical knowledge have evolved with useful application of plants. The age-old practices of rural communities and aboriginals or tribal peoples, in particular, on plant resources for herbal medicines, food, forage, household constructions and implements, beds and sleeping mats, and firewood are the basis of the knowledge for application of plants. Thus, the informations about the ethnomedicinal potential of plants were derived from the gathered knowledge from the aboriginals. In addition, most of the amassed knowledge has been transferred by verbal practice from one generation to another. Prior to the concept of ethnobotany, practicing of plants by humans for different need were depicted in many references of Egyptian, Greek, Roman, Chinese and classical Indian culture like citations of Rig-Veda and Atharva-Veda, 2000 B.C. oldest Vedic resources of India. Various literatures/plant-based pharmacopoeias of ancient civilisation like Papyrus (1515 BC), Hippocrates (462–372 BC), Aristotle (384–322 BC), Theophrastus (372–287 BC) and Galen (131–200 AD) have enlisted utility of plants or plant products in a systematic way (Prioreschi, 1996; Glesinger, 1954; Petrovska, 2012).

Ethnobotany word is derived from the word ethnic, 'ethno' and botany, coined by John William Harshberger in the 1895 (Harshberger, 1986). Ethnobotany can be referred as study of people and plants, emphasizing more on the good relationship between wild plants and humans. Schultes (1962) and Jain (1986) explained 'ethnobotany' as a science of interaction between primal communities with the plants of their ecological niche. In other words, it's an anthropological approach to botany to gather information about utility of diverse plants. The physical and chemical properties of many plant species, apart from phenological and ecological features, will be established by the through investigations into traditional use and management of local flora. According to reports of WHO, 80% of the world populations depend on indigenous medicine and that the majority of traditional therapies involve the use of plant extracts or of their active constituents (Cragg and Newman, 2013). Similarly, various ancient pharmacopoeia of India like Ayurveda, Unani, Siddha and Homoeopathy have listed many uses of plants as medicines (Swargiary et al., 2013). In recent past, National Medicinal Plants Board, Govt. of India has identified 6,000 to 7,000 plant species from the 17,000 to 18,000 species of flowering plants that are used for medicinal purpose (Tripathi et al., 2017).

The development of new and better medicinal or healthcare products, the domestication of new crop plants for food and industrial purposes, understandings of interaction due to plant biodiversity and environmental conservation are propelled by the knowledge that are acquired by the ethnobotanical investigations. On the contrary, due to lack of the specific research and awareness, the plants at the site are facing threat of loss due to ignorance and various mismanagement activities of human beings. Thus, the documentation of ethnobotanical knowledge is not only important for species conservation but also sustainable use of resources. This chapter will try to compile and summarise the ethnomedicinal plants that are still in practice by the local peoples of Mayurbhanj, Keonjhar, and Baragarh districts of Odisha, India, covered with large and deep forest. Furthermore, such studies are often informative about locally available important plant species and sometimes that may lead to the discovery of newer drugs.

14.2 BIODIVERSITY AND DISTRIBUTION OF MEDICINAL PLANTS OF ODISHA

The nature, especially plants, represents an enormous pool of natural resources for various products or chemicals and these are being utilized for various purposes, especially as medicines by ethnic people. Ethnobotanical studies reveal that plants are in use to mitigate a wide variety of human needs, both in past and present times (Uprety et al., 2010; Bhushan and Kumar, 2013). This indigenous utility knowledge have matured and accumulated with long-standing traditions and practices of certain regional communities. Recent ethnobotanical studies became basis and interlink disciplines of anthropology, botany, linguistics, nutrition, conservation and pharmacology, enriching the human knowledge (Balick, 1996). Various UNESCO programs now emphasize on simple treatments and effective uses of herbal plants that are commonly found in Asia-Pacific regions. The knowledge of plant diversity is a pre-requisite for assessment and sustainment of ecosystem and this in turn helps us to understand the overall structure and function of an ecosystem (Gairola et al., 2010). Plant biodiversity is the significant part of nature which includes diverse genetic make-up of individuals with richness of variety and this precise information is highly essential for their utilization and conservation. The knowledge on biodiversity allows us to present and to avoid the potential chances of biodiversity loss and often making of future policy for the protection of our environment.

The diverse and rich tribal medicinal flora of India is due to the different climatic conditions and regions. The increasing popularity of tribal medicine among the rural and urban areas of India can easily be observed. The modern systematic studies of Indian ethnobotany began during middle of the 19th century, emphasizing on tribal systems of medicine and cultures. The works of Dr. E.K. Janaki Ammal (1956) has initiated and triggered ethnobotanical studies in India (Janaki Ammal, 1956). However, earlier works like 'A Catalogue of Indian Medicinal Plants and Drugs' by John Fleming (Fleming, 1812) and Indigenous Drugs of India by R.N. Chopra (Chopra and Chopra, 1933) and K.L. Dey (Dey and Mair, 1973) mainly focused on the plants and drugs which had been established in traditional medicines of India. Understanding the potential, the Botanical Survey of India has intensified the ethnobotanical research (Jain and Srivastava, 2001). Studies pursued under All India Coordinated Research Project on Ethnobiology (AICRPE 1992–1998) have established the fact that 8,000 out of 10,000 wild plant species are in practice by tribals for medicinal purpose (Kumar et al., 2011).

Physiographically the forest cover of the Odisha, India is divided into four regions northern plateau, Eastern-Ghats, central and coastal plains. Studies by Saxena and Brahmam (1998) have reported 2,727 indigenous plant species under 228 families. Various studies like plants used in touch therapy at Bargarh district (Sen and Behera, 2008), plants with ethnopaediatric potential observed in the district Koraput (Srivastava and Rout, 1994), plants used for medicinal purpose by Bondo tribe of Malkangri (Prusti and Behera, 2007), identification of 19 diversified plants as edible plants from the Similipal Biosphere Reserve (Rout, 2007) have established the importance of ethnobotany and biodiversity.

Mayurbhanj district being the largest district in Odisha, possess 38% of its area under forest cover and has a great diversity of plant resources especially in medicinal plants (Rout and Panda, 2010). It is surrounded by Rairangpur Forest Division in north, Keonjhar Wildlife Division in south, Similipal Forest in east and Keonjhar Division in west. As per the classification of Champion and Seth (Champion and Seth, 1968), the forests of Mayurbhanj belong to two broad categories namely "Northern Tropical Moist Deciduous Forests" and "Northern Tropical Semi-evergreen Forests." The flora of the district is considerably diverse in terms of taxa. There are 460 polypetalous plant species distributed within 242 genera under 72 families (Mandal et al., 2012). The family *Fabaceae* exhibits highest number of species i.e., 112 belongs to 46 genera followed by *Malvaceae* (24 species of 8 genera), *Cucurbitaceae* (22 species of 16 genera) and *Caesalpiniaceae* (20 species of 6 genera). Similarly, tribal peoples of Kandhamal district, located in the central and girdled high-ranges Odisha use more than 100 plants for therapeutic purpose. Studies of Behera et al. (2006) have reported 98 plant species under 59 families and 93 genera are used for various diseases like asthma, fissure to tuberculosis and some odd 40 plants are for 62 ailments (Panigrahy, 2016). The *Euphorbiaceae* family contribute maximum number of species followed by *Zingiberaceae*, *Fabaceae*, and that of *Caesalpiniaceae, Liliaceae, Lamiaceae, Apocynaceae*, *Asclepiadaceae* family contribute few (one/two) species. Maximum plants are used by the tribal peoples for the curing of major diseases like skin disease, gastrointestinal disease, while fewer number of species for treatment of cold and cough (5 plants), fever (4 plants), head-ache, dysuria and sexual disorder (4 plants) (Panigrahy, 2016).

The western part of Odisha (Bargarh, Balangir and Nawarangpur) is dominated by large forestation and indigenous tribes. Among the herbal recipes prepared from the plant of families *Euphorbiaceae, Lamiaceae, Verbenaceae, Liliaceae* and *Fabaceae,* the members of the family *Fabaceae* are more commonly used (Panda et al., 2014). However, the methods of crude preparation from plants vary with respect to disease and type of plants. The southern-most districts of Odisha (Koraput and Malkangiri) are contiguous to the mainland of Eastern Ghats and southeastern Ghat zone. According to studies of Pattanaik et al. (2006), about 39 plant species belonging to 30 families distributed in 37 genera were used for total of 23 diseases ranging from simple cuts and wounds to severe disease conditions like female disorders. Plants of *Combretaceae* and *Euphorbiaceae* families are more commonly used (3 species each), either singly or in combination with other plant extract.

14.3 BOTANICALS—NATURAL SOURCES FOR ETHNOMEDICINE

A phenomenal number of people around the globe rely on traditional medicines, basically herbal, for their primary need of healthcare (Farnsworth and Soejarto, 1991; Pei, 2002). The global demand for herbal medicine is not only large, but also growing (Srivastava, 2000). The insufficient and sometimes inefficient provisions of western-allopathic medicine, especially in developing countries, lead to revisiting of traditional herbal medicine. This is one of the reasons of market growth of Ayurvedic medicines which is estimated to be expanding at 20% annually in India (Subrat et al., 2002). Plants being of pre-historic origin have been an excellent source of novel natural bioactive compounds. The complex organic compounds producing plants are loosely termed as 'medicinal and aromatic plants' (MAPs) that have many commercial purposes ranging from medicinal utility to foods, condiments and cosmetics (Schippmann et al., 2002). Therefore, the term 'botanicals' is becoming commonly used for a wide range of plant-based products. Moreover, from pharmaceutical point of view, medicinal plants have contributed largely to drug discovery by providing ingredients for drugs. Even some drugs are still extracted directly from plants. Even if the rigorous investigations on terrestrial flora, only 6% of higher plant species (300,000 species approx.) have been systematically evaluated for pharmacological potential while some 15% of species for phytochemical characteristics (Cragg and Newman, 2013; Fabricant and Farnsworth, 2001; Verpoorte, 2000).

Bioactive compounds in plants are complex molecules produced by plants having pharmacological or toxicological effects on humans or other organisms. Phylogenetically, the ability of producing typical bioactive compounds, loosely termed as secondary metabolites, can vary. The secondary bioactive compounds in plants are synthesized non-constitutively but hold important function in plants. For example, flavonoids can protect against free radicals generated during photosynthesis. Terpenoids may attract pollinators or seed dispersers, or inhibit competing plants. Alkaloids usually ward off herbivore animals or insect attacks (phytoalexins). Surveys of plant-derived pure compounds used as drugs conducted by countries with WHO-Traditional Medicine Centers have identified 122 compounds. Of these, 80% were used for the same or related medicinal purposes and were derived from only 94 plant species (Farnsworth et al., 1985). Some notable examples of bioactive compounds in drug formulations are galegine, derived from *Galega officinalis* L., served as a model for metformin and bisguanidine-type in antidiabetic drugs synthesis and papaverine derived from *Papaver somniferum* is an active constituent of hypertension drugs like verapamil (Fabricant and Farnsworth, 2001). In addition, the latter plant is the best source of painkillers such as morphine and codeine (Buss et al., 2003), but probably the best examples of ethnomedicine's role in guiding drug discovery and development is that of the antimalarial drugs, particularly quinine and artemisinin.

14.3.1 PLANTS USED AS ETHNOMEDICINE

Since antiquity, human beings have used remedies from nature to improve their health or to cure illnesses. Increase in consciousness on herbal formulations all over the world has particularly doubled the consumption in the west. Literature shows approx. 20–25% of drugs mentioned in different pharmacopeia are from natural sources, employed in treatment of diseases directly, without modifications, e.g., vincristine from *Catharanthus roseus* (L.) G. Don and silymarin from *Silybum marianum* (L.) Gaertn., or with minor chemical alternation, e.g., aspirin, isolated from *Salix* spp. (Newman et al., 2000). Even in India various studies had emphasized potential of plants in treating various diseases like diabetes (Khurnbongmayum et al., 2005), dysentery and diarrhea (Das and Choudhury, 2012), skin disease (Maruthi et al., 2000), jaundice (Bhatt et al., 2001), reproductive disorders (Bhogaonkar and Kadam, 2006), rheumatoid arthritis (Ramarao Naidu et al., 2008), snake bite (Reddy et al., 1997), antidotes (Thangadurai, 1998) sexually transmitted diseases and urinary complaints, malaria and dental disorder.

Similipal Biosphere Reserve in Mayurbhanj, Odisha is worldwide acknowledged for its plethora of biodiversity. Out of 1076 floral species (approx.) investigated from this sanctuary, more than 200 species are having medicinal uses (Mohanta et al., 2006). Various medicinal plants and its parts are in use against different ailments by tribal peoples of Mayurbhanj district, ranging from gastrointestinal disorders like constipation, diarrhoea, dysentery, gastric, stomachache (16 plant species), skin diseases like blood purification, scabies and skeletal diseases (10 plant species), gynaecological disorder (4 plants), neurological diseases and snakebite (Table 14.1). This emboldens the need of in-depth ethnomedicinal investigations and utility in context with biochemical and pharmaceutical aspects (Rout et al., 2009). Similarly about more than 500 species of medicinal plants are used by different tribes and practitioner of Gandhamardan hills (Bargarh and Bolangir). Some important examples are *Ailanthus excelsa*, *Abrus precatorius* for skin disease, *Abutilon Indicum* for jaundice, *Barleria Prionitis*, *Celastrus Paniculata* and *Aristolochia Indica* for urinary and gynaecological disorder, *Tinospora cordifolia*, *Ichnocarpus frutescens* for diabetics and *Trichodesma indicium, Gymnema sylvestre* for snake bite (Gomes et al., 2010; Behera et al., 2016). The traditional healers of Kandhamal district, Odisha use different ailments or therapeutic agents from plants to treat asthma, arthritis, eczema, glossitis, haematemesis, kidney stone, leucoderma, miliary, nasal bleeding, opthalmia and increase memory power etc. More number of plants is utilized for

TABLE 14.1 Comprehensive List of Ethnomedicinal Plants Used to Treat Various Human Diseases/Ailments in Mayurbhanj, Keonjhar and Baragarh Districts of Odisha, India

Sl	Scientific Name	Family	Common Name	Parts Used	Role/Bioactivity
			District *Mayurbhanj*		
1	*Abrus pracatorius*	Fabaceae	Kaincha	Leaf/Seed	Gonorrhea/Muscle Contusion/Bronchitis, Asthma
2	*Acampe ochracea*	Orchidaceae	Rasna	Leaves	Headache
3	*Aegle marmelos*	Rutaceae	Bael	Fruit	Digestive Disorder
4	*Ageratum conyzoides*	Asteraceae	Pokasungha	Leaves	Wound, Pneumonia
5	*Aglaia elaeagnoidea*	Meliaceae	Pingu	Fruits	Memory Enhanced
6	*Aloe barbadensis*	Asphodelaceae	Aloe vera	Leaf	Skin Diseases
7	*Alstonia scholaris*	Apocynaceae	Chhatiana	Latex	Spermatorrhoea
8	*Andrographis paniculata*	Acanthaceae	Bhuineem	Whole Plant/Leaf/ Shoot	Warts, Eye Problem, Piles, Jaundice/Malaria/ Stomach Trouble
9	*Anogeissus latifolia*	Combretaceae	Dhaura	Bark	Diarrhea
10	*Aristolochia indica*	Aristolochiaceae	Iswarmula	Root Tuber	Abdominal Colic, Constipation, Snake-Bite
11	*Asparagus racemosus*	Liliaceae	Satawari	Tuber	Gastro-Intestinal Problem, Jaundice
12	*Buchanania lanzan*	Anacardiaceae	Chara/Chironji	Leaves/Flower/ Whole Plant	Mouth and Sour Laxatives, Fever, Excessive Thirst
13	*Butea superba Roxb*	Fabaceae	Naipalaso	Shoots	Piles
14	*Calotropis gigantean L.*	Asclepiadaceae	Ark	Leaf	Snake-Bite Antidote
15	*Cassia fistula*	Caesalpiniaceae	Sonari	Leaf	Jaundice
16	*Cassia tora*	Caesalpiniaceae	Chakunda	Leaves, Root	Skin Diseases, Malaria, Ring Worm, Intestinal Disorder, Skin Itching
17	*Catunaregam spinosa*	Rubiaceae	Putua	Root	Rheumatic Fever and Pain in Limbs

TABLE 14.1 *(Continued)*

Sl	Scientific Name	Family	Common Name	Parts Used	Role/Bioactivity
18	*Cissus quadrangularis*	Vitaceae	Hada Jodi	Leaves	Bone Healing
19	*Coccinia grandis L.*	Cucurbitaceae	Bankundri	Root	Jaundice
20	*Croton roxburghii*	Euphorbiaceae	Putudi	Leaves	Cancer, Diabetes
21	*Curculigo orchioides*	Amaryllidaceae	Talmuli	Root	Piles
22	*Curcuma longa L.*	Zingiberaceae	Haldi	Fresh Rhizome	Piles, Cure Blood
23	*Cuscuta reflexa*	Convolvulaceae	Nirmuli	Leaves	Constipation, Spleen Disease
24	*Cycas cricinalis*	Cycadaceae	Arguna	Shoots	Astringent, Diuretic
25	*Cynodon dactylon*	Poaceae	Dub	Leaves	Nose Bleeding
26	*Dendrophthoe falcata*	Loranthaceae	Malang	Leaves	Wound
27	*Derris indica*	Fabaceae	Karonj	Seed	Skin Diseases
28	*Desmodium heterocarpon*	Fabaceae	Salparni	Root	Menstrual Problems
29	*Desmodium polycaroum*	Fabaceae	Krushnapurni	Whole Plant	Cough, Fever
30	*Enhydra fluctuans*	Asteraceae	Hidmicha	Leaf	Gonorrhea
31	*Ficus benghalensis L.*	Moraceae	Boro	Bark	Fissure, Piles
32	*Ficus racemosa L.*	Moraceae	Dimiri	Stem	Gastro-Intestinal Problem
33	*Haldinia cordifolia*	Rubiaceae	Koim	Bark	Redness In Eye
34	*Hemidesmus indicus*	Apocynaceae	Anantamula	Root	Coolant, Blood Purifier
35	*Hemidesmus indicus L.*	Periocaceae	Antamula	Root/Leaf	Blood Purification, Gout and Joint Pain/Piles
36	*Holarrhena antidysenterica*	Apocynaceae	Indrajal	Leaves	Gut Mortality Disorder
37	*Holarrhena pubescens*	Apocynaceae	Kuluchi	Bark	Skin Diseases, Gastro-Intestinal Problem

TABLE 14.1 *(Continued)*

Sl	Scientific Name	Family	Common Name	Parts Used	Role/Bioactivity
38	*Hygrophila auriculata*	Acanthaceae	Koilekha	Leaf	Bronchitis, Cough and Epidemic Fever.
39	*Ichnocarpus frutescens L.*	Apocynaceae	Dudhilata	Root	Scabies
40	*Ixora pavetta*	Rubiaceae	Telakurmi	Leaves	Acidity
41	*Lawsomnia inermis L.*	Lythraceae	Manjuati	Root	Jaundice, Tonic for Liver
42	*Lawsonia inermis*	Lythraceae	Mehndi	Roots	Jaundice
43	*Litsea monopetale*	Lauraceae	Pojo	Bark	Gastro-Intestinal Problem
44	*Madhuca longifolia*	Sapotaceae	Mahula	Flowers	Cough, Cold, Headache, Diabetes
45	*Marsdenia tenacissina*	Asclepiadaceae	Ahirigada	*Root*	*Abdominal Colic, Constipation, Snake-Bite*
46	*Melastoma malabathari-cum L.*	Melastomaceae	Koroli	Leaf	Skin Diseases, Tumor, Colic Disease
47	*Millettia extensa*	Fabaceae	Gaudhuni	Roots	Skin Diseases
48	*Mimosa himalayana Gamble*	Mimosaceae	Khirkichi	Root	Disorders of Lungs and Inflammation of Liver
49	*Mimosa pudica L.*	Mimosaceae	Lajkoli	Whole Plant	Bleeding Piles
50	*Mitragyna parvifolia*	Rubiaceae	Gudikoim	Root	Pimples
51	*Morinda citrifolia L.*	Rubiaceae	Pindra	Leaf	Rheumatic Fever, Joint Pain
52	*Morinda pubescens*	Rubiaceae	Achu	Root	Dysentery
53	*Mucuna nigricans*	Fabaceae	Boidonko	Seeds	Ulcer of Genital Organs
54	*Nyctanthes arbortristis*	Oleaceae	Gangasiuli	Leaves, Shoot	Fever, Malaria
55	*Opuntia dillenii*	Cactaceae	Nagapheni	Pulp	Eyes Disorder
56	*Oroxylum indicum*	Bignoniaceae	Phenphena	Bark	Rheumatic Pain, Stomach Upset
57	*Oxystelma secamone L.*	Asclepiadaceae	Dudhilata	Root	Jaundice, Chronic Liver Problems

TABLE 14.1 *(Continued)*

Sl	Scientific Name	Family	Common Name	Parts Used	Role/Bioactivity
58	*Phaseolus calcaratus*	Fabaceae	Banmungo	Whole Plant	Fever and Gout
59	*Phyllanthus fraternus*	Euphorbiaceae	Badiamla	Whole Plant	Liver Cirrhosis, Hepatitis
60	*Polyalthia cerasoides*	Annonaceae	Champati	Bark	Diabetes
61	*Pterocarpus marsupium*	Fabaceae	Piasala	Bark	Diabetes, Stomach Pain, Mouth Ulcers
62	*Pterospermum xylocarpum*	Sterculiaceae	Mucchkundo	Bark	General Debility or Weakness
63	*Pueraria tuberosa*	Fabaceae	Bhuikakharu	Root	Treat Rheumatism
64	*Rauvolfia serpentina*	Apocynaceae	Patalgaruda	Root	Snake-Bite Antidote, High Blood Pressure
65	*Rubia cordifolia*	Rubiaceae	Chireita	Root	Intoxication
66	*Sansevieria roxburghiana*	Asparagaceae	Murga	Leaves	Antiseptic Fast Aid
67	*Saraca asoca*	Caesalpiniaceae	Asoka	Leaf	Pain Reliever, Gynecological Benefits, Diabetes
68	*Schleichera oleosa*	Sapindaceae	Kusum	Fruit/Stem /Seed	Skin Itching/Treatment of Gout, Scabies
69	*Semecarpus anacardium*	Anacardiaceae	Bhalia	Seed/Gum	Body Pain/Neurological Diseases
70	*Shorea robusta*	Dipterocarpaceae	Sal	Seeds	Stomach Pain
71	*Smilax perfoliata*	Smilacaceae	Ramdatuni	Root	Gastric Problem
72	*Soymida febrifuge*	Meliaceae	Ruhini	Bark	Dysentery
73	*Stereospermum suaveolens*	Bignoniaceae	Patoli	Root	Nervous Disorders
74	*Syzygium cerasoides*	Myrtaceae	Poijam	Bark	Diabetes
75	*Syzygium cumini*	Myrtaceae	Jamu	Seed	Diabetes
76	*Tephrosia purpurea L.*	Fabaceae	Bankultha	Root	Stomach Upset
77	*Terminalia chebula*	Combretaceae	Harida	Fruit/Seed	Cardiovascular Diseases/Bleeding Piles
78	*Terminalia alata*	Combretaceae	Asan	Bark	Chest Pain

TABLE 14.1 *(Continued)*

Sl	Scientific Name	Family	Common Name	Parts Used	Role/Bioactivity
79	*Tragia involucrate L.*	Euphorbiaceae	Bichhuati	Whole Plant	Diarrhea Alternating Constipation
80	*Vanda tessellata*	Orchidaceae	Kankata	Leaves	Earache, Galactogogue
81	*Ventilago maderaspatana*	Rhamnaceae	Rakta Pichula	Bark	Snake-Bite Antidote
82	*Woodfordia fruticosa*	Lythraceae	Dhatuki	Leaves/Flower	Stimulant, Dysentery/Leucorrhoea
			District Keonjhar		
83	*Adhotoda vasica*	Acanthaceae	Basanga	Leaves	Piles, Leprosy, Cough, Tuberculosis
84	*Aegele marmelos*	Rutaceae	Bela	Stem Bark/Fruit	Gastrics, Diarrhea, Dysentery, Diabetes
85	*Butea monospema*	Eabaceae	Palasi	Seed	Diarrhea
86	*Dioscorea wallichii*	Dioscoreaceae	Pita Alu	Root	Stomach Pain
87	*Diospyros melanoxylon*	Ebenaceae	Kendu	Leaf/Bark	Night Blindness, Constipation, Diarrhea
89	*Emblica officinalis*	Euphorbiaceae	Anla	Fruit/Leaves	Fertility Enhancement, Brain Functioning
90	*Madhuca indica*	Sapotaceae	Mahua Flower	Seed Oil	Rheumatic Pain, Dhuda
91	*Pongamia pinnata*	Fabaceae	Karanja	Stem Bark	Antimalaria, Skin Diseases, Piles, Rheumatic
92	*Rauwolfia serpentine*	Apocynaceae	Patalgarud	Root	Snake-Bite Antidote, High Blood Pressure
93	*Shorea robusta*	Dipterocarpaceae	Sal	Leaves	Carminative/Stomachic/Astringent Properties.
94	*Terminalia chebula*	Combretaceae	Harida	Fruit/Leaf	Dysentery, Cough/Anemia
95	*Withania somnifera*	Solanaceae	Ashwagandha	Root/Flower	Seminal Weakness /Spermatorrhea
			District Baragarh		
96	*Abelmoschus esculentus*	Malvaceae	Bhendi	Unripe Fruit	Pulmonary Tuberculosis
97	*Abutilon indicum*	Malvaceae	Kuthelchitra	Leaf, Stem Bark	Jaundice, Diarrhea, Anemia,
98	*Acacia catechu*	Mimosaceae	Khair	Stem Bark	Night Blindness, Facture of Knee and Bone

TABLE 14.1 *(Continued)*

Sl	Scientific Name	Family	Common Name	Parts Used	Role/Bioactivity
100	*Acacia nilotica*	Mimosaceae	Bamur	Leaf/Bark Paste	Diabetes/Increase Male Fertility
101	*Acalypha indica*	Euphorbiaceae	Indramaris	Areal Parts	Skin Diseases, Ulcers, Bronchitis
102	*Achyranthes aspera*	Amaranthaceae	Aphamarga	Leaf/Root	Scorpion Bite, Dental
104	*Acorus calamus*	Araceae	Bhutnashan	Rhizome	Bronchitis, Snake Bite,
105	*Adhatoda vasica*	Acanthaceae	Basong	Leaf	Piles, Leprosy, Cough, TB
106	*Aerva lanata*	Amaranthaceae	Chaul Dhua	Dried Plant	Gall Bladder, Kidney and Uteral Stone
107	*Agava americana*	Agavaceae	Hatipazar	Whole Plant	Ulcer, Tuberculosis, Jaundice
108	*Alangium salvifolium*	Alangiaceae	Aankel	Leaf, Root, Bark	Redness of Eye /Anthelmintic
109	*Albizia lebbek*	Fabaceae	Sesua	Bark/Leaves	Dysentery, Diarrhea
100	*Bambusa vulgaris*	Poaceae	Katang	Root, Bark	Piles, Promote Flow of Urines
101	*Bauhinia variegata*	Caesalpiniaceae	Kuler	Root Bark	Glandular TB, Reducing Bulkiness of the Body
102	*Boerhavia diffusa L.*	Nyctaginaceae	Gadhapurni	Whole Plant	Bronchitis, Leucorrhoea, Asthma, Anemia
103	*Bryophyllum pinnatum*	Crassulaceae	Patargaja	Leaves	Boils, Wound, Sores, Insect Bites
104	*Butea monosperma*	Fabaceae	Palasa	Seed, Bark	Dysentery, Piles, Hydrocele, Menstrual Disorder
105	*Caesalpinia pulcherrima*	Caesalpiniaceae	Radhachuda	Root, Leaf, Bark	Cholera, Malaria, Puragative, Absortion
106	*Cajanus cajan*	Fabaceae	Kandul	Leaf Juice, Seed	Leprosy, Bronchitis, Mouth Ulcers, Jaundice
107	*Careya arborea*	Barringtoniaceae	Khumbhi	Root Paste, Bark	Body Pain/Piles
108	*Desmodium heterocarpon*	Fabaceae	Krishnapani	Aerial Parts	Tonic for Fattening of Body
109	*Dillenia aurea*	Dilleniaceae	Kermetta	Stem Bark, Leaf	Restoration of Health After Child Birth, Griping Pain
110	*Dillenia indica*	Dilleniaceae	Awoo	Leave	Treatment of Carbuncle

TABLE 14.1 *(Continued)*

Sl	Scientific Name	Family	Common Name	Parts Used	Role/Bioactivity
111	*Dolichos lablab*	Fabaceae	Shemi	Leaves, Seeds	Abdomen Pains, Inflammation of Urethra, Snake Bite,
112	*Enydra fluctuans*	Asteraceae	Hidimichi	Leaves	Headache, Eye Disease, Hook Worm Infection
113	*Eucalyptus rostrata*	Myteraceae	Eucalyptus	Leaves	Asthma, Dysentery, Tb, Diabetes, Malaria, Cold
114	*Euphorbia ligularia*	Euphorbiaceae	Thua	Latex	Purgative, Paralysis, Rheumatism, Earache
115	*Geranium wallichianum*	Poaceae	Rattenjot	Leaf, Root	Blood Purification, Jaundice
116	*Heliotropium indicum*	Boraginaceae	Hatisundha	Root	Iron Deficiency Against Anemia During Pregnancy
117	*Hemidusmus indicus*	Asclepiadaceae	Ananthmula	Root	Leucoderma, Blood Purification, Skin Diseases
118	*Holoptelea integrifolia*	Ulmaceae	Dharanja	Stem/Bark	Rheumatic Pains
119	*Impatiens kleinii*	Balsaminaceae	Haragaura	Leaves	Externally for Burn Places
120	*Jatropha curcas*	Euphorbiaceae	Ramjada	Seed Oil/Leaf	Pneumonia, Body Pain/Jaundice
121	*Lantana camara*	Verbenaceae	Naguari	Leaf	Malaria
122	*Lawsonia innermis*	Lytheraceae	Benjati	Roots	Anemia, Jaundice
123	*Leucas aspera*	Lamiaceae	Bhutmari	Leaf	Psoriasis, Skin Diseases
124	*Nymphaea pubescens*	Nymphaceae	*Kai*	Rhizome	Goitre
125	*Nymphoea stellate*	Nymphaceae	*Nila Kai*	Flower	Head Ache, Night Blindness, Redness of Eye, Acne
126	*Ocimum gratissimum*	Lamiaceae	*Gandha Tulsi*	Leaf	Diaphoretic, Cold & Cough, Anthelmintic
127	*Opuntia stricta*	Cactaceae	Nagphani	Phylloclade	For Swelling of Joints
128	*Oroxylum indicum*	Bignoniaceae	*Phenphena*	Stem Bark	Digestive Aid, Diarrhea, Menorrhagia, Puragative
129	*Phyla nodiflora*	Verbenaceae	Gosingi	Root	To Promote Sexual Desire of Women
130	*Pistica stratiotes*	Araceae	*Borajhanj*	Leaf	Diabetes, Blood In Urine, Anemia, Dysentery

TABLE 14.1 *(Continued)*

Sl	Scientific Name	Family	Common Name	Parts Used	Role/Bioactivity
131	*Plumbago indica*	*Plumbaginaceae*	*Raktachitaparu*	Root, Latex	Dyspepsia, Ophthalmic, Scabies, Leucoderma
132	*Psidium guajava*	Myteraceae	*Maya*	Flower/Fruit	Headache/Jaundice, Diabetes
133	*Psoralia corylifolia*	Fabaceae	*Bakuchi*	Leaf, Seed	Diarrhea, Leucoderma, Elephantiasis
134	*Pterocarpus marsupium*	Fabaceae	*Bija*	Stem Bark, Gum	Fertility, Blood Dysentery
135	*Swertia affinis*	Gentianaceae	Chirayetta	Leaf	Inflammations, Antipyretic, Leucoderma, Piles,
136	*Syzygium cumini*	Myteraceae	Jam	Seed/Bark	Diabetes/Dysentery/Hemorrhages, Leukorrhoea
137	*Tagetes erecta*	Asteraceae	Ganja	Whole Plant	Chest Pain, Anxiety, Muscle Pain, Colds, Ulcer
138	*Tagetes minuta*	Asteraceae	Katki Ganja	Leaves	Piles, Kidney Troubles, Muscular Pains
139	*Terminalia arjuna*	Combretaceae	*Kau*	Stem /Bark	Malaria, Heart Disease
140	*Terminalia bellirica*	Combretaceae	*Behera*	Stem/Bark	Leucoderma
141	*Tinospora cordifolia*	Menispermaceae	*Baikhujen*	Stem/Root	Jaundice, Acidity, Leprosy
142	*Toona ciliata*	Meliaceae	Mahalim	Whole Plant	Tumors, Worms, Leprosy, Tuberculosis
143	*Vigna trilobata*	Fabaceae	Sanobiri	Leaf Paste	Scorpion Sting
144	*Vitex negundo*	Verbenaceae	Nirgundi	Root Bark/Leaf	Typhoid Fever/Arthritis
145	*Ziziphus oenoplia*	Rhamnaceae	*Kathaukoli*	Stem Bark	Dysentery
146	*Ziziphus rugosav*	Rhamnaceae	*Chunkoli*	Leaf/Bark	Anti Pox/Dyspepsia

Papers cited: Rout et al. (2009), Routray and Nayak (2017), Kandari et al. (2012), Sahu et al. (2010).

diarrhea (12 plants), followed by rheumatism, itches, body ache (11 plants), dysentery and jaundice (10 plants) and seven species for migraine and acidity (Behera et al., 2006). Plants used in the treatment by the diverse races of Koraput district are dominated by the members of *Combretaceae* and *Euphorbiaceae* family are used by indigenous peoples. Apart from curing diseases, plants of families *Rubiaceae* and *Euphorbiaceae* (Makhija and Khamar, 2010), *Acanthaceae, Amaranthaceae and Mimosaceae* (Mitra and Mukherjee, 2012) and *Fabaceae* (Jain et al., 2011) are commonly used to treat snakebite, a major problem in forest region.

14.3.2 PLANT DERIVED BIOACTIVE COMPOUNDS

Since beginning of mankind, natural products and similar/related moieties have been served as guiding molecules for diverse therapeutic agents. With remarkable improvements in human health care on one hand and environmental deterioration on the other, a growing demand for natural products and phytomedicine has shifted research and development works into new drug discovery. In early 1900's before synthetic era 80% of all medicines were obtained from plant source (Harvey, 2008). As it is evident from pharmacopeias of Indian medicinal system which has a long history and one of the oldest organized systems of medicine. An analysis into the sources of new drug from 1981 to 2007 reveals that almost half of the drug approved since 1994 were based on natural product (Harvey, 2008; Butler, 2008). Ethnobotanical information is crucial for researchers in selecting the plants for onward screening and various chemical analysis (Balandrin et al., 1993). Usually the plant materials were collected and identified with reference to ethnobotanical information and subsequently screened for bioactive compounds by phytochemical analysis, performed in consultation with local users of the herbal medicine. The identified lead compounds of herbal medicines are tested through phytochemistry, molecular and animal experiments, and clinical trials. Basically, the natural products (secondary metabolites) have been more successful as a lead source of potential drug. According survey of WHO, at least 122 chemical compounds, isolated/ derived from 94 floral species, are considered as viable drugs by many countries. Of these 119 drugs, 74% were discovered as a result of chemical studies directed at the isolation of the active substances from plants used in traditional medicine (Farnsworth et al., 1985; Cragg and Newman, 2013). Thus use of ethnobotanical information for drug development through various approaches constitutes an essential part of the methodology.

Bioactive compounds of plant origin plays significant role in immune-modulation, anti-diabetic, anti-inflammatory conditions and more commonly used as anti-malarial, anti-microbial, anti-obesity and anti-neoplastic agents. The compounds withanolides from *Withania somnifera*, berberine from *Berberis aristata*, guggulsterone from *Commiphora mukul* (guggul) (Shishodia and Aggarwal, 2004) and nimbidin from neem (*Azadirachta indica*) (Gupta et al., 1977) act as anti-inflammatory while thevetin A, B, peruvoside from *Thevetia neriifolia* are potent cardiac glycoside (Bose et al., 1999). The reserpine from *Rauwolfia serpentine* was first tested in India for anti-hypertensive activity. The *Terminalia arjuna* bark has been used for treatment of angina. *Garcinia cambogia* have hydroxycitric acid which is used as an anti-obesity agent (Heymsfield et al., 1998). Naphthylisoquinoline alkaloids isolated from leaves of *Anastrocladus heyneanus* particularly anastrocladidine, ancistrocladidine, ancistrocladinium B and ancistrotanzanine have been shown to exhibit significant antiplasmodial activity (Bringmann et al., 2004). *Withania somnifera* and the steroidal alkaloid solasodine from *Solanum nigrum,* are used as immunomodulators (Sarma and Khosa, 1994). Plant-derived khellin compound from *Ammi visnaga* (toothpick plant) act as a lead molecule in asthma and allergy drugs. Galantamine is mainly used against Alzheimer‘s disease was derived from *Galanthus caucasicus* (commonly called Snowdrop) (Tsakadze et al., 1969). *Strychnos nux-vomica* L., an important wild medicinal plant from Odisha possesses a wide range of antimicrobial activity against potential human pathogens (Dwibedy et al., 2015).

14.4 PLANTS WITH ECONOMIC VALUES

Non-medicinal plants are also an important source of income. Their sale and barter contribute to the economic development of rural communities and support modern industrial development. Economic important plants produce fibres, gums, resins, starch, sugar, and the countless other materials used by man. Tribal communities from Odisha are inevitable part of forest ecosystem with their own socio-cultural pattern, tradition and quintessential food practices. Their diet mostly comprises a variety of unconventional foods, as edible flowers, fruits, leaves, seeds, stems, tubers and wild mushrooms (Saxena, 1996). According to an estimate, 80% of forest dwellers in Jharkhand (part of erstwhile Bihar), Odisha (erstwhile Orissa), Chhattisgarh- Madhya Pradesh, West-Bengal and Himachal Pradesh depend on forest for 25 to 50% of their annual food requirements (Sinha and Lakra, 2005). Among the tribes of Mayurbhanj, Kheonjhar, wild fruits like *Badru*

(*Olax scandens* Roxb.), *Bankundri* (*Melothria heterophylla* (Lour.) Cogn., *Kankodo* (*Momordica dioica* Roxb.ex Willd.), *Lawa* (*Ficus glomerata* Roxb.), *Karmata* (*Dillenia aurea* Sm.), *Korkotta* (*Dillenia indica* Linn.), *Oserwa* (*Capparis zeylanica* Linn.) and *Pakare* (*Ficus lucescens* Blume) are cooked and eaten as vegetable. Roots and tubers are excellent food source for man and livestock as these are readily digested and have a high-energy content. The fiber yielding plants are second to the food plants with economic importance for human society. Many fiber yielding plants, including *Boehmeria nivea* Gaud. (Ramie), *Crotolaria juncea* L. (Sunhemp), *Corchorus capsularis* L. (Jute), *Gossypium arboreum* L. (Cotton), *Hibiscus cannabinus* L. (Kenaf), *Linum usitatissimum* L. (Flax) (Table 14.2) are the best known commercial plants which provide durable and flexible fibre.

14.5 SIGNIFICANCE OF BIODIVERSITY

Biodiversity encompasses all species of plants, animals & microorganisms, the ecosystems they live in and the ecological processes of which they are a part (McNeely et al., 1990). Biodiversity is clearly a fundamental component of life on earth. It can be defined as functional natural resources on which humans are dependent for their livelihood and socioeconomic development from ancient to present times and extending into future generations. The utilization and conservation of biodiversity involves intrinsic interactions between species, genetic populations, communities, landscape, and natural ecosystems, in one hand, and culture, technology, social, economic and indigenous knowledge, institutions and information on the other. It thus creates and maintains an ecological system. Functioning biodiversity maintains and accelerates ecosystem productivity where each species weather very small microbes or humans have an important role to play (Hoiling et al., 1997). This is evident from the fact that a larger number of plant species represents variety of crops as food or economic importance. This also ensures sustained diversity of species of all life forms. This diversity and species richness withstand and restore from a variety of calamities. Healthy biodiversity is not only responsible for protection of soil, water and recycling of nutrients (Chivian, 2002) but also act as a reservoir for food, medicinal bioactive compounds, timber and non-timber products, ornamental plants and diverse gene pool. Thus the ability and cost of replacing or replicating such system is beyond any imagination. So, any destruction reduces and isolates habitats, thereby reducing the interaction of species with a large gene pool.

TABLE 14.2 List of Economic Plants Used by Locals of Mayurbhanj, Keonjhar and Baragarh Districts of Odisha, India

Sl	Scientific Name	Family	Common Name	Parts Used	Role/Bioactivity
1	*Abelmoschus moschatus*	Malvaceae	Jangli Bhindi	Stem	Cotton Fiber, Jute
2	*Abrus precatorius*	Malvaceae	*Bano Bhindi*	Bark	Fiber
3	*Acorus calamus*	Poaceae	Sabai Grass	Grass Part	Paper Manufacturing, Weaving, Furniture's
4	*Alstonia scholaris*	Malvaceae	Pedi-Pedica	Stem	Cordage and also Mixed with Jute
5	*Alternanthera philoxeroides*	*Fabaceae*	Sola	Plant	Fodder
6	*Alternanthera sessilis*	Araceae	Bacha	Rhizome	Mohuli
7	*Bacopa monnieri*	Amaranthaceae	Madaranga Saga	Leaves	Vegetable
8	*Bauhinia purpurea*	Apocynaceae	Sangu	Stem	Cordage, Bowstrings
9	*Bauhinia racemosa*	Caesalpiniaceae	Barada	Bark	Cordage
10	*Bauhinia tomentosa*	Caesalpiniaceae	Ambalota	Bast	Cordage
11	*Bauhinia vahlii*	Dipterocarpaceae	Sal Tree	Seeds and Fruit	Lamp Oil, Vegetable Fat
12	*Bauhinia vahlii*	Caesalpiniaceae	Kanchan	Bark	Fiber
13	*Bauhinia variegata*	Caesalpiniaceae	Siali	Shoot, Bark	*Rope, Thread*
14	*Boehmeria macrophylla*	Caesalpiniaceae	Kanchano	Bark	Fiber
15	*Boehmeria nivea*	Urticaceae	Kankura	Bark	Cordage and Fishing-Nets
16	*Boerhavia chinensis*	Scrophulariaceae	Brahmi	Leaves	Vegetable
17	*Borassus flabellifer*	Bombacaceae	Simili	Fruit	Mattresses, Pillow
18	*Calotropis procera*	Asclepiadaceae	Akaona	Bark	Bow Strings, Twine, Fishing-Nets
19	*Canna indica*	Asclepiadaceae	Arakh	Stem	Fishing Nets, Twine
20	*Cinnamomum tamala*	Lauraceae	Tejpatta	Leaves/Seeds	Spices/Grind Seeds for Cooking
21	*Colocasia esculenta*	Commelinaceae	Kanissera	Leaves, Shoot	Vegetable
22	*Commelina benghalensis*	Apiaceae	Thalkudi	Leaves	Vegetable
23	*Croton lucidas*	Pepper Rod	Euphorbiaceae	Leave	Insecticidal Activity

TABLE 14.2 *(Continued)*

Sl	Scientific Name	Family	Common Name	Parts Used	Role/Bioactivity
24	*Croton sp.*	Basket Hoop	Euphorbiaceae	Leave	Cytotoxicity
25	*Cymbopogoncitratus*	Various	Euphorbiaceae	Leave, Seed	Insecticidal Activities
26	*Echinochloa crusgalli*	*Amaranthaceae*	Ghodamadaranga	Stem, Leaves	Pig Feed
27	*Echinochloa stagnina*	Poaceae	Dhera	Vegetative Parts	Fodder
28	*Fagara martinicensis*	Coffee Rose	Apocynaceae	Seed, Leaves	Insecticidal Activity
29	*Glinus oppositifolius*	Poaceae	Dhera	Seeds	Grains
30	*Hydrilla verticillata*	Cyperaceae	Kanaka	Tuber	Pig Feed
31	*Hygrophila auriculata*	Hydrophyllaceae	Langulia	Whole Plant	Supplementary Food
32	*Ipomoea aquatica*	Acanthaceae	Koelekha	Leaves	Vegetable
33	*Marsilea quadrifolia*	Scrophulariaceae	Keralata	Leaves	Vegetable
34	*Nelumbo nucifera*	Pontederiaceae	*Kaupana*	*Inflorescence*	Vegetable
35	*Neptunia oleracea*	Nymphaeaceae	Padma/Kamala	Rhizome	Vegetable
36	*Nymphaea nouchali*	Mimosaceae	Panilajakuli	Leaves	Vegetable
37	*Nymphaea pubescens*	Nymphaeaceae	Nilakain	Rhizome	Vegetable
38	*Oryza rufipogon*	Nymphaeaceae	Rangakain	Rhizome, Seeds	Roasted Food
39	*Oryza sativa*	Poaceae	Balunga	Plant	Fodder
40	*Phyllanthus emblica*	Fabaceae	Siali Tree	Leaves, Fruits	Food, Fiber, Tannin
41	*Shorea robusta*	Combretaceae	Sal Tree	Leaves	Leaves for Sericulture
42	*Terminalia elliptica*	Combretaceae	Asan Tree	Timber, Fruit	Making Furnitures, Construction

Papers cited: Panda and Misra (2011), Kandari et al. (2012), Sahu et al. (2013).

14.5.1 ROLE OF ETHNOBOTANY IN CONSERVATION OF BIODIVERSITY

Biodiversity is declining in present scenarios due to involvement of diverse factors coupled with excessive land use for farming or industry, climatic change, heightened in explosion and exploitation of natural resources, and other anthropological pollutions. Whether natural or human-induced, in most of the scenarios, factors tend to interact and affect reciprocally biodiversity and humans. A clear relationship of alteration of biodiversity with respect to direct drivers like deforestation led habitat loss, establishment of invasive alien species, overexploitation of natural resources ranging from domestic firewood to farming, and excessive use of synthetic fertilizer pollutants was acknowledged worldwide. In addition, excessive nutrient can allow a particular species to dominate above others, out competing them altogether. On the other hand, indirect drivers that are accountable for the changes in biodiversity include economic and technological activity, human population, and socio-political and cultural practices. Thus, replicating such complex biodiverse system by humans is impossible to imagine. Modern life, including access to modern allopathic medicines have transformed the consumption patterns of medicinal plants among local communities, from homemade small quantities to massive harvesting for marketing in large quantities. Overharvesting of medicinal plants has not only resulted in resource degradation and loss of biodiversity but also loss of indigenous medical knowledge and traditions. This ultimately diminishes the knowledge on traditional medical systems. The in-depth value of that biodiversity, both intrinsically and to mankind, is immeasurable, and thus must be protected and preserved.

Numerous studies have reported the magnitude of ethnobotany in the management and conservation of vegetation resources. One of the most important roles that ethnobotany could perform to help conserve biodiversity would be to propose realistic and functional models for natural resource usage and management. Being an interdisciplinary science, the onus of preserving the intellectual wealth of traditional knowledge that indigenous people possess pertaining to natural interactive environment including the medicinal plants resides on ethnobotany. Ethnobotanical studies helps in management of resources at the local level by understanding the interactions between people and biotic entities. It also helps to transfer indigenous techniques, practices and knowledge to replicate in other areas of the same ecosystem for resource management. It facilitates local people's participation in the collection, documentation, analysis, and assessment of indigenous knowledge through field studies. Thereby, it increases the inventory of plants used by humans ranging from medicinal, food, and fodder to non-timber forest products and religious utility. This includes their knowledge on the utilization and maintenance of different types of plant resources on a long-term basis without damaging or destroying their habitats. Hence, maximum efforts should be made to document and integrate indigenous knowledge on land-use, vegetation and forest management, non-timber forest products, medicinal plants, agro-forestry, and home-gardens. Understanding the indigenous knowledge of rural people in relation to biodiversity resource management and cultural traditions is important for the development of the region. In reciprocation, the traditional practice helps maintaining and taking care of plants, results in restoration of biodiversity.

14.6 FUTURE PROSPECTS

Traditional knowledge is confined and acquiescent only to a specified locality. Ethnomedicinal studies not only try to collect information regarding utilities of plants but also bring it to the public domain. It becomes highly significant when the collected medico folklore data guides in discovering novel drugs. It also provides information about biodiversity in pursuing conservation strategies and other biological utilization of the species. In many cases today, over-harvesting of medicinal plants, degradation of medicinal resources, and loss of traditional medical knowledge in local communities where these resources are found are common problems. However, during policies formation for continued use and conservation of plant wealth as well as standardization of appropriate drugs or dose-illness relationship, the knowledge of community-based ethnobotany is immensely utilized (Poonam and Singh, 2009). This traditional Econo-medical knowledge is important not only for its potential contribution to drug development and health care but also for development of socio-economic values of people. Thus, the documentation and inventory can be considered the information bank of traditional medicine.

KEYWORDS

- **biodiversity**
- **ethnobotanical**
- **medicinal plants**
- **traditional medicine**

REFERENCES

AICRPE (All India Coordinated Project on Ethnobiology). Final Technical Report, (Ministry of Environment and Forests, Government of India, New Delhi), 1992–1998.

Balandrin, M. F.; Kinghorn, A. D.; Farnsworth, N. R. Plant-Derived Natural Products in Drug Discovery and Development: An Overview; *ACS Publications*, **1993**.

Balick, M.J. Transforming Ethnobotany for the New Millennium. *Annals of the Missouri Botanical Garden*, **1996**, 58–66.

Behera, B. C.; Behera, B.; Nanda, B. K.; Sahoo, R. K.; Meher, A. Ethnomedicinal Plants of Gandhamardan Hills (Odisha): A Review. *The Pharm student* **2016**, *27*, 01–06.

Behera, S. K.; Panda, A.; Behera, S. K.; Misra, M. K. Medicinal Plants Used by the Kandhas of Kandhamal District of Orissa. *Indian Journal of Traditional Knowledge* **2006**, *5* (4), 519–528.

Bhatt, D. C.; Mitaliya, K. D.; Pandya, N. A.; Baxi, U. S. Herbal Therapy for Jaundice. *Advances in Plant Sciences.* **2001**, *14 (1)*, 123–126.

Bhogaonkar, P. Y.; Kadam, V. N. Ethnopharmacology of Banjara Tribe of Umarkhed Taluka, District Yavatmal, Maharashtra for Reproductive Disorders. *Indian J. Tradit. Know* **2006**, *5*, 336–341.

Bhushan, B.; Kumar, M. Ethnobotanically Important Medicinal Plants of Tehsil Billawar, District Kathua, J&K, India. *Journal of Pharmacognosy and Phytochemistry* **2013**, *2* (4), 14–21.

Bose, T. K.; Basu, R. K.; Biswas, B.; De, J. N.; Majumdar, B. C.; Datta, S. Cardiovascular Effects of Yellow Oleander Ingestion. *Journal of the Indian Medical Association* **1999**, *97* (10), 407–410.

Bringmann, G.; Dreyer, M.; Michel, M.; Tayman, F. S.; Brun, R. Ancistroheynine B and Two Further 7, 3′-Coupled Naphthylisoquinoline Alkaloids from Ancistrocladus Heyneanus Wall. *Phytochemistry* **2004**, *65* (21), 2903–2907.

Buss, A. D.; Cox, B.; Waigh, R. D. Natural Products as Leads for New Pharmaceuticals. *Burger's medicinal chemistry and drug discovery* **2003**.

Butler, M. S. Natural Products to Drugs: Natural Product-Derived Compounds in Clinical Trials. *Natural product reports* **2008**, *25* (3), 475–516.

Champion, S. H.; Seth, S. K. A Revised Survey of the Forest Types of India. *A revised survey of the forest types of India.* **1968**.

Chivian, E. *Biodiversity: Its Importance to Human Health*; Center for Health and the Global Environment, Harvard Medical School, Cambridge, MA., **2002**.

Chopra, R. N.; Chopra, I. C. *Indigenous Drugs of India*; Academic publishers, 1933.

Cragg, G. M.; Newman, D. J. Natural Products: A Continuing Source of Novel Drug Leads. *Biochimica et Biophysica Acta (BBA)-General Subjects* **2013**, *1830* (6), 3670–3695.

Das, S.; Choudhury, M. D. Ethnomedicinal Uses of Some Traditional Medicinal Plants Found in Tripura, India. *Journal of Medicinal Plants Research.* **2012**, *6* (35), 4908–4914.

Dey, K. L.; Mair, W. *The Indigenous Drugs of India,* 2nd Ed., Pama Primlane, Chronica Botanica: New Delhi, India, **1973**, 186–187.

Dwibedy, A. S.; Moharana, A.; Kumar, S.; Naik, S. K.; BarikΨ, D. P. Qualitative Estimation of Bioactive Compounds and Evaluation of Antimicrobial Activity of Strychnos Nux-Vomica L. Leaf Extracts. *Plant Science Research* **2015**, *37* (1&2), 65–70.

Fabricant, D. S.; Farnsworth, N. R. The Value of Plants Used in Traditional Medicine for Drug Discovery. *Environmental health perspectives.* **2001**, *109* (1): 69.

Farnsworth, N. R.; Akerele, R. O.; Bingel, A. S.; Soejarto, D. D.; Guo, Z. Medicinal Plants in Therapy. *Bulletin of the World Health Organization.* **1985**, *63,* 965–981.

Farnsworth, N.R.; Soejarto, D.D. Global Importance of Medicinal Plants. *The conservation of medicinal plants.* **1991**, 25–51.

Fleming, J. *A Catalogue of Indian Medicinal Plants and Drugs, with Their Names in the Hindusta'ni and Sanscrit Languages.* J. Cuthell Pub, **1812**.

Gairola, S.; Sharma, C. M.; Rana, C. S.; Ghildiyal, S. K.; Suyal, S. Phytodiversity (Angiosperms and Gymnosperms) in Mandal-Chopta Forest of Garhwal Himalaya, Uttarakhand, India. *Nature and Science.* **2010**, *8* (1), 1–17.

Glesinger, L. Medicine through Centuries. *Zagreb: Zora* **1954**, 21–38.

Gomes, A.; Das, R.; Sarkhel, S.; Mishra, R.; Mukherjee, S.; Bhattacharya, S.; Gomes, A. Herbs and Herbal Constituents Active against Snake Bite. *Indian journal of experimental biology* **2010**, *48* (9), 865.

Gupta, O. P.; Ali, M. M.; Ray, B. G.; Atal, C. K. Some Pharmacological Investigations of Embelin and Its Semisynthetic Derivatives. *Indian journal of physiology and pharmacology* **1977**, *21* (1), 31–39.

Harshberger, J.W. The purpose of ethnobotany. *Bot. Gaz.* **1896**, *21*, 146–158.

Harvey, A. L. Natural Products in Drug Discovery. *Drug discovery today* **2008**, *13* (19–20), 894–901.

Heymsfield, S. B.; Allison, D. B.; Vasselli, J. R.; Pietrobelli, A.; Greenfield, D.; Nunez, C. Garcinia Cambogia (Hydroxycitric Acid) as a Potential Antiobesity Agent: A Randomized Controlled Trial. *Jama* **1998**, *280* (18), 1596–1600.

Hoiling, C.; Schindler, D.; Walker, B. W.; Roughgarden, J. Biodiversity in the Functioning of Ecosystems: An Ecological Synthesis. *Biodiversity loss: economic and ecological issues* **1997**, 44.

Jain, A.; Katewa, S. S.; Sharma, S. K.; Galav, P.; Jain, V. Snakelore and Indigenous Snakebite Remedies Practiced by Some Tribals of Rajasthan. *Indian Journal of Traditional Knowledge* **2011**, *10* (2), 258–268.

Jain, S. K.; Srivastava, S. Indian Ethnobotanical Literature in Last Two Decades-A Graphic Review and Future Directions. *Ethnobotany,* **2001**, *13*, 1–8.

Jain, S.K. Ethnobotany. *Interdisciplinary Science Reviews* **1986**, *11* (3), 285–292.

Janaki Ammal, E. K. Introduction to the Subsistence Economy of India. *Man's role in changing face of the earth (ed. William LT Jr), University of Chicago Press, Chicago* **1956**, 324–335.

Kandari, L. S.; Gharai, A. K.; Negi, T.; Phondani, P. C. Ethnobotanical Knowledge of Medicinal Plants among Tribal Communities in Orissa. *India. J Forest Res Open Access* **2012**, *1* (104), 2.

Khurnbongmayum, A. O.; Khan, M. L.; Tripathi, R. S. Ethnomedicinal Plants in the Sacred Groves of Manipur. *Indian Journal of Traditional Knowledge* **2005**, *4* (1), 21–32.

Kumar, M.; Bussmann, R. W.; Mukesh, J.; Kumar, P. Ethnomedicinal Uses of Plants Close to Rural Habitation in Garhwal Himalaya, India. *Journal of Medicinal Plants Research* **2011**, *5* (11), 2252–2260.

Makhija, I. K.; Khamar, D. Anti-Snake Venom Properties of Medicinal Plants. *Der Pharmacia Lettre* **2010**, *2* (5), 399–411.

Mandal, K. K.; Kar, T.; Reddy, C. S.; Biswal, A. K. Polypetalous plant diversity of Mayurbhanj with special reference to ret species. *Threats and Concerns to Biodiversity*. **2012**, pp. 20–25.

Maruthi, K. R.; Krishna, V.; Manjunatha, B. K.; Nagaraja, Y. P. Traditional Medicinal Plants of Davanagere District, Karnataka with Reference to Cure of Skin Diseases. *Environment and Ecology* **2000**, *18* (2), 441–446.

McNeely, J. A.; Miller, K. R.; Reid, W. V.; Mittermeier, R. A.; Werner, T. B. *Conserving the World's Biological Diversity.* International Union for Conservation of Nature and Natural Resources, **1990**.

Mitra, S.; Mukherjee, S. K. Some Plants Used as Antidote to Snake Bite in West Bengal, India. *Diversity and Conservation of Plants and Traditional Knowledge* **2012**, 515–537.

Mohanta, R. K.; Rout, S. D.; Sahu, H. K. Ethnomedicinal Plant Resources of Similipal Biosphere Reserve, Orissa, India. *Zoos Print J.* **2006**, *21* (8), 2372–2374.

Newman, D. J.; Cragg, G. M.; Snader, K. M. The Influence of Natural Products upon Drug Discovery. *Natural product reports* **2000**, *17* (3), 215–234.

Panda, A.; Misra, M. K. Ethnomedicinal Survey of Some Wetland Plants of South Orissa and Their Conservation. *Indian Journal of Traditional Knowledge* **2011**, *10* (2), 296–303.

Panda, S. S.; Dhal, N. K.; Muduli, S. D. Ethnobotanical Studies in Nawarangpur District, Odisha, India. *International Journal of Medical Research and Review* **2014**, *2* (2), 257–276.

Panigrahy, J.; Behera, S. K.; Venugopal, A.; Leelaveni, A. Ethnomedicinal Study of Some Medicinal Plants from Kandhamal District, Odisha. *Int. J. Herbal Med* **2016**, *4* (5), 36–40.

Pattanaik, C.; Reddy, C. S.; Murthy, M. S. R.; Reddy, P. M. Ethno-Medicinal Observation among the Tribal People of Koraput District Orissa, India. *Research Journal of Botany* **2006**, *1* (3), 125–128.

Pei, S. Ethnobotany and Modernisation of Traditional Chinese Medicine. In *Paper at a Workshop on Wise Practices and Experiential Learning in the Conservation and Management of Himalayan Medicinal Plants, Kathmandu, Nepal*; **2002**; pp 15–20.

Petrovska, B. B. Historical Review of Medicinal Plants' Usage. *Pharmacognosy reviews* **2012**, *6* (11), 1.

Poonam, K.; Singh, G. S. Ethnobotanical Study of Medicinal Plants Used by the Taungya Community in Terai Arc Landscape, India. *Journal of Ethnopharmacology* **2009**, *123* (1), 167–176.

Prioreschi, P. *A History of Medicine: Roman Medicine*; Edwin Mellen Press. **1996**, Vol. 3.

Prusti, A. B.; Behera, K. K. Ethnobotanical Exploration of Malkangiri District of Orissa, India. *Ethnobotanical Leaflets **2007*** (1), 14.

Ramarao Naidu, B. V. A.; Seetharami Reddi, T. V. V.; Prasanthi, S. Folk Herbal Remedies for Rheumatoid Arthritis in Srikakulam District of Andhra Pradesh. *Ethnobotany* **2008**, *20*, 76–79.

Reddy, M. H.; Vijayalakshmi, K.; Venkataraju, R. R. Native Phytotherapy for Snakebite in Nallamalai's Eastern Ghats, India. *J Econ Taxon Bot* **1997**, *12*, 214–217.

Rout, S. D. Ethnobotany of Diversified Wild Edible Fruit Plants in Similipal Biosphere Reserve, Orissa. *Ethnobotany* **2007**, *19*, 137–139.

Rout, S. D.; Panda, T.; Mishra, N. Ethno-Medicinal Plants Used to Cure Different Diseases by Tribals of Mayurbhanj District of North Orissa. *Studies on Ethno-Medicine* **2009**, *3* (1), 27–32.

Rout, S. D.; Panda, S. K. Ethnomedicinal plant resources of Mayurbhanj district, Orissa. *Indian J Traditional Knowledge* **2010**, *9* (1), 68–72.

Routray, A.; Nayak, S. Ethnomedicinal Plants Resource of Stakosia RF of Karanjia Division, Mayurbhanj, Odisha. *Imperial Journal of Interdisciplinary Research* **2017**, *3* (4).

Sahu, A. R.; Behera, N.; Mishra, S. P. Use of Ethnomedicinal Plants by Natives of Bargarh District of Orissa, India. **2010**, *14*, 889–910.

Sahu, S. C.; Pattnaik, S. K.; Dash, S. S.; Dhal, N. K. Fibre-Yielding Plant Resources of Odisha and Traditional Fibre Preparation Knowledge– An Overview. *IJNPR* **2013**, *4* (4), 339–347.

Sarma, D. N. K.; Khosa, R. L. Immunomodulators of Plant Origin–a Review. *Ancient science of Life* **1994**, *13* (3–4), 326.

Saxena, H. O.; Brahmam, M. Eastern-Ghat flora of Orissa: An epitome of biodiversity worth conserving, in: *Proc. National Seminar on Conservation of Eastern Ghats.* **1998**, pp. 128–134.

Saxena, R. Tribes of central India. *Nutrition.* **1996**, 30 (2), 14.

Schippmann, U.; Leaman, D. J.; Cunningham, A. B. Impact of Cultivation and Gathering of Medicinal Plants on Biodiversity: Global Trends and Issues. *Biodiversity and the ecosystem approach in agriculture, forestry and fisheries* **2002**.

Schultes, R. E. The Role of the Ethnobotanist in the Search for New Medicinal Plants. *Lloydia* **1962**, *25*, 257–266.

Sen, S. K.; Behera, L. M. Ethnomedicinal Plants Used by the Tribals of Bargarh District to Cure Diarrhoea and Dysentery. *Indian Journal Of Traditional Knowledge* **2008**, *7* (3), 425–428.

Shishodia, S.; Aggarwal, B. B. Guggulsterone Inhibits NF-KB and IκBα Kinase Activation, Suppresses Expression of Anti-Apoptotic Gene Products, and Enhances Apoptosis. *Journal of Biological Chemistry* **2004**, *279* (45), 47148–47158.

Sinha, R.; Lakra, V. Wild Tribal Food Plants of Orissa. *Indian Journal of Traditional Knowledge* **2005**, *4* (3), 246–252.

Srivastava, R. Studying the Information Needs of Medicinal Plant Stakeholders in Europe. *Traffic Dispatches* **2000**, *15* (5), 13.

Srivastava, S. C.; Rout, N. Some Plants of Ethnopaediatric Importance in District Koraput, Orissa. *Nelumbo* **1994**, *36* (1–4), 166–168.

Subrat, N.; Iyer, M.; Prasad, R. The Ayurvedic Medicine Industry: Current Status and Sustainability. *Ecotech Services (India) Pvt. Ltd* **2002**, *63*.

Swargiary, A.; Boro, H.; Brahma, B. K.; Rahman, S. Ethno-Botanical Study of Anti-Diabetic Medicinal Plants Used by the Local People of Kokrajhar District of Bodoland Territorial Council, India. *Journal of Medicinal Plants Studies* **2013**, *1* (5), 51–58.

Thangadurai, D. Ethnobotanical Plants Used as Antidote for Poisonous Bites among the Tribals South Western Ghats, India. In *National conference on recent trends in spices & medicinal plant research, Calcutta, WB, India*; **1998**.

Tripathi, J.; Singh, R.; Ahirwar, R. P. Ethnomedicinal Study of Plants Used by Tribal Person for Diarrhoea Diseases in Tikamgarh District MP. *Journal of Medicinal Plants* **2017**, *5* (1), 248–253.

Tsakadze, D. M.; Abdusamatov, A.; Yunusov, S. Y. Alkaloids of Galanthus Caucasicus. *Chemistry of Natural Compounds* **1969**, *5* (4), 281–282.

Uprety, Y.; Asselin, H.; Boon, E. K.; Yadav, S.; Shrestha, K. K. Indigenous Use and Bio-Efficacy of Medicinal Plants in the Rasuwa District, Central Nepal. *Journal of Ethnobiology and Ethnomedicine* **2010**, *6* (1), 3.

Verpoorte, R. Pharmacognosy in the New Millennium: Lead finding and Biotechnology. *Journal of pharmacy and pharmacology* **2000**, *52* (3), 253–262.

CHAPTER 15

Structural Profiling of Bioactive Compounds with Medicinal Potential from Traditional Indian Medicinal Plants

C. SAREENA, A. ANJU SURESH, SWETHA SUNIL, and T. V. SUCHITHRA*

School of Biotechnology, National Institute of Technology Calicut (NITC), P.O. NIT Campus, Kozhikode–673601, Kerala, India,
**E-mail: drsuchithratv@nitc.ac.in*

ABSTRACT

There are many traditional medicinal plants containing potential bioactive phytocompounds which can be developed as better leads in future drug discovery process. However, studies on plants are very limited and marginal due to complexity in their chemical constituents and lack of knowledge about their mechanism of action. This chapter presents details of 100 major phytocompounds reported from Indian traditional plants with structure, bioactivities, and sources. An extensive literature survey was carried out in this regard; the data were collected from research publications (PubMed-NCBI) and also from multiple online databases (Dr. Dukes phytochemical and ethnobotanical database, MAPS Database, PubChem, ChemSpider, STITCH and Super Target). The study can be concluded that many Indian plants have been investigated for their potential use in different types of ailments and many of the compounds studied can be developed as future drugs.

15.1 INTRODUCTION

India has a unique position in the world, where a number of recognized indigenous systems of medicine namely Ayurveda, Siddha, Unani, Naturopathy, etc. are being used for health care of people (Tambekar and Dahikar, 2010). A number of polyherbal formulations from these ethnomedicinal practices have been proved as effective to prevent and to treat many diseases. Herbal medicines are considered to be less toxic, possess fewer side-effects than synthetic drugs and easily available at affordable cost (Pari and Umamaheswari, 2000). According to WHO, herbal medicines serve health needs of about 80 percent of the world's population; especially for millions of people in the vast rural areas of developing countries (Njeru et al., 2013). Approximately one-third of top 20 drugs available in the current market are derived from natural products, with a majority being obtained from plants (Ruiz-Torres et al., 2017).

Despite of well-known potentials of phytocompounds, advanced research is extremely lacking in this field. This can be attributed to the following reasons (i) undocumentation of knowledge in successful tribal herbal medicine causes its unavailability to scientific community. Moreover, impact of deforestation, urbanization, and modernization of the tribals also result in decline of the quantity of information acquired through generations, (ii) So far, only about one-third of the existing higher plants have been identified and named scientifically. Among them, majority of plants are unexplored yet. Out of estimated 250,000 to 350,000 plant species identified so far, only about 35,000 species are used worldwide for medicinal purposes (Kong et al., 2003), (iii) Most of plant species associated with rural people is on the verge of extinction. So linking of indigenous knowledge of medicinal plants to modern research activities is necessary to speed up rate of drug discovery programs (Kong et al., 2003). This chapter provides an overview of major phytocompounds from Indian traditional medicinal plants with their structure and source plants. It also gives concise information about their different bioactivities.

15.2 PLANTS USED IN INDIAN TRADITIONAL MEDICINE

The most widely used formulations in Indian traditional medicine are polyherbal in nature (Parasuraman, Thing, and Dhanaraj, 2014). The ingredient plants used in their preparation are usually selected based on experiences of traditional healers and information from ancient manuscripts. These experienced based claims should be replaced by evidence-based claims with scientific support. Based on multifaceted medicinal applications, 100 major phytocompounds of Indian traditional plants are considered here. Certain phytocompounds have been isolated and identified from many plants. For instance, β-sitosterol is reported from plants such as *Azadirachta indica, Pongamia pinnata, Acorus calamus, Lawsonia inermis, Camellia sinensis, Myristica fragrans, Stereospermum suaveolens, Mirabilis jalapa, Allamanda cathartica, Citrus limon, Lawsonia inermis, Bacopa monnieri, Santalum album,* etc. An extensive literature survey was done in this regard (Table 15.1); the data were collected from research publications (PubMed-NCBI) and also from multiple online databases (Dr. Dukes phytochemical and ethnobotanical database (https://phytochem.nal.usda.gov/phytochem/search), MAPS Database (http://www.mapsdatabase.com), PubChem (https://pubchem.ncbi.nlm.nih.gov/search/search.cgi), and ChemSpider (http://www.chemspider.com). Table 15.1 shows list of 100 phytocompounds with multiple medicinal properties and their plant sources. This information regarding plant sources and bioactivities of each compound will be helpful to select the most suitable plants with known active principles for preparing better formulations. Figure 15.1 contains structures of these 100 phytocompounds. This data helps us to find out related structures having a particular bioactivity for studies like Quantitative structure-activity relationships (QSAR). Taxonomical details of source plants are collected from International Plant Naming Index (IPNI) as shown in Table 15.2.

15.3 BIOACTIVITY STUDIES OF PHYTOCOMPOUNDS

This section highlights various bioactivities shown by 100 major phytocompounds from Indian traditional medicinal plants such as antibacterial, antifungal, antiviral, anticancerous, anti-inflammatory, antioxidant, anti-adiposity, antiparasitic, anthelminthic, antidiabetic, antifibrotic, antiarthritic, analgesic, antipyretic, antithrombotic, antiatherogenic, antihyperalgesic, antidepressant, cardioprotective, chemoprotective, pancreas protective, gastroprotective, nephroprotective, neuroprotective, hepatoprotective, lung protective, anesthetic, wound healing, etc.

15.3.1 ANTIMICROBIAL ACTIVITIES

Among 100 phytocompounds with multiple activities considered here, the most studied property is antimicrobial activity, and 75 compounds have been found as antimicrobial in nature. Out of them, 51 compounds show antibacterial properties while 25 compounds are antifungal in nature and only 2 compounds are known for their antiviral properties. Antibacterial compounds are acetic acid (Fraise et al., 2013), (Halstead et al., 2015), allicin (Ankri and Mirelman, 1999), azadirachtin (Sharma, Verma, and Ramteke, 2009), benzaldehyde (Jeyadevi et al., 2013), berberine (Domadia et al., 2008),

TABLE 15.1 Bioactivities of Major Phytocompounds From Indian Traditional Medicinal Plants

No	Name of compound	Source plants	Activity	References)
1.	Acetic acid	*Citrus limon* (L.) Burm. f., *Tamarindus indica* (L.), *Allium cepa* (L.), *Zingiber officinale* Roscoe, *Camellia sinensis* (L.) Kuntze	Antibacterial, Antifungal, Anti-inflammatory, Anti adiposity	Fraise et al. (2013); Halstead et al. (2015); Kumbhare and Sivakumar (2011); Bounihi et al. (2017)
2.	Acetophenone	*Camellia sinensis* (L.) Kuntze, *Cichorium intybus* (L.), *Syzygium aromaticum* (L.) Merr. & L.M. Perry	Anticancerous, Anti-inflammatory, Anti-alzeimers	Street, Sidana, and Prinsloo (2013); Street et al. (2013); Raghav and Singh (2017)
3.	Allicin	*Allium cepa* (L.), *Allium sativum* (L.)	Anticancerous, Antibacterial, Antifungal, Protective against cancer chemotherapy	Abdel-Daim et al. (2017); Ankri and Mirelman (1999); Aggarwal et al. (2011)
4.	Aloe emodin	*Aloe vera* (L.), *Cassia Tora* (L.), *Cassia occidentalis* (L.)	Anticancerous, Antiparasitic, Pancreas protective	Tamokou et al. (2013)
5.	α-phellandrene	*Elettaria cardamomum* Maton, *Trachyspermum ammi Sprague*, *Curcuma longa* (L.), *Lantana camara* (L.), *Mangifera indica* (L.), *Zingiber officinale* Roscoe, *Acorus calamus* (L.), *Camellia sinensis* (L.) Kuntze, *Foeniculum vulgare* Mill., *Citrus limon* (L.) Burm. f.	Anticancerous, Anti-inflammatory, Antibacterial, Antiadiposity	Hosseinzadeh et al. (2015); Zorica et al. (2014); Wang et al. (2017)
6.	α-pinene	*Piper nigrum* (L.), *Nigella sativa* (L.), *Anethum graveolens* (L.), *Myristica fragrans* Houtt., *Rhynchosia Minima* (L.), *Eucalyptus cinerea* F. Muell. ex Benth., *Terminalia chebula* Retz., *Citrus limon* (L.) Burm. f., *Elettaria cardamomum* Maton, *Trachyspermum ammi* Sprague, *Zingiber officinale* Roscoe	Anticancerous, Antibacterial	Chirathaworn and Kongcharoensuntorn (2007); Kovac et al. (2015)
7.	α-terpineol	*Citrus limon* (L.) Burm. f., *Piper nigrum* (L.), *Elettaria cardamomum* Maton, *Myristica fragrans* Houtt., *Curcuma longa* (L.), *Tamarindus indica* (L.), *Zingiber officinale* Roscoe, *Acorus calamus* (L.), *Camellia sinensis* (L.) Kuntze, *Foeniculum vulgare* Mill., *Syzygium aromaticum* (L.) Merr. & L.M. Perry, *Santalum album* (L.)	Anticancerous, Antibacterial	Taha and Eldahshan (2017)

TABLE 15.1 *(Continued)*

No	Name of compound	Source plants	Activity	References)
8.	Anthocyanin	*Prunus domestica* (L.)	Antioxidant, Anticancerous, Anti-inflammatory	Bowen-forbes et al. (2010); Oke-altuntas et al. (2017); Rahmatullah and Jahan, et al. (2010)
9.	Apigenin	*Eclipta alba* Hassk., *Stereospermum suaveolens* DC., *Jatropha gossypifolia* (L.), *Nyctanthes arbor-tristis* (L.), *Cardiospermum halicacabum* (L.), *Allium sativum* (L.), *Camellia sinensis* (L.) Kuntze, *Cichorium intybus* (L.), *Punica granatum* (L.)	Anticancerous	Horinaka et al. (2006)
10.	Ascorbic acid	*Colocasia esculenta* (L.) Schott, *Acacia nilotica* (L.) Delile, *Piper nigrum* (L.), *Nigella sativa* (L.), *Ocimum sanctum* (L.), *Hibiscus rosa-sinensis* (L.), *Terminalia chebula* Retz., *Curcuma longa* (L.), *Trigonella foenum-graecum* (L.), *Aloe vera* (L.), *Centella asiatica* (L.) Urb., *Citrus limon* (L.) Burm. f., *Anethum graveolens* (L.), *Tamarindus indica* (L.), *Aegle marmelos* (L.) Corrêa, *Allium cepa* (L.), *Mangifera indica* (L.), *Phyllanthus emblica* (L.), *Allium sativum* (L.), *Zingiber officinale* Roscoe, *Centella asiatica* (L.) Urb., *Bacopa monnieri* (L.) Wettst., *Acorus calamus* (L.), *Camellia sinensis* (L.) Kuntze, *Cichorium intybus* (L.), *Foeniculum vulgare* Mill., *Punica granatum* (L.), *Syzygium aromaticum* (L.) Merr. & L.M. Perry	Antioxidant, Anticancerous, Detoxifying	Jena et al. (2017); Thirupathi et al. (2016); Moricz et al. (2007)
11.	Azadirachtin	*Azadirachta indica* A. Juss.	Anti bacterial, Anti-cariogenic, Anti-helminthic, Anti-diabetic, Anti-oxidant, Astringent, Anti-viral, Cytotoxic, Anti inflammatory	Sharma et al. (2009); Lakshmi, Krishnan, and Rajendran (2015); Bhat (2008)

TABLE 15.1 *(Continued)*

No	Name of compound	Source plants	Activity	References)
12.	Azuline	*Acorus calamus* (L.), *Curcuma longa* (L.)	Antioxidant, Antiviral	Baqueiro-peña and Guerrero-beltrán (2017); Peet, Selyutina, and Bredihhin (2016)
13.	Benzaldehyde	*Cardiospermum halicacabum* (L.), *Camellia sinensis* (L.) Kuntze, *Hemidesmus indicus* (L.) R. Br., *Syzygium aromaticum* (L.) Merr. & L.M. Perry	Anticancerous, Antibacterial	Salvador et al. (2017); Jeyadevi et al. (2013)
14.	Berberine	*Berberis aristata* DC.	Anticancerous, Anti-alzheimers, Antidiabetic, Antibacterial	Ruan et al. (2017); Jiang and Gao (2017); Yuzhen Jia (2016); Domadia et al. (2008)
15.	B-amyrin	*Sapindus emarginatus* Vahl, *Nigella sativa* (L.)	Antibacterial, Anti-fibrotic, Anti-inflammatory, Anticancerous	Perumal Samy and Gopalakrishnakone (2010); Bandaranayake (2002); Sah and Verma (2012)
16.	β -caryophyllene	*Curcuma longa* (L.), *Trachyspermum ammi* Sprague, *Lantana camara* (L.), *Zingiber officinale* Roscoe, *Syzygium aromaticum* (L.) Merr. & L.M. Perry, *Terminalia chebula* Retz., *Punica granatum* (L.), *Piper betle* (L.)	Anti-inlammatory, Antioxidant, Antibacterial, Anticancerous, Antifungal, Antiarthritic, Analgesicantipyretic	Shen et al. (2017); Tepe et al. (2005); Legault (2007); Tepe et al. (2005); Maroon and Bost (2010); Bandaranayake (2002)
17.	β-elemene	*Piper betle* (L.), *Picrorhiza kurroa* Royle ex Benth.	Anticancerous, Antioxidant	Yu et al. (2017); Bayala et al. (2014); Kumar et al. (2015); Kelly et al. (2016)
18.	β-sitosterol	*Azadirachta indica* A. Juss., *Pongamia pinnata* (L.) Pierre, *Acorus calamus* (L.), *Lawsonia inermis* (L.), *Camellia sinensis* (L.) Kuntze, *Myristica fragrans* Houtt., *Stereospermum suaveolens* DC., *Mirabilis jalapa* (L.), *Allamanda cathartica* (L.), *Citrus limon* (L.) Burm. f., *Bacopa Monnieri, Santalum album* (L.), *Ocimum sanctum* (L.), *Trigonella foenum-graecum* (L.), *Aloe vera* (L.), *Centella asiatica* (L.) Urb., *Cassia Tora* (L.), *Terminalia*	Antibacterial, Anticancerous, Antinflammatory cardioprotective	Zhang et al. (2014); Oloche et al. (2016); Prasad, Kumar, and Kodidhela (2015)

TABLE 15.1 *(Continued)*

No	Name of compound	Source plants	Activity	References)
		bellirica (Gaertn.) Roxb., *Withania somnifera* (L.) Dunal, *Paeonia lactiflora* Pall., *Lantana camara* (L.), *Thevetia neriifolia* Juss. ex A. DC., *Jatropha gossypifolia* (L.), *Allium cepa* (L.), *Phyllanthus emblica* (L.) (*Emblica Officinalis*), *Allium sativum* (L.), *Zingiber officinale* Roscoe, *Centella asiatica* (L.) Urb., *Terminalia arjuna* (Roxb. ex DC.) Wight & Arn., *Caesalpinia pulcherrima* (L.) Sw., *Cichorium intybus* (L.)		
19.	Betulinic acid	*Centella asiatica* (L.) Urb., *Tecomella undulata* (Sm.) Seem., *Ziziphus mauritiana* Lam., *Lantana camara* L., *Centella asiatica* (L.) Urb., *Punica granatum* (L.), *Syzygium aromaticum* (L.) Merr. & L.M. Perry, *Chlorophytum Tuberosum, Trachyspermum ammi* Sprague	Antibacterial, Anicancerous, Antiparasite, Anti-psoriasis	Cowan (1999); Tamokou et al. (2013); Rahmatullah and Samarrai, et al. (2010)
20.	Borneol	*Kaempferia galanga* (L.), *Curcuma longa* (L.), *Annona squamosa* (L.), *Acorus calamus* (L.), *Santalum album* (L.) *Zingiber officinale* Roscoe, *Trachyspermum ammi* Sprague	Antibacterial, Antioxidant, Antithrombotic, Drug delivery	Tepe et al. (2005); Fabricant and Farnsworth (2001)
21.	Caffeic acid	*Citrus limon* (L.) Burm. f., *Allium cepa* (L.), *Allium sativum* (L.), *Zingiber officinale* Roscoe, *Camellia sinensis* (L.) Kuntze, *Cichorium intybus* (L.), *Foeniculum vulgare* Mill., *Punica granatum* (L.), *Nigella sativa* (L.), *Trachyspermum ammi* Sprague, *Myristica fragrans* Houtt.	Antioxidant, Anticancerous, Antiatherogenic	YJ et al. (2001); Tyszka-czochara et al. (2017); Mwonjoria et al. (2014)
22.	Camphor	*Trachyspermum ammi* Sprague	Anti-inflammatory, Analgesic, Anti-infective	Maroon and Bost (2010); Maroon and Bost (2010); Hamidpour et al. (2013)
23.	Capsaicin	*Capsicum frutescens* (L.)	Antibacterial, Antinflammatory, Analgesic	Kalia et al. (2012); Maroon and Bost (2010)

TABLE 15.1 *(Continued)*

No	Name of compound	Source plants	Activity	References)
24.	Carvacrol	*Nigella sativa* (L.), *Ocimum sanctum* (L.), *Piper betle* (L.), *Piper nigrum* (L.), *Camellia sinensis* (L.) Kuntze, *Trachyspermum ammi* Sprague, *Anethum graveolens* (L.)	Anticancerous, Anesthetic, Antibacterial	Kai Fana, Xiaolei Lib et al. (2015); Bianchini et al. (2017); Vasconcelos et al. (2017)
25.	Chlorogenic acid	*Withania somnifera* (L.) Dunal, *Allium sativum* (L.), *Zingiber officinale* Roscoe, *Camellia sinensis* (L.) Kuntze, *Cichorium intybus* (L.), *Foeniculum vulgare* Mill., *Punica granatum* (L.), *Allium sativum* (L.), *Camellia sinensis* (L.) Kuntze, *Pongamia pinnata* (L.) Pierre	Anti-inflammatory, Antioxidant, Antimicrobial, Anticancerous	Rahmatullah and Samarrai, et al. (2010); Chandramohan et al. (2014); Rosa et al. (2016)
26.	Chrysophanol	*Allium cepa* (L.), *Tamarindus indica* (L.), *Syzygium aromaticum* (L.) Merr. & L.M. Perry	Antifungal, Hepatoprotective, Lipid lowering, Lung protective	Duraipandiyan and Ignacimuthu (2010); Danish et al. (2011)
27.	1, 8-cineole	*Eucalyptus cinerea* F. Muell. ex Benth., *Citrus limon* (L.) Burm. f., *Piper nigrum* (L.), *Elettaria cardamomum* Maton, *Zingiber officinale* Roscoe, *Curcuma longa* (L.), *Acorus calamus* (L.), *Foeniculum vulgare* Mill.	Antioxidant, Antimicrobial, Anticancerous	Mann and Markham (1998); Satyal et al. (2011)
28.	Cinnamaldehyde	*Curcuma longa* (L.), *Aloe vera* (L.)	Anticancerous, Antifungal, Antibacterial	Rahmatullah and Jahan, et al. (2010); Domadia et al. (2007)
29.	Cinnamic acid	*Camellia sinensis* (L.) Kuntze, *Foeniculum vulgare* Mill., *Punica granatum* (L.), *Piper nigrum* (L.), *Citrus limon* (L.) Burm. f.	Anticancerous, Anti-alzheimers, Antibacterial, Antiinflammatory	Gabriele et al. (2017); Lan et al. (2017); Cowan (1999); Shen et al. (2013)
30.	Citric acid	*Mangifera indica* (L.), *Cichorium intybus* (L.), *Euphorbia tirucalli* (L.), *Foeniculum vulgare* Mill., *Punica granatum* (L.), *Citrus limon* (L.) Burm. f., *Piper nigrum* (L.), *Elettaria cardamomum* Maton	Antibacterial, Anticancerous	Papetti et al. (2013); Bandaranayake (2002)
31.	Citronellal	*Elettaria cardamomum* Maton, *Aegle marmelos* (L.) Corrêa, *Zingiber officinale* Roscoe, *Foeniculum vulgare* Mill., *Citrus limon* (L.) Burm. f.	Antibacterial, Antifungal, Antihyperalgesic	Nascimento et al. (2000); Bardaweel et al. (2014)
32.	Corilagin	*Picrorhiza kurroa* Royle ex Benth., *Aegle marmelos* (L.) Corrêa	Antibacterial, Antiinfammatory, Nematicidal, Liver fibrosis	Samy and Gopalakrishnakone (2010); Bandaranayake (2002)

TABLE 15.1 *(Continued)*

No	Name of compound	Source plants	Activity	References)
33.	Cuminaldehyde	*Curcuma longa* (L.), *Zingiber officinale* Roscoe	Antibacterial, Anticancerous	Skariyachan, Mahajanakatti, et al. (2011); Yangui et al. (2017)
34.	Curcumin	*Curcuma longa* (L.), *Pongamia pinnata* (L.) Pierre, *Abrus precatorius*	Antibacterial, Anticancer, Antioxidant, Anti-inflammatory, Wound healing	Dipti Rai, Jay Kumar Singh, Nilanjan Roy (2008); Rooney and Ryan (2005); Cowan (1999); Shen et al. (2017),
35.	Cycloartenol	*Allium cepa* (L.), *Curcuma longa* (L.), *Cardiospermum halicacabum* (L.)	Anti-inflammatory, Antidiabetic	Forouzanfar et al. (2014)
36.	δ-cadinene	*Piper nigrum* (L.), *Syzygium aromaticum* (L.) Merr. & L.M. Perry, *Camellia sinensis* (L.) Kuntze, *Vitex Peduncularis* (W.), *Myristica fragrans* Houtt., *Allium cepa* (L.)	Antiparasitic, Antibacterial, Antioxidant	Guo et al. (2017); Zatelli et al. (2016); Jena et al. (2017)
37.	Diallyl sulfide	*Allium sativum* (L.), *Allium cepa* (L.)	Anticancerous, Antidepressant	Wei et al. (2017); Hosseinzadeh et al. (2017).
38.	Diallyl disulfide	*Allium sativum* (L.), *Allium cepa* (L.)	Anticancerous, Antioxidant, Hepatoprotective	Wei et al. (2017); Sumit and Khan (2016)
39.	Dichamanetin	*Piper sarmentosum* Stars	Anticancerous, Antibacterial, Antioxidant	Yong et al. (2013); Lock and Harry (2008); Yong et al. (2013)
40.	Dillapiole	*Stereospermum suaveolens* DC., *Trachyspermum ammi* Sprague	Insecticidal	George et al. (2014); Domingos et al. (2014)
41.	Ellagic acid	*Terminalia chebula* Retz., *Syzygium aromaticum* (L.) Merr. & L.M. Perry, *Terminalia bellirica* (Gaertn.) Roxb., *Phyllanthus emblica* (L.) (*Emblica Officinalis*), *Terminalia arjuna* Roxb. ex DC. Wight & Arn., *Caesalpinia pulcherrima* (L.) Sw., *Euphorbia tirucalli* (L.), *Punica granatum* (L.), *Aloe vera* (L.), *Cassia Tora* (L.), *Cassia occidentalis* (L.)	Antimicrobial, Anticancerous, Neuroprotective, Cardioprotective	Perumal Samy and Gopalakrishnakone (2010); Atanasov et al. (2015)
42.	Emodin	*Allium sativum* (L.), *Punica granatum* (L.), *Vitex Peduncularis* (W.)	Antibacterial, Antiparasitic, Antioxidant	Ji et al. (2017); Danish et al. (2011); Cai et al. (2004)

TABLE 15.1 *(Continued)*

No	Name of compound	Source plants	Activity	References)
43.	Epicatechin	*Delonix elata* Gamble	Antioxidant	Guo et al. (2017); Grzesik et al. (2018)
44.	Epigallocatechin	*Zingiber officinale* Roscoe	Antioxidant	Grzesik et al. (2018)
45.	Epigallocatechin galate	*Camellia sinensis* (L.) Kuntze	Anticancerous, Antioxidant	Chao et al. (2017); Wang et al. (2018); Grzesik et al. (2018)
46.	Esculetin	*Foeniculum vulgare* Mill., *Nigella sativa* (L.), *Foeniculum vulgare* Mill.	Anticancerous, Antioxidant	Liang et al. (2017); Han et al. (2017)
47.	Eugenol	*Ocimum sanctum* (L.), *Curcuma longa* (L.), *Piper betle* (L.) *Annona squamosa* (L.), *Anethum graveolens* (L.), *Lantana camara* (L.), *Acorus calamus* (L.), *Camellia sinensis* (L.) Kuntze, *Foeniculum vulgare* Mill., *Syzygium aromaticum* (L.) Merr. & L.M. Perry, *Santalum album* (L.), *Viola odorata* (L.), *Myristica fragrans* Houtt.	Antipyretic, Antibacterial, Antiparasitic, Anti-inflammatory	Jiaix Feng (1987); Charan Raja et al. (2017); Devi et al. (2010)
48.	Ferulic acid	*Terminalia chebula* Retz., *Anethum graveolens* (L.), *Allium cepa* (L.), *Allium sativum* (L.), *Zingiber officinale* Roscoe, *Cichorium intybus* (L.), *Foeniculum vulgare* Mill., *Punica granatum* (L.), *Viola odorata* (L.), *Cassia fistula* (L.)	Neroprotective, Antioxidant	Tańska et al. (2018)
49.	Gallic acid	*Abrus precatorius* (L.), *Casuarina equisetifolia* (L.), *Euphorbia Hirta* (L.), *Lawsonia inermis* (L.), *Abutilon indicum* (L.) Sweet, *Terminalia chebula* Retz., *Picrorhiza kurroa* Royle ex Benth., *Nymphaea stellata* Willd., *Punica granatum* (L.), *Terminalia bellirica* (Gaertn.) Roxb., *Paeonia lactiflora* Pall., *Mangifera indica* (L.), *Phyllanthus emblica* (L.) (*Emblica Officinalis*), *Terminalia arjuna* Roxb. ex DC. Wight & Arn., *Caesalpinia pulcherrima* (L.) Sw., *Camellia sinensis* (L.) Kuntze, *Syzygium aromaticum* (L.) Merr. & L.M. Perry	Gastroprotective, Anticancerous, Antioxidant, Antibacterial	Hussain et al. (2012); Ranilla et al. (2012); Stapleton et al. (2004)
50.	Genistein	*Prunus puddum* Roxb. ex Wall., *Swertia chirata* C. B. Clarke	Anticancerous, Antibacterial	Li et al. (2017); Abreu et al. (2012)

TABLE 15.1 *(Continued)*

No	Name of compound	Source plants	Activity	References)
51.	Geranial (citral, neral)	*Tamarindus indica* (L.), *Zingiber officinale* Roscoe, *Camellia sinensis* (L.) Kuntze, *Syzygium aromaticum* (L.) Merr. & L.M. *Perry, Citrus limon* (L.) Burm. f.	Antibacterial	Gupta et al. (2017)
52.	Germacrene D	*Piper betle* (L.), *Pongamia pinnata* (L.) Pierre	Antioxidant, Lipase inhibitory, Antibacterial	Mendes et al. (2017); Wang et al. (2017); Zorica et al. (2014)
53.	Glycyrrhizin	*Glycyrrhiza glabra* (L.)	Antibacterial, Anti-inflammatory	Naidu et al. (2009); Asl and Hosseinzadeh (2008)
54.	Guaiacol	*Camellia sinensis* (L.) Kuntze, *Santalum album* (L.), *Peganum harmala* (L.)	Antibacterial, Antioxidant	Asl and Hosseinzadeh (2008)
55.	Hederagenin	*Prunus domestica* (L.), *Nigella sativa* (L.)	Anticancerous, Antibacterial	Liu et al. (2014)
56.	Hyperoside (hyperin)	*Hemidesmus indicus* (L.) R. Br., *Azadirachta indica* A. Juss., *Camellia sinensis* (L.) Kuntze, *Cichorium intybus* (L.), *Syzygium aromaticum* (L.) Merr. & L.M. Perry	Antioxidant, Antiinflammatory hepatoprotective	Rahmatullah and Jahan, et al. (2010); Niu et al. (2017)
57.	Isoquercitrin	*Solanum incanum* (L.), *Stereospermum suaveolens* DC., *Camellia sinensis* (L.) Kuntze, *Foeniculum vulgare* Mill., *Punica granatum* (L.), *Syzygium aromaticum* (L.) Merr. & L.M. *Perry, Acacia nilotica* (L.) Delile, *Citrus limon* (L.) Burm. f.	Anticancerous	Amado et al. (2009)
58.	Isorhamnetin	*Solanum incanum* (L.), *Anethum graveolens* (L.), *Calotropis procera* (Aiton) W. T. Aiton	Antioxidant, Anticancerous, Cardioprotective	R. Guo et al. (2017); Mwonjoria et al. (2014)
59.	Kaempferol	*Allamanda cathartica* (L.), *Solanum incanum* (L.), *Stereospermum suaveolens* DC., *Azadirachta indica* A. Juss. *Mangifera indica* (L.), *Sapindus emarginatus* Vahl, *Trigonella foenum-graecum* (L.), *Centella asiatica* (L.) Urb. *Anethum graveolens* (L.), *Bombax ceiba* (L.), *Cassia fistula* (L.), *Casuarina equisetifolia* (L.), *Nymphaea stellata* Willd. *Paeonia lactiflora* Pall., Thevetia *neriifolia Juss*. ex A. DC. (*Cascabela Thevetia, Thevetia Peruviana*), *Allium cepa* (L.), *Phyllanthus emblica* (L.) (*Emblica Officinalis*), *Allium*	Anticancerous, Anti-inflammatory	Guo et al. (2017)

TABLE 15.1 *(Continued)*

No	Name of compound	Source plants	Activity	References)
		sativum (L.), *Zingiber officinale* Roscoe *Centella asiatica* (L.) Urb., *Camellia sinensis* (L.) Kuntze, *Cichorium intybus* (L.), *Foeniculum vulgare* Mill. *Punica granatum* (L.), *Syzygium aromaticum* (L.) Merr. & L.M. *Perry, Viola odorata* (L.), *Calotropis procera* (Aiton) W. T. Aiton		
60.	Lapachol	*Tecomella undulata* (Sm.) Seem., *Stereospermum suaveolens* DC., *Piper nigrum* (L.)	Antifungal, Anticancerous	Silva et al. (2016); Epifano and Genovese (2013)
61.	Lauric acid	*Mangifera indica* (L.), *Zingiber officinale* Roscoe, *Camellia sinensis* (L.) Kuntze, *Lawsonia inermis* (L.), *Myristica fragrans* Houtt.	Antioxidant, Blood pressure lowering	Alves et al. (2017)
62.	Licoricidin	*Acacia nilotica* (L.) Delile	Antibacterial, Antioxidant, Antifungal, Anti-inflammatory	Nomura et al. (2002); Saxena (2005)
63.	Limonene	*Premna Tomentosa* Willd., *Trachyspermum ammi* Sprague, *Myristica fragrans* Houtt., *Nigella sativa* (L.), *Curcuma longa* (L.) *Anethum graveolens* (L.), *Elettaria cardamomum* Maton, *Rhynchosia Minima* (L.), *Tamarindus indica* (L.) *Mangifera indica* (L.), *Zingiber officinale* Roscoe, *Acorus calamus* (L.), *Camellia sinensis* (L.) Kuntze, *Foeniculum vulgare* Mill. *Annona squamosa* (L.)	Anticancerous, Antinflammatory, Antioxidant, Anti diabetic, Anti-alzeimers	Suh et al. (2017); Gundidza et al. (2009); Suh, Chon, and Choi (2017)
64.	Linalool	*Nyctanthes arbor-tristis* (L.), *Elettaria cardamomum* Maton, *Trachyspermum ammi* Sprague, *Myristica fragrans* Houtt., *Allium sativum* (L.), *Acorus calamus* (L.), *Camellia sinensis* (L.) Kuntze, *Foeniculum vulgare* Mill., *Syzygium aromaticum* (L.) Merr. & L.M. Perry, *Anethum graveolens* (L.)	Anti-inflammatory, Antimicrobial, Antioxidant, Anticancerous	Aytac et al. (2017); Rajeswara Rao et al. (2002); Kumar et al. (2015)
65.	Linoleic acid	*Morinda Tinctoria* Roxb, *Ocimum sanctum* (L.), *Pongamia pinnata* (L.) Pierre, *Glycyrrhiza glabra* (L.)	Antioxidant, Anticholinesterase, Antibacterial.	Shabir et al. (2011); Ji, Li, and Zhang (2009); Ji et al. (2005)

TABLE 15.1 *(Continued)*

No	Name of compound	Source plants	Activity	References)
66.	Linolenic acid	*Glycyrrhiza glabra* (L.)	Cardioprotective, Hypertension	Ji et al. (2005)
67.	Luteolin	*Lawsonia inermis* (L.), *Solanum incanum* (L.), *Stereospermum suaveolens* DC., *Trigonella foenum-graecum* (L.), *Terminalia arjuna* (Roxb. ex DC.) Wight & Arn., *Abutilon indicum* (L.) Sweet, *Punica granatum* (L.), *Cardiospermum halicacabum* (L.), *Stereospermum suaveolens* DC.	Anticancerous, Antiepileptic	Tamokou et al. (2013); Mwonjoria et al. (2014)
68.	Mukurozioside iib	*Citrus limon* (L.) Burm. f. *Azadirachta indica* A. Juss.	Spermicide	Bhat (2008)
69.	Myricetin	*Azadirachta indica* A. Juss., *Allium sativum* (L.), *Zingiber officinale* Roscoe, *Caesalpinia pulcherrima* (L.) Sw., *Camellia sinensis* (L.) Kuntze, *Punica granatum* (L.), *Syzygium aromaticum* (L.) Merr. & L.M. Perry, *Euphorbia hirta* (L.), *Myristica fragrans* Houtt.	Anticancerous, Nephroprotective, Anesthetic	Jayaraman et al. (2010); Street et al. (2013)
70.	1, 4-Naphthalenedione	*Holoptelea integrifolia* (Roxb.) Planch.	Antibacterial, Antioxidant	Vinod et al. (2010)
71.	Nerol (geraniol)	*Annona squamosa* (L.), *Elettaria cardamomum Maton, Lantana camara* (L.), *Tamarindus indica* (L.), *Anethum graveolens* (L.), *Mangifera indica* (L.), *Allium sativum* (L.), *Zingiber officinale* Roscoe, *Camellia sinensis* (L.) *Kuntze, Foeniculum vulgare* Mill., *Santalum album* (L.), *Ocimum sanctum* (L.), *Piper nigrum* (L.)	Hepatoprotective, Anticancerous, Antibacterial, Antihyperglycemic	Ceyhan and Canbek (2017); Queiroz et al. (2017); Rajeswara Rao et al. (2002); Satyal et al. (2011)
72.	Nerolidol	*Zingiber officinale* Roscoe, *Camellia sinensis* (L.) Kuntze, *Nigella sativa* (L.), *Nyctanthes arbor-tristis* (L.)	Antinflammatory, Antioxidant	Javed et al. (2016); Nogueira et al. (2013)
73.	O-coumaric acid	*Foeniculum vulgare* Mill., *Nyctanthes arbor-tristis* (L.), *Curcuma longa* (L.)	Antioxidant, Anticancerous	Sen et al. (2015)

TABLE 15.1 *(Continued)*

No	Name of compound	Source plants	Activity	References)
74.	Oleanolic acid	*Sapindus emarginatus* Vahl, *Nyctanthes arbor-tristis* (L.), *Nymphaea stellata* Willd., *Pergularia Daemia, Punica granatum* (L.), *Allium cepa* (L.), *Allium sativum* (L.), *Terminalia arjuna* (Roxb. ex DC.) Wight & Arn., *Syzygium aromaticum* (L.) Merr. & L.M. Perry, *Trachyspermum ammi* Sprague	Anti-inflammatory, Antioxidant, Anticancerous,	Bogacz et al. (2016); Borchardt et al. (2009); Gao et al. (2016)
75.	P-coumaric acid	*Terminalia chebula* Retz., *Curcuma longa* (L.), *Trigonella foenum-graecum* (L.), *Aloe vera* (L.), *Punica granatum* (L.), *Allium cepa* (L.), *Mangifera indica* (L.), *Allium sativum* (L.), *Zingiber officinale* Roscoe, *Foeniculum vulgare* Mill., *Citrus limon* (L.) Burm. f., *Piper nigrum* (L.), *Prunus puddum* Roxb. ex Wall.	Antioxidant, Anticancerous	Alves et al. (2013); Kong et al. (2013)
76.	P-cymene	*Elettaria cardamomum* Maton, *Trachyspermum ammi* Sprague, *Myristica fragrans* Houtt., *Nyctanthes arbor-tristis* (L.), *Curcuma longa* (L.), *Anethum graveolens* (L.), *Lantana camara* (L.), *Aegle marmelos* (L.) Corrêa, *Zingiber officinale* Roscoe *Acorus calamus* (L.), *Foeniculum vulgare* Mill., *Terminalia chebula* Retz., *Citrus limon* (L.) Burm. f.	Antiparasitic, Neuroprotective, Antibacterial, Antifungal	Souhaiel et al. (2017); Silva et al. (2017); Barbosa et al. (2017); Yangui et al. (2017)
77.	Pectin	*Abrus precatorius* (L.), *Allium cepa* (L.), *Phyllanthus emblica* (L.)*Centella asiatica* (L.) Urb., *Myristica fragrans* Houtt., *Tamarindus indica* (L.), *Allium sativum* (L.), *Centella asiatica* (L.) Urb., *Cichorium intybus* (L.), *Foeniculum vulgare* Mill., *Punica granatum* (L.)*Piper longum* (L.), *Rhoeo spathacea* (Sw.) Stearn	Antidiarrheal, Hepatoprotective, Antioxidant, Antibacterial	Xu et al. (2015); Atanasov et al. (2015)
78.	Phenol	*Acorus calamus* (L.), *Santalum album* (L.), *Prunus domestica* (L.), *Picrorhiza kurroa* Royle ex Benth., *Pongamia pinnata* (L.) Pierre	Analgesic, Antibacterial	Rahmatullah and Jahan, et al. (2010); Cai et al. (2004)

TABLE 15.1 *(Continued)*

No	Name of compound	Source plants	Activity	References)
79.	Piperine	*Piper longum* (L.), *Plumbago zeylanica* (L.)	Antibacterial, Hepatoprotective	Mirza et al. (2011); Bedada and Boga (2017)
80.	Protocatechuic acid	*Allium cepa* (L.), *Foeniculum vulgare* Mill., *Punica granatum* (L.), *Casuarina equisetifolia* (L.), *Prunus puddum* Roxb. ex Wall.	Cardioprotective, Anticancerous, Antioxidant	Mwonjoria et al. (2014); Rahmatullah and Samarrai, et al. (2010)
81.	Punicalagin	*Terminalia chebula* Retz., *Acacia nilotica* (L.) Delile	Antiglycemic, Anti-inflammatory, Antioxidant, Antibacterial, Anticancerous	Bandaranayake (2002); Xu et al. (2017); Adaramoye et al. (2017)
82.	Pyrogallol	*Tephrosia purpurea* (L.) Pers., *Allamanda cathartica* (L.)	Antimicrobial, Antioxidant	Bai et al. (2015); Cowan (1999)
83.	Quercetin	*Citrus limon* (L.) Burm. f., *Acacia nilotica* (L.) Delile, *Anethum graveolens* (L.) *Bombax ceiba* (L.), *Casuarina equisetifolia* (L.), *Solanum incanum* (L.), *Stereospermum suaveolens* DC. *Saraca asoca* (Roxb.) W. J. de Wilde, *Mangifera indica* (L.), *Abutilon indicum* (L.) Sweet, *Viola odorata* (L.) *Sapindus emarginatus* Vahl, *Nyctanthes arbor-tristis* (L.), *Trigonella foenum-graecum* (L.), *Azadirachta indica* A. Juss., *Paeonia lactiflora* Pall., *Thevetia neriifolia* Juss. ex A. DC., *Phyllanthus emblica* (L.), *Allium sativum* (L.), *Zingiber officinale* Roscoe, *Cassia occidentalis* (L.), *Caesalpinia pulcherrima* (L.) Sw. *Camellia sinensis* (L.) Kuntze, *Cichorium intybus* (L.), *Foeniculum vulgare* Mill., *Litsea glutinosa* (Lour.) C. B. Rob., *Punica granatum* (L.), *Syzygium aromaticum* (L.) Merr. & L.M. Perry, *Tecomella undulata* (Sm.) Seem., *Nymphaea stellata* Willd., *Pithecellobium dulce* (Roxb.) Benth.	Antioxidant, Anti-inflammatory, Antibacterial, Antiviral, Gastroprotective	Guo et al. (2017); Suriyanarayanan et al. (2013); Rahmatullah and Jahan, et al. (2010)
84.	Quercitrin	*Euphorbia Hirta* (L.), *Camellia sinensis* (L.) Kuntze, *Anethum graveolens* (L.), *Cassia Tora* (L.), *Azadirachta indica* A. Juss., *Cichorium intybus* (L.)us, *Hibiscus rosa-sinensis* (L.), *Nyctanthes arbor-tristis* (L.)	Antioxidant, Antibacterial, Gastroprotective	Erkekoglou et al. (2017); Rahmatullah and Samarrai, et al. (2010)

TABLE 15.1 *(Continued)*

No	Name of compound	Source plants	Activity	References)
85.	Rhein	*Aloe vera* (L.), *Cassia Tora* (L.), *Cassia occidentalis* (L.) *Rhoeo discolor* (L'Hér.) Hance in Walp., *Rhoeo spathacea* (Sw.) Stearn	Anticancer, Alzheimers, Antifungal	Pérez-Areales et al. (2017); Abu-Eittah et al. (2014)
86.	Rubiadin	*Trigonella foenum-graecum* (L.), *Tephrosia purpurea* (L.) Pers.	Antifungal, Anticancer	Siddiqui et al. (2011)
87.	Rutin	*Citrus limon* (L.) Burm. f., *Ficus benghalensis* (L.), *Stereospermum suaveolens* DC., *Allium sativum* (L.), *Punica granatum* (L.), *Rhoeo spathacea* (Sw.) Stearn, *Morinda Tinctoria* Roxb, *Azadirachta indica* A. Juss., *Allium cepa* (L.), *Phyllanthus emblica* (L.), *Caesalpinia pulcherrima* (L.) Sw., *Cichorium intybus* (L.)us, *Foeniculum vulgare* Mill., *Hemidesmus indicus* (L.) R. Br., *Tecomella undulata* (Sm.) Seem., *Viola odorata* (L.), *Piper nigrum* (L.)	Antiallergic, Anti-inflammatory, Antioxidant, Antibacterial, Anticancerous	R. Guo et al. (2017); Jayaraman et al. (2010); Aggarwal et al. (2011)
88.	Sabinene	*Elettaria cardamomum* Maton, *Myristica fragrans* Houtt. *Acorus calamus* (L.), *Anethum graveolens* (L.), *Foeniculum vulgare* Mill., *Zingiber officinale* Roscoe, *Piper betle* (L.), *Piper nigrum* (L.)	Antibacterial	Alizadeh and Abdollahzadeh (2017)
89.	Sapogenin	*Morinda Tinctoria* Roxb	Reduction of plasma cholesterol and bile salt concentration	Bandaranayake (2002)
90.	Scopoletin	*Morinda Tinctoria* Roxb, *Anethum graveolens* (L.), *Withania somnifera* (L.) Dunal, *Cichorium intybus* (L.), *Foeniculum vulgare* Mill., *Viola odorata* (L.), *Azadirachta indica* A. Juss., *Stereospermum suaveolens* DC., *Cassia fistula* (L.)	Neuroprotective, Anticancer	Arunachalam et al. (2015); Shi et al. (2017)
91.	Sinapic acid	*Cichorium intybus* (L.), *Foeniculum vulgare* Mill. *Ichnocarpus Frutescens*, *Stereospermum suaveolens* DC.	Hypertension, Antioxidants	Tańska et al. (2018)
92.	Tannic acid	*Acacia leucophloea* Wild., *Acorus calamus* (L.), *Caesalpinia pulcherrima* (L.) Sw., *Litsea glutinosa* (Lour.)	Antibacterial, Gastroprotective	Jacob and Khan (2007)

TABLE 15.1 *(Continued)*

No	Name of compound	Source plants	Activity	References)
		C. B. Rob., *Ocimum sanctum* (L.), *Lantana camara* (L.), *Aegle marmelos* (L.) Corrêa, *Cichorium intybus* (L.), *Punica granatum* (L.), *Santalum album* (L.), *Tamarindus indica* (L.), *Mangifera indica* (L.), *Phyllanthus emblica* (L.), *Paeonia lactiflora* Pall., *Terminalia arjuna* (Roxb. ex DC.) Wight & Arn., *Cassia occidentalis* (L.), *Terminalia bellirica* (Gaertn.) Roxb., *Jatropha gossypifolia* (L.), *Pongamia pinnata* (L.) Pierre, *Justicia adhatoda* (L.), *Nigella sativa* (L.), *Terminalia chebula* Retz.		
93.	Terpinene-4ol	*Piper nigrum* (L.), *Anethum graveolens* (L.), *Elettaria cardamomum* Maton, *Syzygium aromaticum* (L.) Merr. & L.M. Perry, *Myristica fragrans* Houtt., *Foeniculum vulgare* Mill., *Acorus calamus* (L.), *Lantana camara* (L.), *Curcuma longa* (L.)	Antibacterial, Anticancerous	Taha and Eldahshan (2017)
94.	Thymol	*Nigella sativa* (L.), *Trachyspermum ammi* Sprague, *Azadirachta indica* A. Juss.	Antiparasitic, Anesthetic, Neuroprotective	Souhaiel et al. (2017); Barbosa et al. (2017)
95.	Thymoquinone	*Nigella sativa* (L.), *Aloe vera* (L.)	Hepato protective anticancerous antibacterial	Forouzanfar et al. (2014); Hala Gali Muhtasib, Albert Roessner (2006); Ryan et al. (2005)
96.	Totarol	*Podocarpus totara* (L.)	Antibacterial	Jaiswal et al. (2007), Muroi and Himejima (1992)
97.	Trans-anethole	*Rhynchosia Minima* (L.), *Elettaria cardamomum* Maton	Antifungal	Fujita et al. (2017)
98.	Umbelliferone	*Cichorium intybus* (L.)us, *Foeniculum vulgare* Mill., *Tecomella undulata* (Sm.) Seem., *Boerhavia diffusa* (L.)	Antibacterial, Antidepressant, Antioxidant	Kayser and Kolodziej (1999); Cao, Li, and Yang (2016); Asl and Hosseinzadeh (2008)
99.	Ursolic acid	*Punica granatum* (L.), *Thevetia neriifolia* Juss. ex A. DC., *Nigella sativa* (L.), *Azadirachta indica* A. Juss.	Anti-inflammatory, Anticancer, Antioxidant	González-Chávez et al. (2017); Tueller, Harley, and Hancock (2017); Chi et al. (2017); Micota et al. (2014)

TABLE 15.1 *(Continued)*

No	Name of compound	Source plants	Activity	References)
100.	Vanillic acid	*Casuarina equisetifolia* (L.), *Allium cepa* (L.), *Allium sativum* (L.), *Cichorium intybus* (L.)us, *Foeniculum vulgare* Mill., *Zingiber officinale* Roscoe, *Justicia adhatoda* (L.), *Berberis aristata* DC.	Antibacterial	Alves et al. (2013)

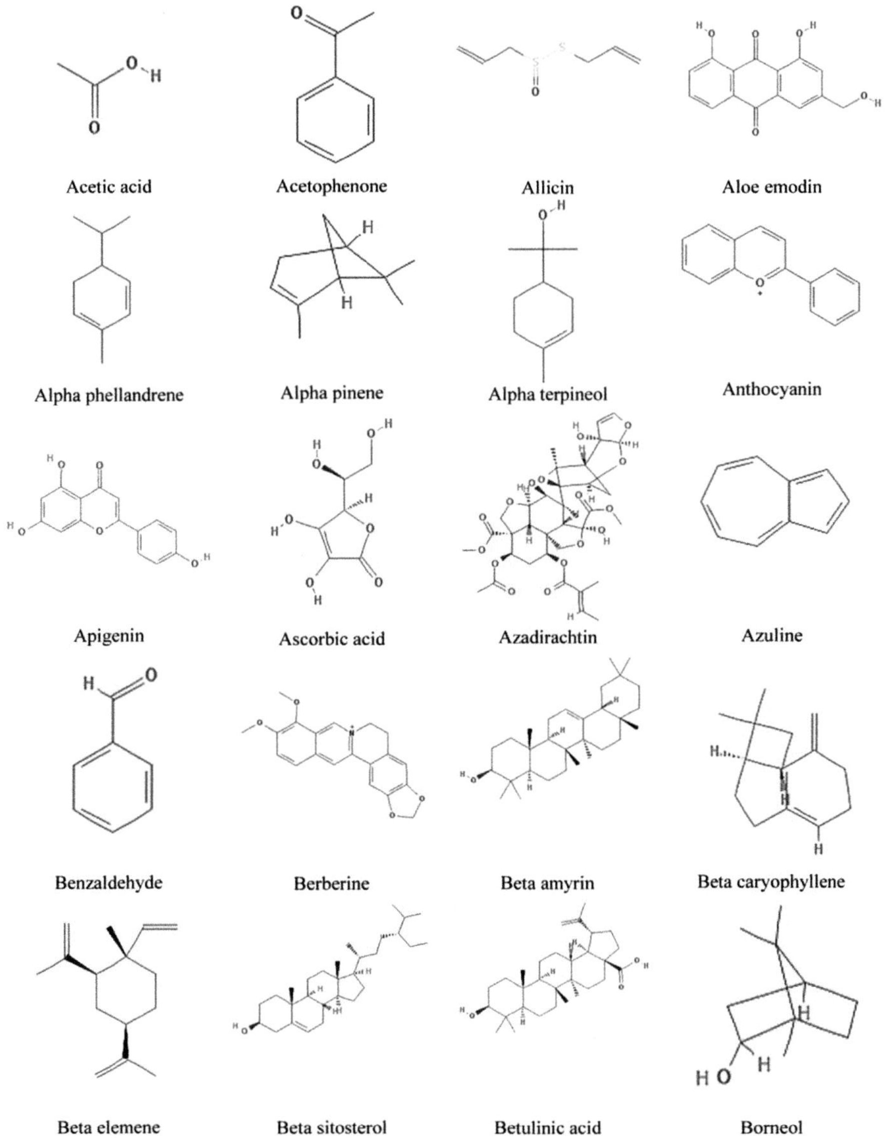

FIGURE 15.1 Structures of major bioactive compounds from Indian traditional medicinal plants.

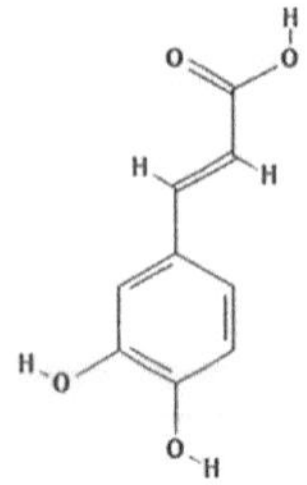
Caffeic acid

Camphor

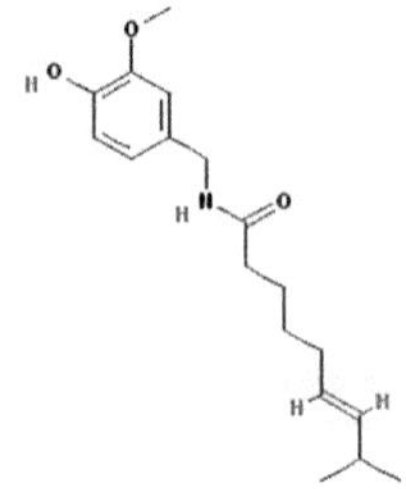
Capsaicin

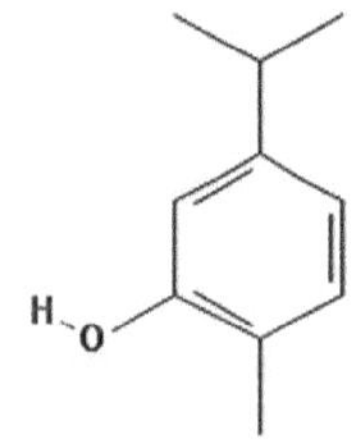
Carvacrol

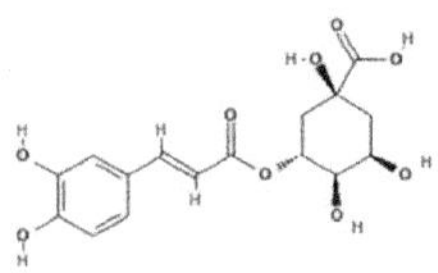
Chlorogenic acid

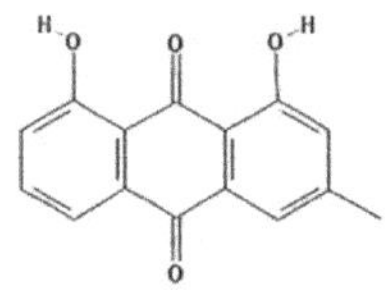
Chrysophanol

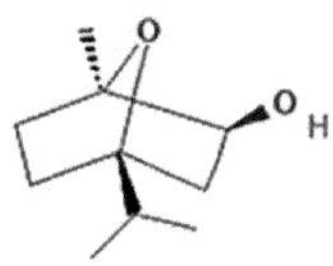
1,8 -Cineole

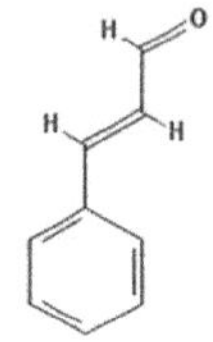
Cinnamaldehyde

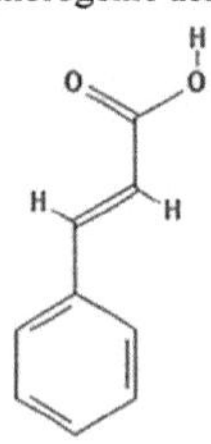
Cinnamic acid

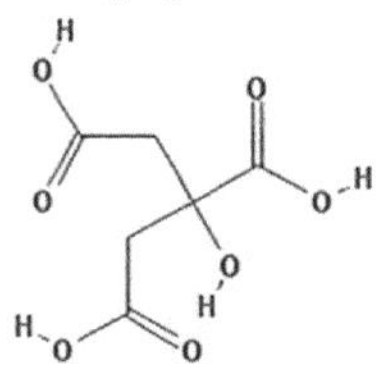
Citric acid

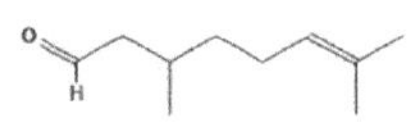
Citronellal

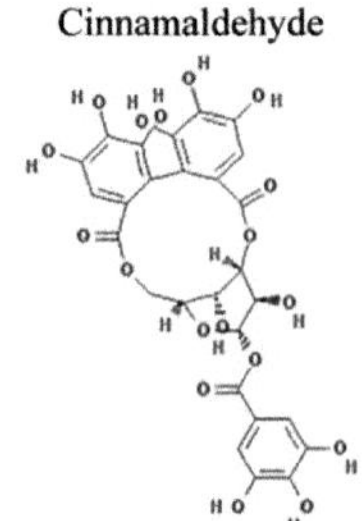
Corilagin

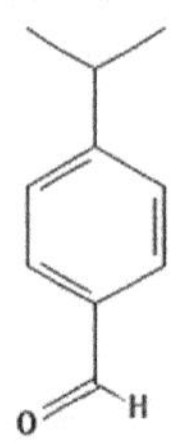
Cuminaldehyde

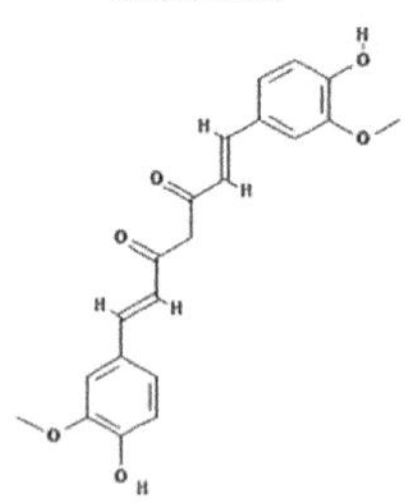
Curcumin

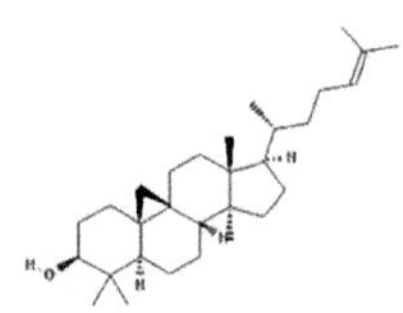
Cycloartenol

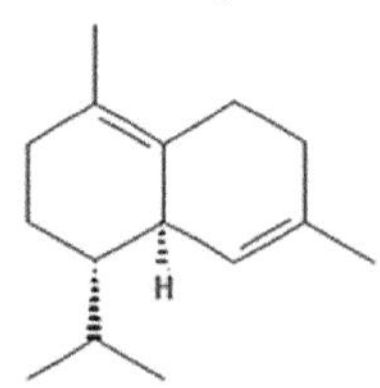
Delta cadinene

Diallyl sulfide

Diallyl disulfide

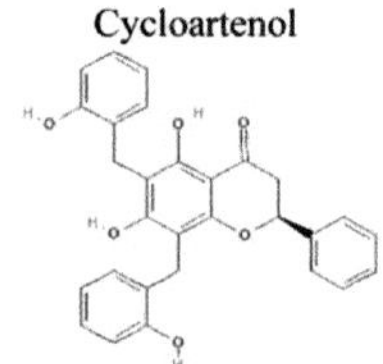
Dichamanetin

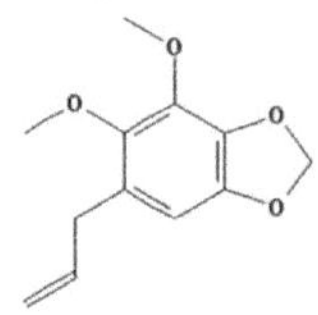
Dillapiole

FIGURE 15.1 *(Continued).*

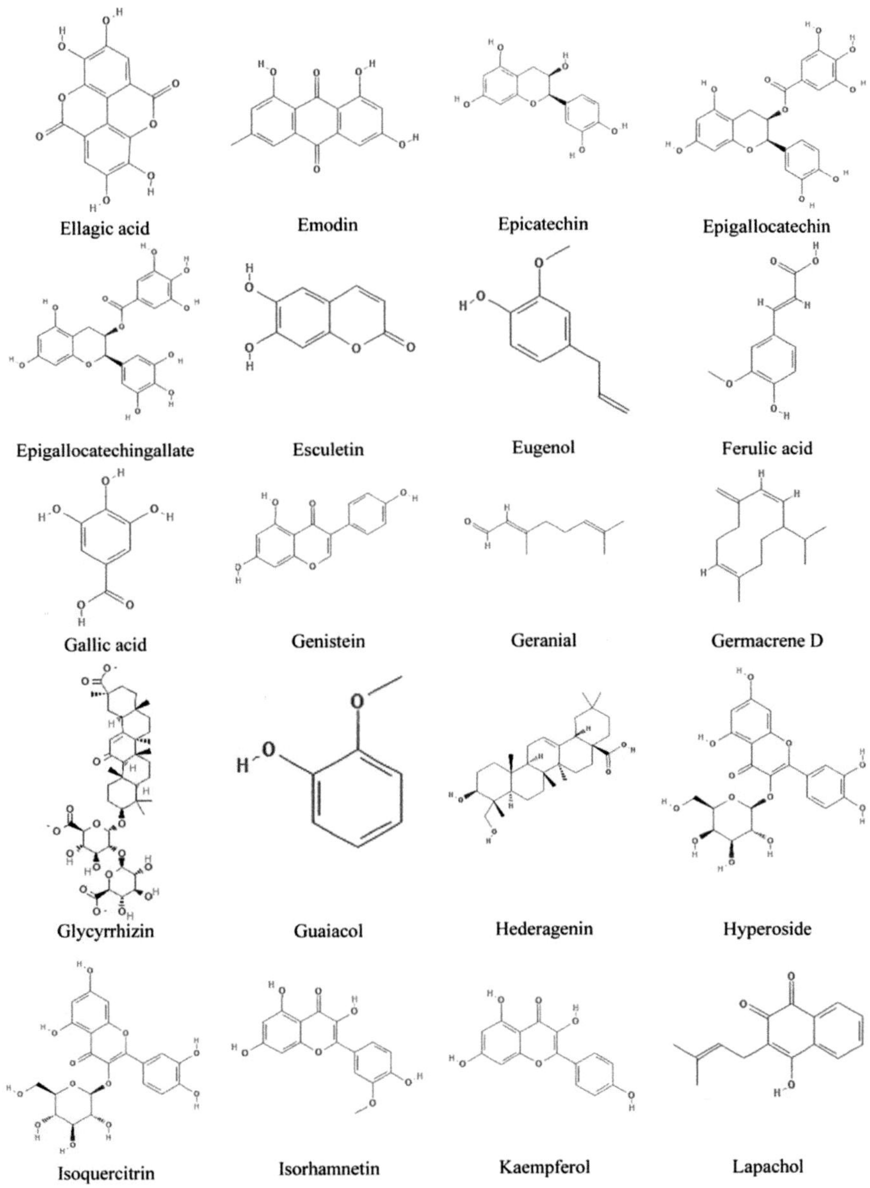

FIGURE 15.1 *(Continued).*

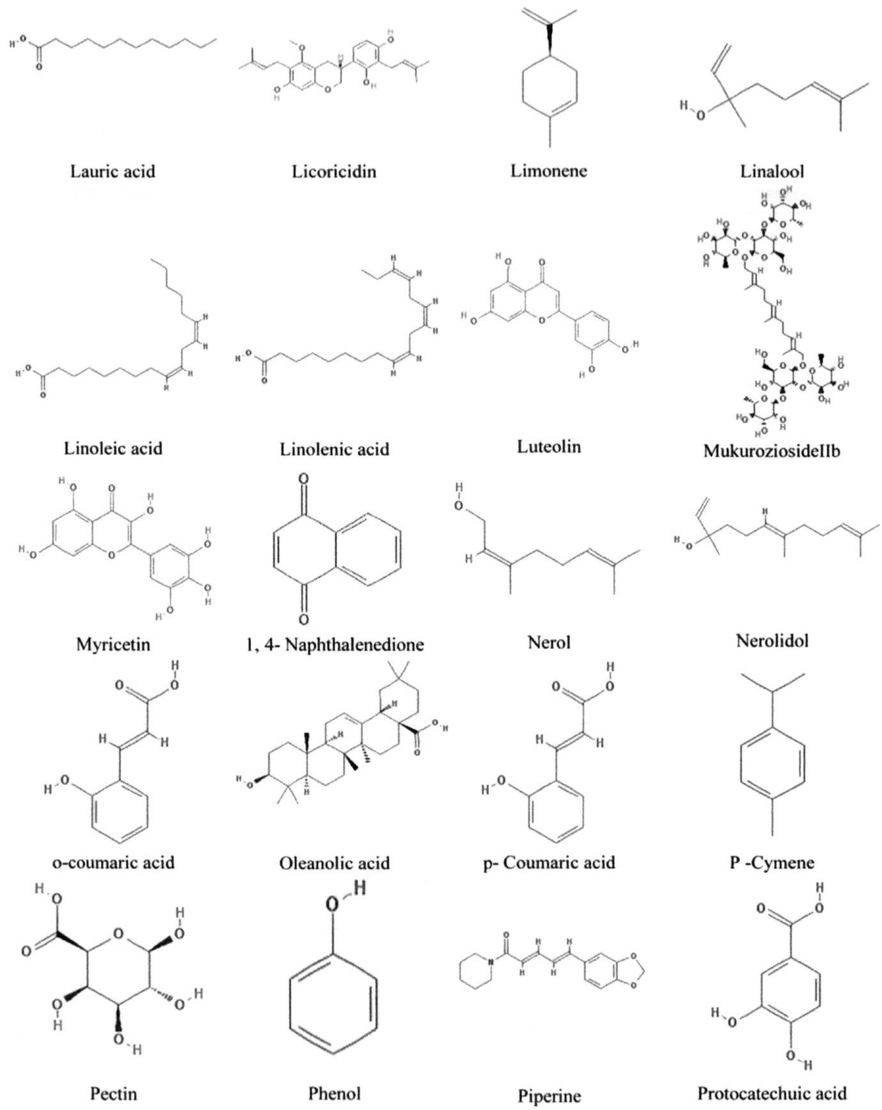

FIGURE 15.1 *(Continued).*

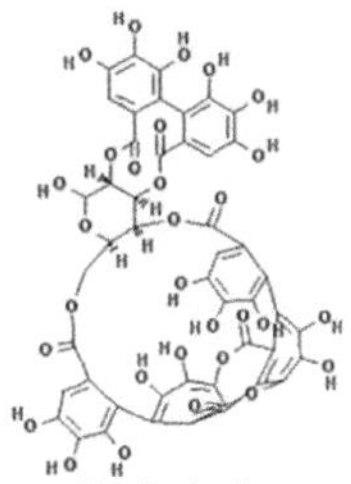
Punicalagin

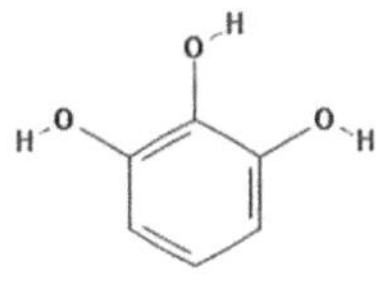
Pyrogallol

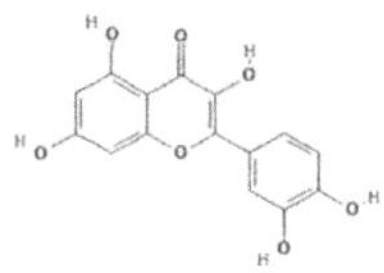
Quercetin

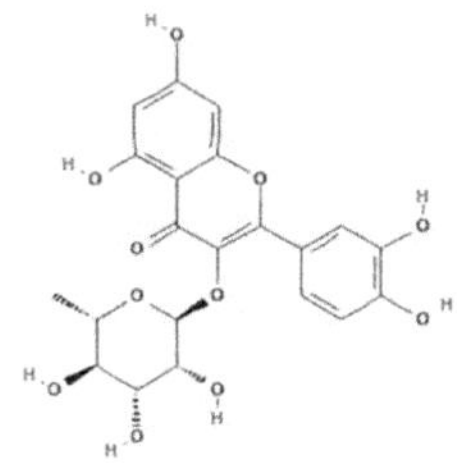
Quercitrin

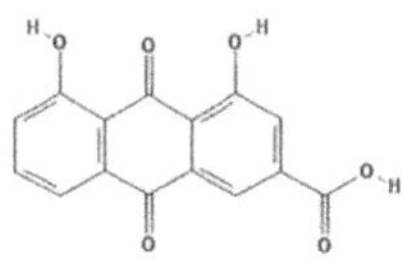
Rhein

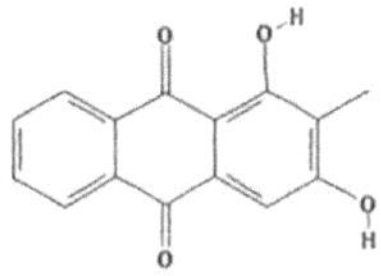
Rubiadin

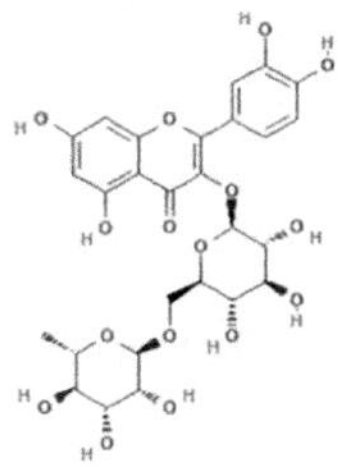
Rutin

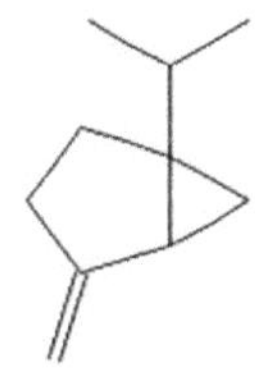
Sabinene

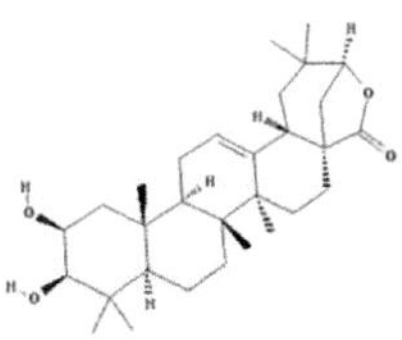
Sapogenin

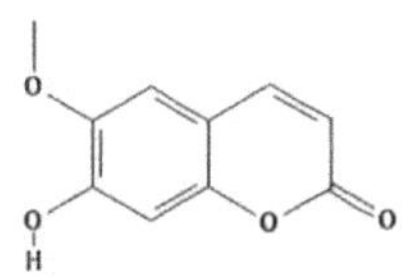
Scopoletin

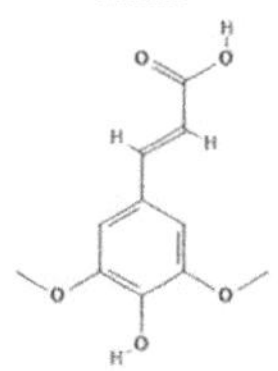
Sinapic acid

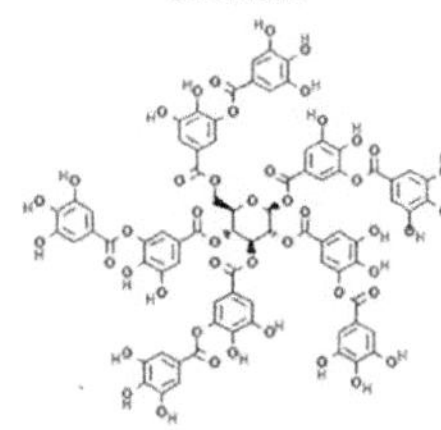
Tannic acid

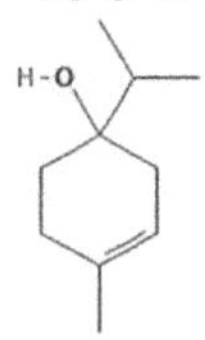
Terpinene-4ol

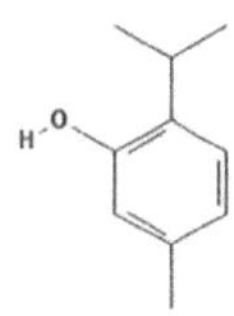
Thymol

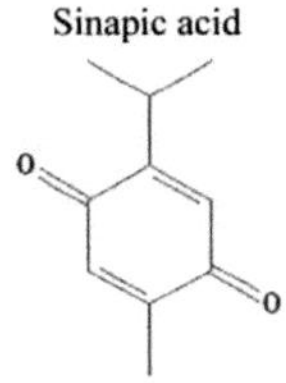
Thymoquinone

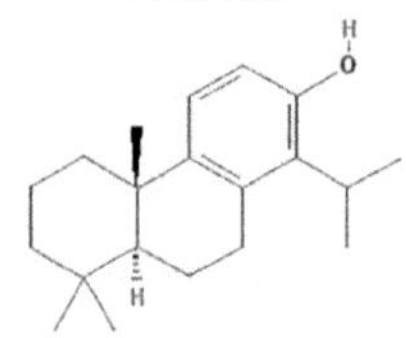
Totarol

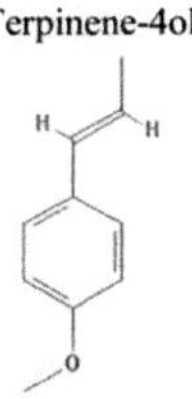
Trans anethole

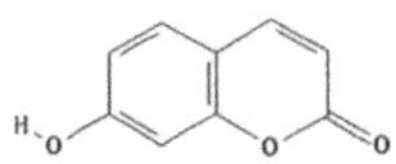
Umbelliferone

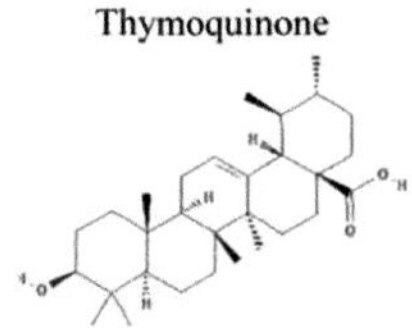
Ursolic acid

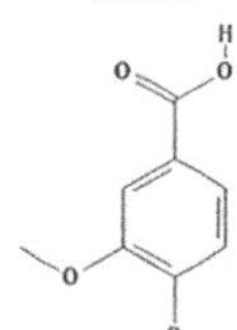
Vanillic acid

FIGURE 15.1 *(Continued).*

TABLE 15.2 Taxonomic Details of Selected Indian Traditional Medicinal Plants

No.	Scientific Name	Common Name	Family
1	*Abrus precatorius* (L.)	Weather vine, Crab's Eye and Precatory Bean	Leguminosae
2	*Abutilon indicum* (L.) Sweet	Indian Mallow, Indian Abutilon	Malvaceae
3	*Acacia leucophloea* Willd.	Pilang Bast, Reonja	Leguminosae
4	*Acacia nilotica* (L.) Delile	Babul	Leguminosae
5	*Acorus calamus* (L.)	Calamus, Sweet Flag	Acoraceae
6	*Aegle marmelos* (L.) Corrêa	Bael	Rutaceae
7	*Allamanda cathartica* (L.)	Golden Trumpet	Apocynaceae
8	*Allium cepa* (L.)	Onion	Alliaceae
9	*Allium sativum* (L.)	Garlic	Alliaceae
10	*Aloe vera* (L.)	Aloe	Aloaceae
11	*Andrographis echioides* Nees	False Water willow	Acanthaceae
12	*Anethum graveolens* (L.)	Garden Dill	Apiaceae
13	*Annona squamosa* (L.)	Custard Apple	Annonaceae
14	*Azadirachta indica* A. Juss.	Nimba, Neem	Meliaceae
15	*Bacopa monnieri* (L.) Wettst.	Water hyssop, Brahmi	Scrophulariaceae
16	*Berberis aristata* DC.	Daruharidra	Berberidaceae
17	*Boerhavia diffusa* (L.)	Punarnava	Nyctaginaceae
18	*Bombax ceiba* (L.)	Red Silk Cotton Tree	Bombacaceae
19	*Caesalpinia pulcherrima* (L.) Sw.	Poinciana, Peacock Flower, Settimandaram	Leguminosae
20	*Calotropis procera* (Aiton) W. T. Aiton	Ushar, Giant Milkweed	Asclepiadaceae
21	*Camellia sinensis* (L.) Kuntze	Tea Plant	Theaceae
22	*Cardiospermum halicacabum* (L.)	Balloon Vine, Heart Vine, Heart Pea	Sapindaceae
23	*Cassia fistula* (L.)	Golden Shower, Semi-Wild Indian Labernum	Leguminosae
24	*Cassia occidentalis* (L.)	Coffee Senna, Rubbish Cassia, Stinking /Coffee Weed	Leguminosae
25	*Cassia tora* (L.)	Sickle Senna	Leguminosae
26	*Casuarina equisetifolia* (L.)	Sheoak, Horsetail Tree	Casuarinaceae
27	*Centella asiatica* (L.) Urb.	Centella or Gotu Kola	Mackinlayaceae
28	*Cichorium intybus* (L.)	Common Chicory	Asteraceae
29	*Citrus limon* (L.) Burm. f.	Lemon	Rutaceae
30	*Cocos nucifera* (L.)	Cocunut	Arecaceae
31	*Colocasia esculenta* (L.) Schott	Taro	Araceae
32	*Colubrina asiatica* (L.) Brongn.	Latherwood, Asian Nakedwood, Asian Snake wood	Rhamnaceae
33	*Curcuma longa* (L.)	Turmeric	Zingiberaceae
34	*Cynodon dactylon* (L.) Pers.	Dūrvā Grass, Dhoob, Bermuda Grass, Dubo	Poaceae

TABLE 15.2 *(Continued)*

No.	Scientific Name	Common Name	Family
35	*Delonix elata* Gamble	White Gulmohar	Leguminosae
36	*Delonix regia* (*Bojer*) Raf.	Royal Poinciana, Gulmohar	Leguminosae
37	*Eclipta alba* Hassk.	Bhringaraja, False Daisy	Asteraceae
38	*Elettaria cardamomum* Maton	Cardamom, Malabar cardamom, Ceylon cardamom	Zingiberaceae
39	*Eucalyptus cinerea* (F.) Muel (L.) ex Benth.	Argyle Apple, Silver Dollar Eucalyptus	Myrtaceae
40	*Euphorbia hirta* (L.)	Snake Weed	Euphorbiaceae
41	*Euphorbia tirucalli* (L.)	Aveloz, Firestick Plants, Indian Tree Spurge	Euphorbiaceae
42	*Ficus benghalensis* (L.)	Banyan Tree	Moraceae
43	*Foeniculum vulgare* Mil (L.)	Fennel, Saunf	Apiaceae (Umbelliferae)
44	*Glycyrrhiza glabra* (L.)	Madhuka, Licorice	Leguminosae
45	*Gmelina asiatica* (L.)	Vikarin	Lamiaceae
46	*Hemidesmus indicus* (L.) R. Br.	Indian Sarsaparilla	Asclepiadaceae
47	*Hibiscus rosa-sinensis* (L.)	Shoe flower	Malvaceae
48	*Holarrhena antidysenterica* (L.) Wal (L.)	Tellicherry Bark, Conessi Bark	Apocynaceae
49	*Holoptelea integrifolia* (Roxb.) Planch.	Indian Elm, Entire-Leaved Elm Tree, Chilbil	Ulmaceae (Elm Family)
50	*Ichnocarpus frutescens* (L.) R. Br.	Sarivaa	Apocynaceae
51	*Jatropha gossypifolia* (L.)	Bellyache Bush, Physicnut	Euphorbiaceae
52	*Justicia adhatoda* (L.)	Malabar Nut	Acanthaceae
53	*Kaempferia galanga* (L.)	Kencur, Aromatic Ginger	Zingiberaceae
54	*Lantana camara* (L.)	Wild Sage	Verbenaceae
55	*Lawsonia inermis* (L.)	Mignonette Tree	Lythraceae
56	*Litsea glutinosa* (Lour.) C. B. Rob.	Soft Bollygum, Bolly Beech, Bollywood	Lauraceae
57	*Mangifera indica* (L.)	Common Mango	Anacardiaceae
58	*Mirabilis jalapa* (L.)	Four-O-clock	Nyctaginaceae
59	*Morinda tinctoria* Roxb.	Togaru	Rubiaceae
60	*Myristica fragrans* Houtt.	Jati	Myristicaceae
61	*Nigella sativa* (L.)	Black cumin	Ranunculaceae
62	*Nyctanthes arbor-tristis* (L.)	Night-flowering jasmine	Oleaceae
63	*Nymphaea stellata* Willd.	Nilotpala	Nymphaeaceae
64	*Ocimum sanctum* (L.)	Tulsi, Basil	Lamiaceae
65	*Paeonia lactiflora* Pal (L.)	Peony	Paeoniaceae
66	*Peganum harmala* (L.)	Wild Rue, Esfand, Harmel	Nitrariaceae
67	*Pelargonium tomentosum* Jacq.	Krishnapalai	Geraniaceae

TABLE 15.2 *(Continued)*

No.	Scientific Name	Common Name	Family
68	*Pergularia daemia* (Forssk.) Chiov.	Trellis-Vine	Asclepiadaceae
69	*Phyllanthus emblica* (L.)	Indian Gooseberry	Phyllanthaceae
70	*Picrorhiza kurroa* Royle ex Benth.	Katurohini	Scrophulariaceae
71	*Piper betle* (L.)	Betel	Piperaceae
72	*Piper longum* (L.)	Black Pepper	Piperaceae
73	*Piper nigrum* (L.)	Black Pepper	Piperaceae
74	*Pithecellobium dulce* (Roxb.) Benth.	Manila Tamarind, Madras Thorn	Leguminosae
75	*Plumbago zeylanica* (L.)	Leadwort, Chitraka	Plumbagonaceae
76	*Pongamia pinnata* (L.) Pierre	Naktamala	Leguminosae
77	*Prunus domestica* (L.)	Common Plum	Rosaceae
78	*Prunus puddum* Roxb. ex Wal (L.)	Padmaka, Padmagandhi	Rosaceae
79	*Punica granatum* (L.)	Pomegranate	Lythraceae
80	*Rhoeo discolor* (L'Hér.) Hance in Walp.	Moses-in-the-Cradle	Commelinaceae
81	*Rhoeo spathacea* (Sw.) Stearn	Oyster plant	Commelinaceae
82	*Rhynchosia minima* (L.) DC.	Least Snout Bean	Leguminosae
83	*Rubia cordifolia* (L.)	Manjishta	Rubiaceae
84	*Santalum album* (L.)	Indian Sandalwood	Santalaceae
85	*Sapindus emarginatus* Vahl	Soapberries, Soapnuts	Sapindaceae
86	*Saraca asoca* (Roxb.) W. J. de Wilde	Ashoka Tree	Leguminosae
87	*Saussurea lappa* (Decne.) C. B. Clarke	Kushta	Asteraceae
88	*Solanum incanum* (L.)	Bitter Apple, Bitter Garden Egg	Solanaceae
89	*Stereospermum suaveolens* DC.	Patola	Bignoniaceae
90	*Streblus asper* Lour.	Shakhotaka, Siamese Rough Bush, Tooth Brush Tree	Moraceae
91	*Swertia chirata* C. B. Clarke	Chirayita	Gentianaceae
92	*Symplocos racemosa* Roxb.	Lodhra	Symplocaceae
93	*Syzygium aromaticum* (L.) Merr. & (L.)M. Perry	Clove	Myrtaceae
94	*Tamarindus indica* (L.)	Tamarind	Leguminosae
95	*Tecomella undulata* (Sm.) Seem.	Rohida	Bignoniaceae
96	*Tephrosia purpurea* (L.) *Pers.*	Wild indigo	Leguminosae
97	*Terminalia arjuna* (Roxb. ex DC.) Wight & Arn.	Arjuna (Arjun Tree)	Combretaceae

TABLE 15.2 *(Continued)*

No.	Scientific Name	Common Name	Family
98	*Terminalia bellirica* (Gaertn.) Roxb.	Bahera	Combretaceae
99	*Terminalia catappa* (L.)	Myrobalan	Combretaceae
100	*Terminalia chebula* Retz.	Abhaya	Combretaceae
101	*Thevetia neriifolia* Juss. ex A. DC. (*Cascabela thevetia* (L.) *Lippold, Thevetia peruviana* K. Schum.)	Yellow Oleander	Apocynaceae
102	*Trachyspermum ammi* Sprague	Ajwan	Apiaceae
103	*Trigonella foenum-graecum* (L.)	Fenugreek	Leguminosae
104	*Viola odorata* (L.)	Sweet Violet, Banafsa	Violaceae
105	*Vitex peduncularis* Wal (L.)	Mayiladi	Lamiaceae
106	*Withania somnifera* (L.) Dunal	Ashwagandha, Indian Ginseng	Solanaceae
107	*Woodfordia fruticosa* Kurz	Fire Flame Bush	Lythraceae
108	*Wrightia tinctoria* R. Br.	Centella, Gotu Kola	Apocynaceae
109	*Zingiber officinale* Roscoe	Ginger	Zingiberaceae
110	*Ziziphus mauritiana* Lam.	Jujube, Indian Plum	Rhamnaceae

β-caryophyllene (Tepe et al., 2005), beta sitosterol (Zhang et al., 2014), betulinic acid (Cowan, 1999), borneol (Tepe et al., 2005), camphor (Hamidpour et al., 2013), capsaicin (Kalia et al., 2012), carvacrol (Vasconcelos et al., 2017), chlorogenic acid (Chandramohan, Divya, and Dhanarajan, 2014), 1,8-cineole (Mann and Markham, 1998), cinnamaldehyde (Domadia et al., 2007), cinnamic acid (Cowan, 1999), citric acid (Papetti et al., 2013), citronellal (Nascimento et al., 2000), corilagin (Perumal Samy and Gopalakrishnakone, 2010), cuminaldehyde (Skariyachan, Mahajanakatti, et al., 2011), curcumin (Dipti Rai, Jay Kumar Singh, Nilanjan Roy, 2008), δ-cadinene (Zatelli et al., 2016), dichamanetin (Lock and Harry, 2008), emodin (Ji et al., 2017), eugenol (Devi et al., 2010), gallic acid (Stapleton et al., 2004), genistein (Abreu, McBain, and Simões, 2012), geranial (citral, neral) (Gupta et al., 2017), germacrene D (Zorica et al., 2014), glycyrrhizin (Naidu, Lalam, and Bobbarala, 2009), guaiacol (Asl and Hosseinzadeh, 2008), hederagenin (Liu et al., 2014), licoricidin (Nomura, Fukai, and Akiyama, 2002), linalool (Rajeswara Rao et al., 2002, linoleic acid. (Ji et al., 2005), 1, 4-naphthalenedione (Vinod, Haridas, and Sadasivan, 2010), nerol (geraniol) (Rajeswara Rao et al., 2002), p-cymene (Yangui et al., 2017), pectin (Atanasov, Waltenberger, and Pferschy-Wenzig,

2015), phenol (Cai et al., 2004), piperine (Mirza et al., 2011), punicalagin (Xu et al., 2017), pyrogallol (Bai et al., 2015), quercetin (Suriyanarayanan, Shanmugam, and Santhosh, 2013), quercitrin (Rahmatullah and Samarrai, et al., 2010), rutin (Jayaraman et al., 2010), sabinene (Alizadeh and Abdollahzadeh, 2017), tannic acid (Jacob and Khan, 2007), terpinene-4ol (Taha and Eldahshan, 2017), thymoquinone (Ryan, Rooney, and Ryan, 2005), totarol (Jaiswal et al., 2007), umbelliferone (Kayser and Kolodziej, 1999) and vanillic acid (Alves et al., 2013). While most of the phytocompounds showing antibacterial activity, comparatively lesser number exhibit antifungal activity namely, acetic acid, allicin, β-caryophyllene, cinnamaldehyde, citronellal, lapachol, licoricidin, p-cymene, rubiadin, trans-anethole. Even though 34 plants were studied for antiviral properties, only Azuline and Quercetin were found as antiviral agents.

15.3.2 ANTICANCEROUS ACTIVITIES

The second most studied bioactivity for phytocompounds is their anticancerous property. There are 50 compounds with reported anticancerous property from Indian traditional medicine namely, acetophenone (Street, Sidana, and Prinsloo, 2013), allicin (Abdel-Daim et al., 2017), aloe emodin (Tamokou et al., 2013), α-phellandrene (Hosseinzadeh et al., 2015), α-pinene (Chirathaworn and Kongcharoensuntorn, 2007), α-terpineol (Taha and Eldahshan, 2017), anthocyanin (Rahmatullah and Jahan, et al., 2010), apigenin (Horinaka et al., 2006), ascorbic acid (Thirupathi et al., 2016), azadirachtin (Bhat, 2008), benzaldehyde (Salvador et al., 2017), berberine (Ruan, et al., 2017), β-caryophyllene (Legault J 2007), β-elemene (Yu et al., 2017), β-sitosterol (Oloche, Okwuasaba, and Obochi, 2016), betulinic acid (Tamokou et al., 2013), caffeic acid (Tyszka-czochara, Bukowska-strakova, and Majka, 2017), carvacrol (Kai Fana, Xiaolei Lib et al., 2015), 1, 8-cineole (Satyal et al., 2011), cinnamaldehyde (Rahmatullah and Jahan, et al., 2010), cinnamic acid (Gabriele et al., 2017), citric acid (Bandaranayake, 2002), cuminaldehyde (Yangui et al., 2017), curcumin (Rooney and Ryan, 2005), diallyl sulfide (Wei et al., 2017), diallyl disulfide (Wei et al., 2017), dichamanetin (Yong et al., 2013), ellagic acid (Atanasov et al., 2015), epigallocatechin galate (Chao et al., 2017), esculetin (Liang et al., 2017), gallic acid (Hussain, Fareed, and Ali, 2012), genistein (Li et al., 2017), hederagenin (Liu et al., 2014), isoquercitrin (Amado et al., 2009), isorhamnetin (Guo et al., 2017), kaempferol (Guo et al., 2017), lapachol (Epifano and Genovese, 2013), limonene (Suh, Chon, and Choi, 2017), linalool (Suneel Kumar & Venkatarathanamma, 2015), luteolin

(Tamokou et al., 2013), nerol (geraniol) (Queiroz et al., 2017), o-coumaric acid (Sen et al., 2015), oleanolic acid (Gao et al., 2016), p-coumaric acid (Kong et al., 2013, protocatechuic acid (Rahmatullah and Samarrai, et al., 2010), punicalagin (Adaramoye et al., 2017), rubiadin (Siddiqui et al., 2011), rutin (Aggarwal et al., 2011), scopoletin (Shi et al., 2017), terpinene-4ol (Taha and Eldahshan, 2017), thymoquinone (Muhtasib and Roessner, 2006) are the major reported anticancerous compounds.

15.3.3 ANTIOXIDANT ACTIVITIES

The third most studied property is antioxidant activity. Anthocyanin (Bowen-forbes, Zhang, and Nair, 2010), ascorbic acid (Jena et al., 2017), azadirachtin (Bhat, 2008), azuline (Baqueiro-peña and Guerrero-beltrán, 2017), beta caryophyllene (Tepe et al., 2005), β-elemene (Suneel Kumar and Venkatarathanamma, 2015), borneol (Tepe et al., 2005), caffeic acid (Shiao, and SY, 2001), chlorogenic acid (Samarrai, et al., 2010), 1,8-cineole (Satyal et al., 2011), curcumin (Cowan, 1999), delta cadinene (Jena et al., 2017), diallyl disulfide (Sumit and Khan, 2016), dichamanetin (Yong et al., 2013), emodin (Cai et al., 2004), epicatechin (Guo et al., 2017), epigallocatechin (Grzesik et al., 2018), epigallocatechin galate (Grzesik et al., 2018), esculetin (Han et al., 2017), ferulic acid (Tańska, Mikołajczak, and Konopka, 2018), gallic acid (Ranilla, Apostolidis, and Shetty, 2012), germacrene D (Mendes et al., 2017), guaiacol (Asl and Hosseinzadeh, 2008), hyperoside (hyperin) (Rahmatullah and Jahan, et al., 2010), isorhamnetin (Guo et al., 2017), lauric acid (Alves et al., 2017), licoricidin (Saxena, 2005), limonene (Gundidza et al., 2009, linalool (Kumar et al., 2015), linoleic acid (Shabir et al., 2011), 1, 4-naphthalenedione (N. V Vinod et al., 2010), nerolidol (Nogueira, Antonia, and Cardoso, 2013), o-coumaric acid (Sen et al., 2015), oleanolic acid Borchardt et al., 2009), p-coumaric acid (Alves et al., 2013), pectin (Xu et al., n.d.), protocatechuic acid (Samarrai, et al., 2010), punicalagin (Bandaranayake, 2002), pyrogallol (Cowan, 1999), quercetin (R. Guo et al., 2017), sinapic acid (Tańska et al., 2018), umbelliferone (Asl and Hosseinzadeh, 2008) and ursoilc acid (Micota et al., 2014) are main antioxidant phytocompounds reported here.

15.3.4 ANTI-INFLAMMATORY ACTIVITIES

Anti-inflammatory property also has a major position in the bioactivities of phytocompounds. Acetic acid (Kumbhare and Sivakumar, 2011), acetophenone (Street et al., 2013), α-phellandrene (Hosseinzadeh et al., 2015), anthocyanin (Rahmatullah and Jahan, et al., 2010), β-amyrin (Sah and Verma, 2012), β-caryophyllene (Shen et al., 2017), β-sitosterol (Oloche et al., 2016), camphor (Maroon and Bost, 2010), capsaicin (Maroon, and Bost, 2010), chlorogenic acid (Rahmatullah and Samarrai, et al., 2010), corilagin (Bandaranayake, 2002), curcumin (Shen et al., 2017), cycloartenol (Forouzanfar et al., 2014), eugenol (Devi et al., 2010), glycyrrhizin (Asl and Hosseinzadeh, 2008), hyperoside (hyperin) (Rahmatullah and Jahan, et al., 2010), kaempferol (Guo et al., 2017), rutin (Guo et al., 2017) and ursolic acid (González-Chávez et al., 2017) are the major antioxidant compounds listed here.

15.3.5 OTHER ACTIVITIES

Other properties reported were antiadiposity (acetic acid (Bounihi et al., 2017), chrysophanol (Danish et al., 2011), germacrene D (Wang et al., 2017), sapogenin (Bandaranayake, 2002)), anti Alzheimers (acetophenone (Raghav and Singh, 2017), berberine (Jiang and Gao, 2017), cinnamic acid (Lan et al., 2017), rhein (Pérez-Areales et al., 2017)), antiparasitic (aloe emodin (Tamokou et al., 2013), azadirachtin (Bhat, 2008), betulinic acid (Samarrai, et al., 2010), δ-cadinene (Guo et al., 2017), emodin (Tamokou et al., 2013), eugenol (Charan Raja et al., 2017), p-cymene (Souhaiel et al., 2017), thymol (Souhaiel et al., 2017)), gastroprotective (ascorbic acid (Moricz, et al., 2007), quercetin (Rahmatullah and Jahan, et al., 2010), quercitrin (Rahmatullah and Samarrai, et al., 2010), tannic acid (Jacob and Khan, 2007)), antidiabetic (azadirachtin (Bhat, 2008), berberine (Yuzhen Jia, 2016), cycloartenol (Forouzanfar et al., 2014), nerol (geraniol) (Satyal et al., 2011), punicalagin (Bandaranayake, 2002)). A very few compounds were reported for their antifibrotic, antiarthritic, analgesic, antipyretic, antithrombotic, antiatherogenic, antihyperalgesic, antidepressant, cardioprotective, chemoprotective, pancreas protective, nephroprotective, neuroprotective, hepatoprotective, lung protective, anesthetic, wound healing, etc.

15.4 TARGET STUDIES OF ANTIBACTERIAL PHYTOCOMPOUNDS

Extensive literature review done in this area reveals that most of the studies on antimicrobial medicinal plants are confined to screening of activities against different pathogens without further scientific exploration up to drug developments. This research scenario points towards indispensable advanced research on the exact mechanism of their antimicrobial activity which is essential for safe incorporation of herbal medicine to modern medicine. Only a limited number of reports are available to reveal identity of active principles involved and its real mechanism of action. Data were collected from published findings, STITCH, and Super Target. Phytocompounds with known mechanism of action identified by *in-silico* and *in-vitro* studies are shown in Table 15.3.

The mode of action of 2-hydroxyl 5-bezylisouvarinol, berberine, curcumin, dichamanetin, kaempferol, p-coumaric acid, quercitrin and sanguinarine were reported against *Bacillus subtilis.* The most common target reported in *B. subtilis* is FtZ. Berberine, 2-hydroxy 5-benzyl isouvarinol, curcumin, dichamanetin and sanguinarine can block bacterial division by binding on the FtZ protein. In addition to FtZ, berberine has two more target namely, Bmr multidrug efflux transporter and BmrR, transcriptional regulator. In the case of kaempferol, targets are ImrA (a transcriptional repressor of lmrAB and yxaGH operons), ImrB Efflux pump protein, Quercetin dioxygenase, YxaF (a transcriptional regulator) and YxaH (a putative integral inner membrane protein). The p-coumaric acid targets histidine ammonia-lyase, phenolic acid decarboxylase, succinyl CoA-3-oxoacid CoA-transferase, and urease γ subunit. While the targets of action of quercitrin are DesR-DesK two-component response regulator, phosphotransferase system (PTS) lichenan-specific enzyme IIA component, polar chromosome segregation protein, and quercetin dioxygenase. They can interfere with membrane lipid fluidity regulation, uptake, and metabolism of lichenan, axial filament formation, and resistance mechanism respectively.

Capsaicin, chlorogenic acid, chrysophaentin A, germacrene D, germacrene D-4-ol, 1, 4-naphthalenedione (Vinod et al., 2010), piperine, terpinen-4-ol, and ursolic acid were found as antibacterial against *S. aureus.* Epicatechin gallate (ECG), epigallocatechin gallate (EGCG), glycerol monolaurate, poli-docanol, corilagin (Stapleton and Taylor, 2007), licoricidin and tellimagrandin can block signal transduction in plasma membrane and cell wall biosynthesis in MRSA strains (Stapleton and Taylor, 2007). Corilagin, ECG (Stapleton et al., 2004), EGCG (Bernal et al., 2010), licoricidin and tellimagrandin are cell wall inhibitors that can bind to pencillin binding protein2A (PBP2A) while glycerol monolaurate and polidocanol which can bind on MecR1 function as signal transduction inhibitors (Stapleton and Taylor, 2007).

The killing effect of benzoic acid on *Burkholderia xenovorans* has been studied, and several targets were reported such as amidase, putative carbon-nitrogen hydrolase protein, 2-hydroxy-6-oxo-6-phenylhexa-2, 4-dienoate

TABLE 15.3 Phytocompounds with Known Mechanism of Action in Bacteria

In vitro studies					
Antibacterial phytocompounds	**Targets identified**	**Role of target**	**Major source plants**	**Bacteria studied**	**References**
1, 4-Naphthalenedione	β-lactamase	Drug metabolism	*Holoptelea integrifolia*	*Staphylococcus aureus*	Vinod et al., (2010)
2-hydroxy, 5-bezylisouvarinol	FtsZ	Bacterial division	*Xylopia africana*	*Bacillus subtilis*	Lock and Harry, (2008)
α-pinene	HspR	Efflux pump protein	*Cuminum cyminum, Cannabis sativa*	*Campylobacter jejuni*	Kovac et al., (2015)
	HrcA	Efflux pump protein		*Campylobacter jejuni*	Kovac et al., (2015)
Azadirachtin	Acetyl-CoA carboxylase-biotin carboxylase	Biosynthesis of fatty acids	*Azadirachta indica*	*Klebsiella pneumoniae*	Stitch 4.0
Benzoic acid	Benzoate-coenzyme A ligase	Drug metabolism	*Fragaria ananassa, Duchesnea indica.*	*Burkholderia xenovorans*	Stitch 4.0
	2-Hydroxy-6-oxo-6-phenyl-hexa-2, 4-dienoate hydrolase (BphD)	Survival in macrophages		*Burkholderia xenovorans*	Stitch 4.0
	Benzoate 1, 2-dioxygenase electron transfer subunit (BenC)	Drug metabolism		*Burkholderia xenovorans*	Stitch 4.0
	BenB, Benzoate 1, 2-dioxygenase subunit β-	Drug metabolism		*Burkholderia xenovorans*	Stitch 4.0
	Benzaldehyde dehydrogenase	Drug metabolism		*Burkholderia xenovorans*	Stitch 4.0
	BenA, Benzoate 1, 2-dioxygenase α-subunit	Drug metabolism		*Burkholderia xenovorans*	Stitch 4.0
	Benzoate-coenzyme A ligase	Drug metabolism		*Burkholderia xenovorans*	Stitch 4.0
	Amidase	Virulence factor		*Burkholderia xenovorans*	Stitch 4.0

TABLE 15.3 *(Continued)*

In vitro studies					
Antibacterial phytocompounds	**Targets identified**	**Role of target**	**Major source plants**	**Bacteria studied**	**References**
	Putative carbon-nitrogen hydrolase protein	Virulence factor		*Burkholderia xenovorans*	Stitch 4.0
Benzyl (6Z, 9Z, 12Z)-6, 9, 12-octadecatrienoate, 3-benzyloxy-1-nitro-butan-2-ol, 1, 3-Cyclohexane dicarbohydrazide	LuxR	Quorum sensing	*Salvadora persica*	*Streptococcus mutans*	Al-Sohaibani and Murugan, (2012)
Berberine	FtsZ	Bacterial division	*Berberis vulgaris, Berberis aristata*	*Escherichia coli*	Domadia et al., (2008)
	BmrR, transcriptional regulator	Regulation of transcription		*Bacillus subtilis*	Stitch 4.0
	Bmr multidrug efflux transporter	Efflux pump protein		*Bacillus subtilis*	Stitch 4.0
	FtsZ	Bacterial division		*Bacillus subtilis*	Stitch 4.0
Borneol	CspA cold shock protein	Resistance mechanism	*Blumea balsamifera, Kaempferia galanga*	*Salmonella enterica*	Stitch 4.0
Caffeic acid	Histidine ammonia lyase	Amino acid metabolism	*Eucalyptus globulus*	*Rhodobacter sphaeroides*	Stitch 4.0
Capsaicin	NorA	Efflux pump protein	*Capsicum frutescens*	*Staphylococcus aureus*	Kalia et al., (2012)
Carvacrol	DnaK molecular chaperone	Folding of nucleic acids	*Origanum vulgare*	*Lactobacillus helveticus*	Stitch 4.0
	GroEL chaperonin	Folding of nucleic acids			
Caryophyllene	O-Methyltransferase	Drug metabolism	*Origanum vulgare, Piper nigrum*	*Streptococcus pneumoniae*	Stitch 4.0

TABLE 15.3 *(Continued)*

In vitro studies					
Antibacterial phytocompounds	**Targets identified**	**Role of target**	**Major source plants**	**Bacteria studied**	**References**
Chlorogenic acid	SortaseA	Covalent anchoring of specific proteins to the peptidoglycan of the cell wall	*Coffea canephora, Coffea arabica*	*Staphylococcus aureus*	Wang et al., (2015)
	FabG	Fatty acid biosynthesis		*Escherichia coli*	Li et al., (2006)
Chrysophaentin A	FtsZ	Bacterial division	*Chrysophaeum taylori*	*Staphylococcus aureus*	Keffer et al., (2013)
Cinnamaldehyde	FtsZ	Bacterial division	*Cinnamomum verum*	*Escherichia coli*	Domadia et al., (2007)
Citronellal	Aldehyde dehydrogenase (NAD)	Removal of aldehydes and alcohols in cells that were under stress	*Eucalyptus citriodora*	*Mycobacterium smegmatis*	Stitch 4.0
	γ-aminobutyraldehyde dehydrogenase	Putrescine Catabolism		*Mycobacterium smegmatis*	Stitch 4.0
Corilagin	PBP2A	Cell wall biosynthesis	*Arctostaphylos uva-ursi*	*MRSA*	Stapleton and Taylor, (2007)
Curcumin	FtsZ	Bacterial division	*Curcuma longa*	*Bacillus subtilis*	Rai et al., (2008)
	Predicted NADP-dependent, Zn-dependent oxidoreductase	Drug metabolism		*Escherichia coli*	Stitch 4.0
	Dihydropteridine reductase	Synthesis of nucleotides and amino acids.		*Escherichia coli*	Stitch 4.0
	FtsZ	Bacterial division		*Escherichia coli*	Stitch 4.0
	Tyrosine kinase	Extracellular polysaccharide, colanic acid synthesis.		*Escherichia coli*	Stitch 4.0

TABLE 15.3 *(Continued)*

In vitro studies					
Antibacterial phytocompounds	**Targets identified**	**Role of target**	**Major source plants**	**Bacteria studied**	**References**
Dichamanetin	FtsZ	Bacterial division	*Uvaria chamae*	*Bacillus subtilis*	Lock and Harry, (2008)
Ellagic acid	Arylamine N-acetyltransferase	Drug metabolism	*Quercus alba, Quercus robur*	*Pseudomonas aeruginosa*	Stitch 4.0
Epicatechin gallate (ECG)	PBP2A	Cell wall biosynthesis	*Camellia sinensis*	*MRSA*	Stapleton et al., (2004)
Epigallocatechin gallate (EGCG)	FabI	Fatty acid biosynthesis	*Camellia sinensis*	*Escherichia coli*	Zhang and Rock, (2004)
	Glucosyltransferase	Extracellular matrix formation of cariogenic biofilms		*Streptococcus mutans*	Kim et al., (2017)
	FabG	Fatty acid synthesis		*Escherichia coli*	Zhang and Rock, (2004)
	PBP2A	Cell wall biosynthesis		*MRSA*	Bernal et al., (2010)
Eugenol	Acetolactate synthase	Synthesis of branched chain amino acids	*Syzygium aromaticum*	*Brucella melitensis*	Stitch 4.0
Gallic acid	Amidohydrolase 2	Drug metabolism	*Caesalpinia mimosides, Boswellin dalzzielli*	*Polaromonas sp.*	Stitch 4.0
	Protocatechuate 3, 4-dioxygenase α-subunit	Drug metabolism		*Polaromonas sp.*	Stitch 4.0
	HlyD Secretion protein	Secretion of proteins		*Polaromonas sp.*	Stitch 4.0
Germacrene D	FtsZ	Bacterial division	*Pinus nigra*	*Staphylococcus aureus*	Zorica et al., (2014)
Germacrene D-4-ol	FtsZ	Bacterial division	*Pinus nigra*	*Staphylococcus aureus*	Zorica et al., (2014)

TABLE 15.3 *(Continued)*

In vitro studies					
Antibacterial phytocompounds	**Targets identified**	**Role of target**	**Major source plants**	**Bacteria studied**	**References**
Glycerol monolaurate	MecR1	Signal transduction in PM	*Cocos nucifera*	*MRSA*	Stapleton and Taylor, (2007)
Glycyrrhizin	β-glucuronidase	Virulent factor	*Glycyrrhiza glabra*	*Thermotoga maritima*	Stitch 4.0
Kaempferol	Quercetin dioxygenase	Drug metabolism	*Aloe vera, Moringa oleifera*	*Bacillus subtilis*	Stitch 4.0
	YxaF, Transcriptional regulator	Regulation of transcription		*Bacillus subtilis*	Stitch 4.0
	YxaH, Putative integral inner membrane protein	Membrane potential		*Bacillus subtilis*	Stitch 4.0
	lmrA, Transcriptional repressor of lmrAB and yxaGH operons	Transcriptional regulation		*Bacillus subtilis*	Stitch 4.0
	lmrB Efflux pump protein	Efflux pump		*Bacillus subtilis*	Stitch 4.0
Licoricidin	PBP2A	Cell wall biosynthesis	*Glycyrrhiza glabra*	*MRSA*	Stapleton and Taylor, (2007)
Linoleic acid	FabI	Fatty acid synthesis	*Helichrysum pedunculatum, Schotia brachypetala*	*Escherichia coli*	Ji et al., (2005)
Myrcene	Aryl-alcohol dehydrogenase	Anaerobic respiration	*Cymbopogon citratus, Mangifera indica*	*Rhodococcus opacus*	Stitch 4.0
Oleic acid	Glucosyl transferase	Glucan formation from sucrose	*Prunus salicina*	*Streptococcus mutans*	Won et al., (2007)
p-Coumaric acid	Phenolic acid decarboxylase	Decarboxylation and detoxification of phenolics	*Gnetum cleistostachyum.*	*Bacillus subtilis*	Stitch 4.0

TABLE 15.3 *(Continued)*

		In vitro studies			
Antibacterial phytocompounds	**Targets identified**	**Role of target**	**Major source plants**	**Bacteria studied**	**References**
	Histidine ammonia-lyase	Aminoacid metabolism		*Bacillus subtilis*	Stitch 4.0
	Urease γsubunit	Vitulent factor		*Bacillus subtilis*	Stitch 4.0
	Succinyl CoA-3-oxoacid CoA-transferase	Synthesis and degradation of ketone bodies and degradation of amino acids		*Bacillus subtilis*	Stitch 4.0
p-Cymene	AmtR, Transcriptional regulator	Virulence factor	*Cuminum cyminum, Thymus vulgaris*	*Corynebacterium glutamicum*	Stitch 4.0
Piperine	MdeA, efflux pump protein	Efflux pump protein	*Piper longum, Piper nigrum*	*Staphylococcus aureus*	Mirza et al., (2011b)
	Rv1258c, efflux pump protein	Efflux pump protein		*Mycobacterium tuberculosis*	Sharma et al., (2010)
	NorA, efflux pump protein	Efflux pump protein		*Staphylococcus aureus*	Khan et al., (2006)
Polidocanol	MecR1	Signal transduction in PM	*Cocos nucifera*	*MRSA*	Stapleton and Taylor, (2007)
Quercetin	ArpR	Transcriptional regulator	*Fragaria ananassa, Spinacia oleracea*	*Pseudomonas putida*	Stitch 4.0
	FabZ	Saturated fatty acids biosynthesis		*Pseudomonas putida*	Stitch 4.0
	DnaK	Molecular chaperone		*Pseudomonas putida*	Stitch 4.0
	Glutathione peroxidase	Virulence factor		*Pseudomonas putida*	Stitch 4.0
	UxpA, Lipoprotein	Adhesion to host cell and translocation of virulence factors into host cells		*Pseudomonas putida*	Stitch 4.0

TABLE 15.3 *(Continued)*

In vitro studies					
Antibacterial phytocompounds	**Targets identified**	**Role of target**	**Major source plants**	**Bacteria studied**	**References**
	F0F1 ATP synthase subunit α-and β-	ATP synthesis		*Pseudomonas putida*	Stitch 4.0
	GyrA, DNA gyrase A	Negative supercoiling of DNA		*Pseudomonas putida*	Stitch 4.0
Quercetin diacylglycoside	GyrB, DNA gyrase B	Negative supercoiling of DNA	*Fragaria ananassa, Spinacia oleracea*	*Escherichia coli*	Suriyanarayanan et al., (2013)
Quercetin diacylglycoside	Topoisomerase IV	Resolution of chromosome dimers at DNA replication	*Fragaria ananassa, Spinacia oleracea*	*Escherichia coli*	Hossion et al., (2011)
Quercitrin	Quercetin dioxygenase	Resistance mechanism	*Fagopyrum tataricum, Nymphaea odorata*	*Bacillus subtilis*	Stitch 4.0
	Polar chromosome segregation protein	Axial filament formation		*Bacillus subtilis*	Stitch 4.0
	Phosphotransferase system (PTS) lichenan-specific enzyme IIA component	Uptake and metabolism of lichenan		*Bacillus subtilis*	Stitch 4.0
	DesR-DesK two-component response regulator	Membrane lipid fluidity regulation		*Bacillus subtilis*	Stitch 4.0
Rutin	Chitinase	Chitin degradation	*Asparagus aethiopicus*	*Lactococcus lactis*	Stitch 4.0
Sabinene	YfcC, Predicted inner membrane protein	Adhesion of bacterial cells to host	*Myristica fragrans, Piper longum*	*Escherichia coli*	Stitch 4.0
Sanguinarine	FtsZ	Bacterial division	*Sanguinaria canadensis*	*Bacillus subtilis*	Beuria, Santra, and Panda, (2005)
Scopoletin	FtsZ	Bacterial division	*Scopolia carniolica, Scopolia japonica*	*Mycobacterium tuberculosis*	Duggirala et al., (2014)

TABLE 15.3 *(Continued)*

		In vitro studies			
Antibacterial phytocompounds	**Targets identified**	**Role of target**	**Major source plants**	**Bacteria studied**	**References**
Tannin	TanC α, βfold family hydrolase	Hydrolysis of complex biomolecules	*Camellia sinensis, Quercus robur.*	*Lactobacillus plantarum*	Stitch 4.0
Tellimagrandin	PBP2A	Cell wall biosynthesis	*Rosa canina L*	*MRSA*	Stapleton and Taylor, (2007)
Terpinen-4-ol	Protein VraX	Cell wall metabolism (a polypeptide associated with cell wall stress)	*Melaleuca alternifolia*	*Staphylococcus aureus*	Stitch 4.0
Thymol	LeuB 3-isopropylmalate dehydrogenase	Biosynthesis of amino acids (leucine)	*Thymus vulgaris*	*Haemophilus influenzae*	Stitch 4.0
Thymoquinone	ATP synthase	ATP synthesis	*Nigella sativa*	*Escherichia coli*	Ahmad, Laughlin, and Kady, (2015)
Totarol	FtsZ	Bacterial division	*Podocarpus totara*	*Mycobacterium tuberculosis*	Jaiswal et al., (2007)
Ursolic acid	Matrix metalloproteinases	Biofilm formation	*Leonurus cardiaca*	*Staphylococcus aureus*	Micota et al., (2014)
(4S)-2-Methyl-2-phenyl-pentane-1, 4-diol	Glucosamine 6 phosphate synthase	Cell wall biosynthesis	*Vitex negundo Linn*	*Escherichia coli*	Wojciechowski et al., (2005)
3-Benzyloxy-1-nitro-butan-2-ol	QS DNA-binding response regulator	Biofilm formation	*Salvadora persica*	*Streptococcus mutans*	Al-Sohaibani and Murugan, (2012)
β-sitosterol	PBP2A	Alteration of target	*Azadirachta indica*	*Staphylococcus aureus*	Skariyachan, Krishnan, et al., (2011)
Butein	RmlA, RmlB, RmlC & RmlD	Rhamnose synthesis	*Butea monosperma*	*Mycobacterium tuberculosis*	Sundarrajan, Lulu, and Arumugam, (2015)

TABLE 15.3 *(Continued)*

In vitro studies					
Antibacterial phytocompounds	**Targets identified**	**Role of target**	**Major source plants**	**Bacteria studied**	**References**
Chlorogenic acid	PBP2A	Alteration of target	*Punica granatum*	*Staphylococcus aureus*	Chandramohan et al., (2014)
Citric acid	PBP2A	Alteration of target	*Punica granatum*	*Staphylococcus aureus*	Chandramohan et al., (2014)
Curcumine	Deltatoxin	Virulence factor (toxin)	*Curcuma longa*	*Clostridium perfringens*	Skariyachan, Mahajanakatti, et al., (2011)
Ellagic acid	PBP2A	Alteration of target	*Punica granatum*	*Staphylococcus aureus*	Chandramohan et al., (2014)
	Proline dehydrogenase	Aminoacid metabolism	*Phyllanthus niruri*	*Helicobacter pylori*	Lin et al., (2005)
Gloriosal	Biotin protein ligase	Biotinylation of proteins	*Gloriosa superba*	*Mycobacterium tuberculosis*	Daisy, Nivedha, and Bakiya, (2013)
Hydroxycinnamic acid	Proline dehydrogenase	Aminoacid metabolism	*Phyllanthus niruri*	*Helicobacter pylori*	Lin et al., (2005)
Isothiocyanate	Fimbrial protein	Bacterial adhesion to uroepithelial cells	*Brassica oleracea*	*Escherichia coli*	Jasmine and Selvakumar, (2012)
Meliantriol	PVL	Virulence factor (toxin)	*Aloe vera*	*Staphylococcus aureus*	Skariyachan, Krishnan, et al., (2011)
Solasodine	β-lactamase	Drug metabolism	*Solanum xanthocarpum*	*Escherichia coli*	Swain and Padhy, (2016)
Stigmasterol glucoside	β-lactamase	Drug metabolism	*Solanum xanthocarpum*	*Escherichia coli*	Swain and Padhy, (2016)

hydrolase (BphD), benzoate 1, 2-dioxygenase (BenA, BenB, and BenC), benzaldehyde dehydrogenase and benzoate-coenzyme A ligase. While amidase and putative carbon-nitrogen hydrolase protein affect virulence factor, others can block drug metabolism (Stitch 4.0.). Another virulence factor such as AmtR, a transcriptional regulator that present in *Corynebacterium glutamicum* can be inhibited by p-cymene. Kovac et al., studied the target action of α-pinene in *Campylobacter jejuni* and found it can bind to HrcA and HspR which are efflux pump proteins of the bacteria (Kovac et al., 2015). The most studied phytocompounds against *E. coli* are berberine, chlorogenic acid, cinnamaldehyde, EGCG, quercetin diacylglycoside, thymoquinone, curcumin, linoleic acid, and sabinene. Among them, curcumin, EGCG and quercetin diacylglycoside possess multiple targets in *E. coli.* The targets of curcumin are dihydropteridine reductase, FtsZ, predicted NADP-dependent, Zn-dependent oxidoreductase and tyrosine kinase which can block synthesis of nucleotides and amino acids, bacterial division, drug metabolism, extra-cellular polysaccharide, and colanic acid synthesis respectively (Stitch 4.0.). While EGCG has only two targets-Fab G and Fab I which function in fatty acid biosynthesis (Zhang and Rock, 2004). Like EGCG, linoleic acid also can target on Fab I. Quercetin diacylglycoside targets on DNA gyrase B (Suriyanarayanan et al., 2013) and topoisomerase IV (Hossion et al., 2011) which are responsible for negative supercoiling of DNA and resolution of chromosome dimers at DNA replication. Sabinene inhibits the adhesion of *E. Coli* to host cells by binding to YfcC, a predicted membrane protein.

Certain studies on effective killing of *Mycobacterium tuberculosis* reveals that scopoletin (Duggirala et al., 2014) and totarol (Jaiswal et al., 2007) can inhibit the cell division by binding on FtZ protein while piperine (Sharma et al., 2010) can effect an efflux pump protein, called Rv1258c. The target of citronellal in *Mycobacterium smegmatis* was also studied as well. It was found that citronellal can bind to aldehyde dehydrogenase and γ-aminobutyraldehyde dehydrogenase and can affect putrescine catabolism and removal of aldehydes and alcohols in cells when cells are under stress. The mechanisms of action of quercetin in *Pseudomonas putida* have been studied very well. The reported targets are TArpR, DnaK, α-and β-subunits of F0 F1 ATP synthase, FabZ, glutathione peroxidase, DNA gyrase A and UxpA lipoprotein. Some phytocompounds are less studied and reported with single target against single pathogen. For instance, Carvacrol can effect the folding of bacterial nucleic acids and found effective against *Lactobacillus helveticus (*Stitch 4.0.). Antibacterial activity of Tannin has also been observed effective against the same genus, but in different species, L. *plantarum* (Stitch 4.0.) and its target molecule found to be TanC α, β fold family hydrolase. Even though some targets were identified by *in-silico* studies, the most of such findings are not validated by wet lab experiments.

15.5 CONCLUSION AND FUTURE PERSPECTIVES

Plants are the potential source for many phytocompounds with single or multiple properties which in turn found to be a tremendous source of new drugs. The experience-based claims from ethnomedicine practitioners must be transformed into evidence-based claims by our scientists, in order to establish credibility for Indian traditional herbal drug use in the modern settings. Moreover, increasing concern about drug resistance even in common pathogens increases the interest in the studied antimicrobial phytocompounds. Of course, this will be an intense area of research in coming future.

KEYWORDS

- **antibacterial target studies**
- **bioactivity**
- **phytocompounds**
- **traditional medicine**

REFERENCES

Abdenour Bounihi; Arezki Bitam; Asma Bouazza; Lyece Yargui; Elhadj Ahmed Koceir. "Fruit Vinegars Attenuate Cardiac Injury via Anti-Inflammatory and Anti-Adiposity Actions in High-Fat Diet-Induced Obese Rats." *Pharmaceutical Biology* **2017,** *55*(1), 43–52.

Abhishek Kumar Sah; Vinod Kumar Verma. "Phytochemicals and Pharmacological Potential of *Nyctanthes arbortristis*: A Comprehensive Review." *International Journal of Research in Pharmaceutical and Biomedical Sciences* **2012,** *3*(1), 420–427.

Abu-Eittah, R. et al., "Antifungal Activity of Rhein Isolated From Cassia Fistula L. Flower." *Webmedcentral Pharmacology* **2014,** *1*(4), 1–8.

Abugafar Hossion, M. L. et al., "Quercetin Diacylglycoside Analogues Showing Dual Inhibition of DNA Gyrase and Topoisomerase IV as Novel Antibacterial Agents." *Journal of Medicinal Chemistry* **2011,** *54*(11), 3686–3703.

Adele Papetti et al., "Identification of Organic Acids in Cichorium Intybus Inhibiting Virulence-Related Properties of Oral Pathogenic Bacteria." *Food Chemistry* **2013,** *138* (2–3), 1706–1712.

Aisha Siddiqui; Amin, K. M. Y.; Zuberi, R. H.; Anwar Jamal. "Standardization of Majith (*Rubia cordifolia* Linn.)." *Indian Journal of Traditional Knowledge* **2011,** *10*(1), 330–33.

Al Shaimaa M. Taha; Omayma A. Eldahshan. "Chemical Characteristics, Antimicrobial, and Cytotoxic Activities of the Essential Oil of Egyptian *Cinnamomum glanduliferum* Bark." *Chemistry & Biodiversity* **2017,** *14*(5), 160–183.

Alaattin Sen et al., "Modulatory Actions of O-Coumaric Acid on Carcinogen-Activating Cytochrome P450 Isozymes and the Potential for Drug Interactions in Human Hepatocarcinoma Cells." *Pharmaceutical Biology* **2015,** *53*(9), 1391–1398.

Alves, M. J. et al., "Antimicrobial Activity of Phenolic Compounds Identified in Wild Mushrooms, SAR Analysis, and Docking Studies." *Journal of Applied Microbiology* **2013,** *115*(2), 346–357.

Ana Cristina Abreu; Andrew J. McBain; Manuel Simões. "Plants as Sources of New Antimicrobials and Resistance-Modifying Agents." *Natural Product Reports* **2012,** *29*(9), 1007–1017.

Anjana Sharma; Rani Verma; Padmini Ramteke. "Antibacterial Activity of Some Medicinal Plants Used by Tribals Against Uti Causing Pathogens." *World Applied Sciences Journal* **2009,** *7*(3), 332–339.

Anna Bogacz; Joanna Bartkowiak-Wieczorek; Lucjusz Zaprutko; Przemyslaw L. Mikolajczak. "Strong and Long-Lasting Antinociceptive and Anti-Inflammatory Conjugate of Naturally Occurring Oleanolic Acid and Aspirin." *Frontiers in Pharmacology* **2016,** *7*(3), 1–18.

Ardalan Alizadeh; Hamid Abdollahzadeh. "Essential Oil Constituents and Antimicrobial Activity of *Pycnocycla bashagardiana* Mozaff. from Iran." *Natural Product Research* **2017,** *31*(17), 2081–2084.

Atanas G. Atanasov; Birgit Waltenberger; Eva-Maria Pferschy-Wenzig. "Discovery and Resupply of Pharmacologically Active Plant-Derived Natural Products: A Review." *Biotechnol Adv* **2015,** *33*(8), 1582–1614.

Azam Hosseinzadeh; Davood Jafari; Tunku Kamarul; Abolfazll Bagheri; Ali M. Sharifi. "Evaluating the Protective Effects and Mechanisms of Diallyl Disulfide on Interleukin-1β-Induced Oxidative Stress and Mitochondrial Apoptotic Signaling Pathways in Cultured Chondrocytes." *Journal of Cellular Biochemistry* **2017,** *118*(7), 1879–1888.

Bagora Bayala; Imaël H. N. Bassole; Riccardo Scifo; Charlemagne Gnoula; Laurent Morel. "Anticancer Activity of Essential Oils and Their Chemical Components -a Review." *Am J Cancer Res* **2014,** *4*(6), 591–607.

Balasubramanian Suriyanarayanan; Karthi Shanmugam; Ramachandran Sarojini Santhosh. "Synthetic Quercetin Inhibits Mycobacterial Growth Possibly by Interacting with DNA Gyrase." *Romanian Biotechnological Letters* **2013,** *18*(5), 8587–93.

Bandaranayake, W. M. "Bioactivities, Bioactive Compounds and Chemical Constituents of Mangrove Plants." *Wetland Ecology and Management* **2002,** *10*(1), 421–452.

Bao-Xin-Zi Liu et al., "Hederagenin from the Leaves of Ivy (Hedera Helix L.) Induces Apoptosis in Human LoVo Colon Cells through the Mitochondrial Pathway." *BMC Complementary and Alternative Medicine* **2014,** *14*(412), 1–10.

Barbosa, R. et al., "Effects of Lippia Sidoides Essential Oil, Thymol, P-Cymene, Myrcene and Caryophyllene on Rat Sciatic Nerve Excitability." *Brazilian Journal of Medical and Biological Research* **2017,** *50*(12), 51–63.

Bartłomiej Micota; Beata Sadowska; Anna Podsędek; Małgorzata Redzynia; Barbara Różalska. "*Leonurus Cardiaca* L. Herb Derived Extract and Ursolic Acid as the Factors Affecting the Adhesion Capacity of *Staphylococcus aureus* in the Context of Infective Endocarditis." *Acta Biochimica Polonica* **2014,** *61*(2), 385–388.

Bektas Tepe; Dimitra Daferera; Atalay Sokmen; Munevver Sokmen; Moschos Polissiou. "Antimicrobial and Antioxidant Activities of the Essential Oil and Various Extracts of *Salvia tomentosa* Miller (Lamiaceae)." *Food Chemistry* **2005,** *90*(3), 333–340.

Bharat B. Aggarwal; Sahdeo Prasad; Simone Reuter; Ramaswamy Kannappan; R. Vivek. "Allicin Protects against Cisplatin-Induced Vestibular Dysfunction by Inhibiting the Apoptotic Pathway." *Curr Drug Targets* **2011,** *12*(11), 1595–1653.

Bianchini, A. E. et al., "Monoterpenoids (Thymol, Carvacrol and S-(+)-Linalool) with Anesthetic Activity in Silver Catfish (*Rhamdia quelen*): Evaluation of Acetylcholinesterase and GABAergic Activity." *Brazilian journal of medical and biological research* **2017,** *50*(12), 46–63

Bing-Hui Li; Xiao-Feng Ma; Xiao-Dong Wu; Wei-Xi Tian. "Inhibitory Activity of Chlorogenic Acid on Enzymes Involved in the Fatty Acid Synthesis in Animals and Bacteria." *IUBMB Life* **2006,** *58*(1), 39–46.

Binsong Han et al., "Comprehensive Characterization and Identification of Antioxidants in Folium Artemisiae Argyi Using High-Resolution Tandem Mass Spectrometry." *Journal of Chromatography B* **2017,** 1063(1), 84–92.

Cabo, M. L.; Braber, A. F. M.; Koenraad, P. M. F. J. "Apparent Antifungal Activity of Several Lactic Acid Bacteria against *Penicillium discolor* Is Due to Acetic Acid in the Medium." *Journal of Food Protection* **2002,** *65*(8), 1309–1316.

Camille S. Bowen-Forbes; Yanjun Zhang; and Muraleedharan G. Nair. "Journal of Food Composition and Analysis Anthocyanin Content, Antioxidant, Anti-Inflammatory and Anticancer Properties of Blackberry and Raspberry Fruits." *Journal of Food Composition and Analysis* **2010,** *23*(1), 554–560.

Chandramohan, A.; Divya, S. R.; Dhanarajan, M. S. "Antimicrobial Potential of Active Molecules from *Punica* granatum Fruit Extract by *in Vitro* and *in Silico* Methods." *International Journal of Bioscience Research* **2014,** *3*(4), 1–8.

Chang Ji et al., "Fatty Acid Synthesis Is a Target for Antibacterial Activity of Unsaturated Fatty Acids." *FEBS Letters* **2005,** *579*(3), 5157–5162.

Mamilla R. Charan Raja; Anand Babu Velappan; Davidraj Chellappan; Joy Debnath; Santanu Kar Mahapatra. "Eugenol Derived Immunomodulatory Molecules against Visceral Leishmaniasis." *European Journal of Medicinal Chemistry* **2017,** *139*(1), 503–518.

Chen, Y. J.; Shiao, M. S.; Wang, S. Y. "The Antioxidant Caffeic Acid Phenethyl Ester Induces Apoptosis Associated with Selective Scavenging of Hydrogen Peroxide in Human Leukemic HL-60 Ce." *Anticancer drugs* **2001,** *12*(2), 143–149.

Chengwei Niu; Man Ma; Xiao Han; Zimin Wang; Hangyan Li. "Hyperion Protects against Cisplatin-Induced Liver Injury in Mice." *Acta Cirurgica Brasileira* **2017,** *32*(8), 633–640.

Chengyuan Liang; Weihui Ju; Shaomeng Pei; Yonghong Tang; Yadong Xiao. "Pharmacological Activities and Synthesis of Esculetin and Its Derivatives: A Mini-Review." *Molecules* **2017,** *22*(3), 387.

Chintana Chirathaworn, and Wisatre Kongcharoensuntorn. "*Myristica fragrans* Houtt. Methanolic Extract Induces Apoptosis in a Human Leukemia Cell Line through SIRT1 mRNA Downregulation." *J Med Assoc Thai* **2007,** *90*(11), 2422–2428.

Chun-Yan Shen; Jian-Guo Jiang; Wei Zhu; Qin Ou-Yang. "Anti-Inflammatory Effect of Essential Oil from *Citrus aurantium* L. Var.amara Engl." *Journal of Agricultural and Food Chemistry* **2017,** *65*(39), 8586–8594.

Damasceno Nogueira; Neto Antonia; Amanda Cardoso. "Antioxidant Effects of Nerolidol in Mice Hippocampus After Open Field Test." *Neurochem Res* **2013,** *38*(1), 1861–1870.

Daniel J. Tueller; Jackson S. Harley; and Chad R. Hancock. "Effects of Curcumin and Ursolic Acid on the Mitochondrial Coupling Efficiency and Hydrogen Peroxide Emission of Intact Skeletal Myoblasts." *Biochemical and Biophysical Research Communications* **2017,** *492*(3), 368–372.

David R. George; Robert D. Finn; Kirsty M. Graham; Olivier A. E. Sparagano. "Present and Future Potential of Plant-Derived Products to Control Arthropods of Veterinary and Medical Significance." *Parasites & Vectors* **2014,** *7*(1), 1–12.

De-Yang Shen et al., "Chemical Constituents from *Andrographis echioides* and Their Anti-Inflammatory Activity." *International Journal of Molecular Science* **2013,** *14*(1), 496–514.

Dipti Rai; Jay Kumar Singh; Nilanjan Roy; Dulal Panda. "Curcumin Inhibits FtsZ Assembly: An Attractive Mechanism for Its Antibacterial Activity." *Biochemical Journal* **2008,** *410*(1), 147–155.

Dipti Rai; Jay Kumar Singh; Nilanjan Roy; Dulal Panda. "Curcumin Inhibits FtsZ Assembly: an Attractive Mechanism for Its Antibacterial Activity" *Biochemical Journal* **2008,** *410*(1), 147–155.

Duraipandiyan, V.; Ignacimuthu, S. "Antifungal Activity of Rhein Isolated from *Cassia fistula* L. Flower." *Webmedcentral Pharmacology* **2010,** *1*(9), 1–8.

Elena Gabriele, et al., "New Sulfurated Derivatives of Cinnamic Acids and Rosmaricine as Inhibitors of STAT3 and NF-κ B Transcription Factors." *Journal of Enzyme Inhibition and Medicinal Chemistry* **2017,** 1012–1028.

Emre Ceyhan; Mediha Canbek. "Determining the Effects of Geraniol on Liver Regeneration Via the Nuclear Factor-kB Pathway After Partial Hepatectomy." *Alternative therapies in health and medicine* **2017,** *23*(3), 38–45.

Fabricant, D. S.; N. R. Farnsworth. "The Value of Plants Used in Traditional Medicine for Drug Discovery." *Environmental Health Perspectives* **2001,** *109*(1), 69–75.

Fatemeh Forouzanfar; Bibi Sedigheh; Fazly Bazzaz; Hossein Hosseinzadeh. "Black Cumin (*Nigella sativa*) and Its Constituent (Thymoquinone): A Review on Antimicrobial Effects." *Irani J Basic Medi Sci* **2014,** *17*(12), 929–38.

Fenella D. Halstead; Maryam Rauf; Naiem S. Moiemen; Amy Bamford. "The Antibacterial Activity of Acetic Acid against Biofilm-Producing Pathogens of Relevance to Burns Patients." *PLoS ONE* **2015,** *12*(3), 1–15.

Feyza Oke-Altuntas; Selma Ipekcioglu; Ayse Sahin Yaglioglu; Lutfi Behcet. "Phytochemical Analysis, Antiproliferative and Antioxidant Activities *of Chrozophora tinctoria* : A Natural Dye Plant." *Pharmaceutical Biology* **2017,** *6*(1), 176–181.

Fraise, A. P.; Wilkinson, M. A. C.; Bradley, C. R.; Oppenheim, B.; Moiemen, N. "The Antibacterial Activity and Stability of Acetic Acid." *Journal of Hospital Infection* **2013,** *84*(4), 329–331.

Francesco Epifano; Salvatore Genovese. "Lapachol and Its Congeners as Anticancer Agents : A Review." *Phytochemistry Reviews* **2013,** *12*(1), 1–13.

Francisco Javier Pérez-Areales, et al., "Design, Synthesis and Multitarget Biological Profiling of Second-Generation Anti-Alzheimer Rhein–huprine Hybrids." *Future Medicinal Chemistry* **2017,** *9*(10), 965–981.

Gabriele Andressa Zatelli et al., "Antimycoplasmic Activity and Seasonal Variation of Essential Oil of *Eugenia hiemalis* Cambess. (Myrtaceae)." *Natural Product Research* **2016,** *30*(17), 1961–1964.

Gao, L. E. I. et al., "Anticancer Effect of SZC017, a Novel Derivative of Oleanolic Acid, on Human Gastric Cancer Cells." *Oncology Reports* **2016,** *35*(1), 1101–1108.

Ghulam Shabir et al., "Antioxidant and Antimicrobial Attributes and Phenolics of Different Solvent Extracts from Leaves, Flowers and Bark of Gold Mohar [*Delonix regia* (Bojer Ex Hook.) Raf.]." *Molecules* **2011,** *16*(1), 7302–7319.

Gislene G. F. Nascimento; Juliana Locatelli; Paulo C. Freitas; Giuliana L. Silva; Universidade Metodista De Piracicaba. "Antibacterial Activity of Plant Extracts and Phytochemicals on Antibiotic-." *Brazilian Journal of Microbiology* **2000,***31,* 247–256.

Gundidza, M. et al., "Phytochemical Composition and Biological Activities of Essential Oil of *Rhynchosia minima* (L) (DC) (Fabaceae)." *African Journal of Biotechnology* **2009,** *8*(5), 721–724.

Hala Gali Muhtasib; Albert Roessner; Regine Schneider Stock. "Thymoquinone_A Promising Anti-Cancer Drug from Natural Sources-ScienceDirect." *The International Journal of Biochemistry and Cell Biology* **2006,** *38*(8), 1249–1253.

Hayate Javed; Sheikh Azimullah; Salema B. Abul Khair; Shreesh Ojha, M.Emdadul Haque. "Neuroprotective Effect of Nerolidol against Neuroinflammation and Oxidative Stress Induced by Rotenone." *BMC Neuroscience* **2016,** *124*(2) 1–12.

Hisae Muroi; Masaki Himejima. "Antibacterial Activity of Totarol and Its Potentiation." *Journal of Natural Products* **1992,** *55*(10), 1436–1440.

Hong-Fang Ji; Xue-Juan Li; Hong-Yu Zhang. "Natural Products and Drug Discovery. Can Thousands of Years of Ancient Medical Knowledge Lead Us to New and Powerful Drug Combinations in the Fight against Cancer and Dementia?" *EMBO reports* **2009,** *10*(3), 194–200.

Honggao Xu; Kedong Tai; Tong Wei; Fang Yuan; Yanxiang Gao. "Physicochemical and in Vitro Antioxidant Properties of Pectin Extracted from Hot Pepper (Capsicum annuum L. Var. Acuminatum (Fingern)) Residues with Accepted Article Hydrochloric and Sulphuric Acids." *J. Food and Agriculture* **2017,** *97*(14), 4953–4960.

Inshad Ali Khan; Zahid Mehmood Mirza; Ashwani Kumar; Vijeshwar Verma; Ghulam Nabi Qazi. "Piperine, a Phytochemical Potentiator of Ciprofloxacin against *Staphylococcus aureus*." *Antimicrobial Agents and Chemotherapy* **2006,** *50*(2), 810–812.

Ioannis Erkekoglou; Nikolaos Nenadis; Efrosini Samara; Fani Th. Mantzouridou. "Functional Teas from the Leaves of Arbutus Unedo: Phenolic Content, Antioxidant Activity, and Detection of Efficient Radical Scavengers." *Plant Foods for Human Nutrition* **2017,** *72*(2), 176–83.

Islem Yangui; Meriem Zouaoui Boutiti; Mohamed Boussaid; Chokri Messaoud. "Essential Oils of Myrtaceae Species Growing Wild in Tunisia: Chemical Variability and Antifungal Activity Against *Biscogniauxia mediterranea*, the Causative Agent of Charcoal Canker." *Chemistry & Biodiversity* **2017,** *14*(7), 58–67.

Itzamná Baqueiro-Peña; José Á. Guerrero-Beltrán. "Physicochemical and Antioxidant Characterization of *Justicia spicigera*." *Food Chemistry* **2017,** *218*(1), 305–312.

Jasmine, R.; Selvakumar, B. N. "Evaluation of the Mechanism of Action of a Safer Nutritional and Therapeutic Drug from *Brassica oleracea* against Urinary Tract Infection in Pregnant Women." *Research Signpost* **2012,** *661*(2), 165–188.

Jasna Kovac et al., "Antibiotic Resistance Modulation and Modes of Action of alpha-Pinene in *Campylobacter jejuni.*" *PLoS ONE* **2015,** *10*(4), 1–14.

Jean De Dieu Tamokou et al., "Anticancer and Antimicrobial Activities of Some Antioxidant-Rich Cameroonian Medicinal Plants." *PLoS ONE* **2013,** *8*(2), 1–14.

Jessica L. Keffer et al., "Chrysophaentins Are Competitive Inhibitors of FtsZ and Inhibit Z-Ring Formation in Live Bacteria." *Bioorganic and Medicinal Chemistry* **2013,** *21*(18), 5673–5678.

Jeyadevi, R.; Sivasudha, T.; Ilavarasi, A.; Thajuddin, N. "Chemical Constituents and Antimicrobial Activity of Indian Green Leafy Vegetable *Cardiospermum halicacabum*." *Indian J MicrobioL.* **2013,** *53*(2), 208–213.

Jiaix Feng, J.; Lipton, M. "Eugenol Antipyretic Activity in Rabbits." *Neuropharmacology* **1987,** *26*(12), 1775–1778.

Jiang, Y.; Gao, H.; Turdu, G. "Traditional Chinese Medicinal Herbs as Potential AChE Inhibitors for Anti-Alzheimer's Disease_ A Review." *Bioorg Chem.* **2017,** *75*(1), 50–61.

Jianzhang Wang; Gene Chi Wai Man; Tak Hang Chan; Joseph Kwong; Chi Chiu Wang. "A Prodrug of Green Tea Polyphenol (–)-Epigallocatechin-3-Gallate (Pro-EGCG) Serves as a Novel Angiogenesis Inhibitor in Endometrial Cancer." *Cancer Letters* **2018,** *412*(1), 10–20.

Jie Zhang et al., "Antiinfective Therapy with a Small Molecule Inhibitor of *Staphylococcus aureus* Sortase." *Proceedings of the National Academy of Sciences* **2014,** *111*(37), 13517–13522.

Jin-Ming Kong; Ngoh-Khang Goh; Lian-Sai Chia; Tet-Fatt Chia. "Recent Advances in Traditional Plant Drugs and Orchids." *Acta pharmacologica Sinica* **2003,***24*(1), 7–21.

Jin-Shuai Lan et al., "Design, Synthesis, and Evaluation of Novel Cinnamic Acid Derivatives Bearing N-Benzyl Pyridinium Moiety as Multifunctional Cholinesterase Inhibitors for Alzheimer's Disease." *Journal of Enzyme Inhibition and Medicinal Chemistry* **2017,** 776–788.

John K. Mwonjoria et al., "Ethno Medicinal, Phytochemical and Pharmacological Aspects of *Solanum incanum* (Lin.)." *International Journal of Pharmacology and Toxicology* **2014,** *2*(2), 17–20.

Jorge A. R. Salvador et al., "Oleanane-, Ursane-, and Quinone Methide Friedelane-Type Triterpenoid Derivatives: Recent Advances in Cancer Treatment Jorge." *European Journal of Medicinal Chemistry* **2017,** *21*(3), 317–328.

Joseph C. Maroon; Jeffrey W. Bost; Adara Maroon. "Natural Anti-Inflammatory Agents for Pain Relief." *Surg Neurol Int* **2010,** *80*(1), 48–56.

Joy R. Borchardt, et al., "Antioxidant and Antimicrobial Activity of Seed from Plants of the Mississippi River Basin." *Medicinal Plant Research* **2009,** *3*(10), 707–718.

Joyce Kelly et al., "Composition and Cytotoxic and Antioxidant Activities of the Oil of *Piper aequale* Vahl." *Lipids in Health and Disease* **2016,** *15*(174), 1–6.

Julia Peet; Anastasia Selyutina; Aleksei Bredihhin. "Bioorganic & Medicinal Chemistry Antiretroviral (HIV-1) Activity of Azulene Derivatives." *Bioorganic & Medicinal Chemistry* **2016,** *11*(1), 1–5.

Juliana Silva, P.; Isabel C. Silva; Fernando R. Pavan, F. Davi; Márcio P. De Araujo. "Bis (diphenylphosphino)amines-Containing Ruthenium Cymene Complexes as Metallodrugs against *Mycobacterium tuberculosis* Juliana." *Journal of Inorganic Biochemistry* **2017,** *13*(1), 56–67.

Kai Fana; Xiaolei Lib; Yonggang Caoc; Hanping Qic; Lei Lid; Qianhui Zhangc; Hongli Sun. "Carvacrol Inhibits Proliferation and Induces Apoptosis in Human Colon Cancer Cells." *Anti-Cancer Drugs* **2015,** *26*(1), 813–823.

Kantha Deivi Arunachalam; Jaya Krishna Kuruva; Shanmugasundaram Hari; Sathesh Kumar Annamalai; Kamesh Viswanathan Baskaran. "HPTLC Finger Print Analysis and Phytochemical Investigation of *Morinda tinctoria* Roxb Leaf Extracts by HPLC and GCMS." *International Journal of Pharmacy and Pharmaceutical Sciences* **2015,** *7*(2), 360–366.

Ke-Qiang Chi, et al., "Design, Synthesis, and Evaluation of Novel Ursolic Acid Derivatives as HIF-1α Inhibitors with Anticancer Potential." *Bioorganic Chemistry* **2017,** *75*(1), 157–169.

Ken-Ichi Fujita, et al., "Anethole Potentiates Dodecanol's Fungicidal Activity by Reducing PDR5 Expression in Budding Yeast." *Biochimica et Biophysica Acta (BBA) -General Subjects* **2017,** *1861*(2), 477–484.

Kong, C. S.; Jeong, C. H.; Choi, J. S.; Kim, K. J.; Jeong, J. W. "Antiangiogenic Effects of P-Coumaric Acid in Human Endothelial Cells." *Phytother Res.* **2013,** *27*(3), 317–323.

Kwang Sik Suh; Suk Chon; Eun Mi Choi. "Limonene Protects Osteoblasts against Methylglyoxal-Derived Adduct Formation by Regulating Glyoxalase, Oxidative Stress, and Mitochondrial Function." *Chemico-Biological Interactions* **2017b,** *278*(1), 15–21.

Lakshmi, T.; Vidya Krishnan, Rajendran, R.; Madhusudhanan, N. "*Azadirachta indica*_ A Herbal panacea in Dentistry–An Update." *Pharmacogn Rev.* **2015,** *9*(17), 41–44.

Legault, J.; Pichette, A. "Potentiating Effect of Beta-Caryophyllene on Anticancer Activity of Alpha-Humulene, Isocaryophyllene, and Paclitaxel." *J Pharm Pharmacol.* **2007,** *59*(12), 1643–1647.

Lena Gálvez Ranilla; Emmanouil Apostolidis; Kalidas Shetty. "Antimicrobial Activity of an Amazon Medicinal Plant (Chancapiedra) (*Phyllanthus niruri* L.) against Helicobacter pylori and Lactic Acid Bacteria." *Phytotherapy Research* **2012,** *26*(6), 791–799.

Lin Wang et al., "The Therapeutic Effect of Chlorogenic Acid against *Staphylococcus aureus* Infection through Sortase A Inhibition." *Frontiers in Microbiology* **2015,** *6*(3), 1–12.

Lin, Y. T.; Kwon, Y. I.; Labbe, R. G.; Shetty, K. "Inhibition of *Helicobacter pylori* and Associated Urease by Oregano and Cranberry Phytochemical Synergies." *Applied and Environmental Microbiology* **2005,** *71*(12), 8558–8564.

Małgorzata Tańska; Natalia Mikołajczak; Iwona Konopka. "Comparison of the Effect of Sinapic and Ferulic Acids Derivatives (4-Vinylsyringol vs. 4-Vinylguaiacol) as Antioxidants of Rapeseed, Flaxseed, and Extra Virgin Olive Oils." *Food Chemistry* **2018,** *240*(1), 679–685.

Malgorzata Tyszka-Czochara; Karolina Bukowska-Strakova; Marcin Majka. "Metformin and Caffeic Acid Regulate Metabolic Reprogramming In Human Cervical Carcinoma SiHa/HTB-35 Cells and Augment Anticancer Activity of Cisplatin via Cell Cycle Regulation. Malgorzata." *Food and Chemical Toxicology* **2017,** *4*(1), 1–20.

Mann, C. M.; Markham, J. L. "A New Method for Determining the Minimum Inhibitory Concentration of Essential Oils." *Journal of Applied Microbiology* **1998,** *84*(1), 538–544.

Mano Horinaka et al., "The Dietary Flavonoid Apigenin Sensitizes Malignant Tumor Cells to Tumor Necrosis Factor-Related Apoptosis-Inducing Ligand." *Mol Cancer Ther* **2006,** *5*(4), 945–952.

Manoj Kumbhare; Thangavel Sivakumar. "Anti-Inflammatory and Antinociceptive Activity of Pods of *Caesalpinia pulcherrima*." *Journal of Applied Pharmaceutical Science* **2011,** *1*(7), 180–184.

Marco Martín González-Chávez, et al., "Anti-Inflammatory Activity of Standardized Dichloromethane Extract of *Salvia connivance* on Macrophages Stimulated by LPS." *Pharmaceutical Biology* **2017,** *55*(1), 1467–1472.

Marek Wojciechowski; Sławomir Milewski; Jan Mazerski; Edward Borowski. "Glucosamine-6-Phosphate Synthase, a Novel Target for Antifungal Agents. Molecular Modeling Studies in Drug Design." *Acta Biochimica Polonica* **2005,** *52*(3), 647–653.

Marjan Nassiri Asl; Hossein Hosseinzadeh. "Review of Pharmacological Effects of *Glycyrrhiza* Sp. and Its Bioactive Compounds." *Phytotherapy Research* **2008,** *22*(1), 709–724.

Marjorie Murphy Cowan. "Plant Products as Antimicrobial Agents." *Clin. Microbiol. Rev.* **1999,** *12*(4), 564–582.

Maruthi Prasad, E.; Nanda Kumar, Y.; Lakshmi Devi Kodidhela. "Molecular Docking Studies of Phytochemicals of *Vitex negundo* (L.) against Adenosine A1 Receptor as Therapeutic Target in Cardiovascular Diseases." *Journal of Bioinformatics and Proteomics Review* **2015,** *1*(2), 1–5.

Melissa R. Jacob; Shabana I. Khan. "Antioxidant, Antimalarial and Antimicrobial Activities of Tannin-Rich Fractions, Ellagitannins and Phenolic Acids from *Punica granatum* L." *Planta Med* **2007,** *7*(1), 1–11.

Miao Wang et al., "Rapid Prediction and Identification of Lipase Inhibitors in Volatile Oil from *Pinus massoniana* L. Needles." *Phytochemistry* **2017,** *141*(1), 114–120.

Michael F. Ryan; Sara Rooney; Ryan, M. F. "Effects of Alpha-Hederin and Thymoquinone, Constituents of *Nigella sativa*, on Human Cancer Cell Lines." *Anticancer Research* **2005,** *25*(4), 2199–2204.

Michalina Grzesik; Katarzyna Naparło; Grzegorz Bartosz; Izabela Sadowska-Bartosz. "Antioxidant Properties of Catechins: Comparison with Other Antioxidants." *Food Chemistry* **2018,** *241*(1), 480–492.

Mohamed M. Abdel-Daim; Omnia E. Kilany; Hesham A. Khalifa; Amal A. M. Ahmed. "Allicin Ameliorates Doxorubicin-Induced Cardiotoxicity in Rats via Suppression of Oxidative Stress, Inflammation, and Apoptosis." *Cancer Chemotherapy and Pharmacology* **2017,** *80*(4), 745–753.

Mohammad Khursheed Siddiqi Parvez Alam Sumit; Kumar Chaturvedi Rizwan Hasan Khan. "Anti-Amyloidogenic Behavior and Interaction of Diallylsulfide with Human Serum Albumin." *International Journal of Biological Macromolecules* **2016,** *92*(1), 1212–1220.

Mohammed Rahmatullah; Rownak Jahan et al., "A Pharmacological Evaluation of Medicinal Plants Used by Folk Medicinal Practitioners of Station Purbo Para Village of Jamalpur Sadar Upazila in Jamalpur District, Bangladesh." *American-Eurasian Journal of Sustainable Agriculture* **2010,** *4*(2), 170–195.

Mohammed Rahmatullah; Walied Samarrai et al., "An Ethnomedicinal, Pharmacological and Phytochemical Review of Some Bignoniaceae Plants in Folk Medicinal Uses in Bangladesh." *Advances in Natural and Applied Sciences* **2010,** *4*(3), 236–253.

Mohd Danish; Pradeep Singh; Garima Mishra; Shruti Srivastava. "*Cassia fistula* Linn. (Amulthus) -An Important Medicinal Plant: A Review of Its Traditional Uses, Phytochemistry and Pharmacological Properties." *J.Nat.Prod.Plant Resour.* **2011,** *1*(1), 101–118.

Moricz, A. M.; Ott, P. G.; Billes, F.; Otta, K. H.; Tyihak, E. "The Influence of L -Ascorbic Acid on the Antibacterial-Toxic Activity of Aflatoxins on Adsorbent Layer." *Journal of Applied Microbiology* **2007,** *103*(1), 2525–2532.

Naiane Ferraz Bandeira Alves; Thyago Moreira de Queiroz; Rafael de Almeida Travassos; Marciane Magnani; Valdir de Andrade Braga. "Acute Treatment with Lauric Acid Reduces Blood Pressure and Oxidative Stress in Spontaneously Hypertensive Rats." *Basic & Clinical Pharmacology & Toxicology* **2017,** *120*(4), 348–353.

Naidu, K. C.; Ramya Lalam; Varaprasad Bobbarala. "Antimicrobial Agents from *Rubia cordifolia* and *Glycyrrhiza glabra* against Phytopathogens of *Gossypium*." *International Journal of PharmTech Research* **2009,** *1*(4), 1512–1518.

Najet Souhaiel et al., "Ammoides Pusilla (Apiaceae) Essential Oil: Activity against *Acanthamoeba castellanii* Neff." *Experimental Parasitology* **2017,** *183*(1), 99–103.

Nathália G. Amado, et al., "Isoquercitrin Isolated from *Hyptis Fasciculata* Reduces Glioblastoma Cell Proliferation and Changes β-Catenin Cellular Localization." *Anti-Cancer Drugs* **2009,** *20*(7), 543–552.

Neera Raghav; Mamta Singh. "SAR Studies of Some Acetophenone Phenylhydrazone Based Pyrazole Derivatives as Anticathepsin Agents." *Bioorganic Chemistry* **2017,** *75*(1), 38–49.

Nitin Pal Kalia et al., "Capsaicin, a Novel Inhibitor of the NorA Efflux Pump, Reduces the Intracellular Invasion of *Staphylococcus aureus*." *Journal of Antimicrobial Chemotherapy* **2012,** *67*(10), 2401–2408.

Oliver Kayser; Herbert Kolodziej. "Antibacterial Activity of Simple Coumarins: Structural Requirements for Biological Activity." *Journal of Biosciences* **1999,** *54*(3), 169–74.

Oloche, J. J.; F. Okwuasaba, Obochi, G. O. "Review of Phytochemical, Pharmacological and Toxicological Profile of *Stereospermum kunthianum*." *Journal of Advances in Medical and Pharmaceutical Sciences* **2016,** *5*(1), 1–10.

Oluwatosin Adaramoye, et al., "Punicalagin, a Polyphenol from Pomegranate Fruit, Induces Growth Inhibition and Apoptosis in Human PC-3 and LNCaP Cells." *Chemical-Biological Interactions* **2017,** *274*(1), 100–106.

Pandima Devi, K.; Arif Nisha, S.; Sakthivel, R.; Karutha Pandian, S. "Eugenol (an Essential Oil of Clove) Acts as an Antibacterial Agent against *Salmonella typhi* by Disrupting the Cellular Membrane." *Journal of Ethnopharmacology* **2010,** *130*(1), 107–115.

Pari, L.; Umamaheswari, J. "Antihyperglycaemic Activity of *Musa sapientum* Flowers: Effect on Lipid Peroxidation in Alloxan Diabetic Rats." *Phytotherapy Research* **2000,** *14*(2), 136–138.

Patricia Bernal, et al., "Insertion of Epicatechin Gallate into the Cytoplasmic Membrane of Methicillin-Resistant *Staphylococcus aureus* Disrupts Penicillin-Binding Protein (PBP) 2a-Mediated β-Lactam Resistance by Delocalizing PBP2." *Journal of Biological Chemistry* **2010,** *285*(31), 24055–24065.

Paul D. Stapleton et al., "Modulation of β-Lactam Resistance in *Staphylococcus aureus* by Catechins and Gallates." *International Journal of Antimicrobial Agents* **2004,** *23*(5), 462–67.

Paul D. Stapleton; Peter W. Taylor. "Methicillin Resistance in *Staphylococcus aureus* : Mechanism and Modulation." *Sci Prog* **2007,** *85*(1), 1–14.

Pedro Rauel Cândido Domingos; Ana Cristina da Silva Pinto; Joselita Maria Mendes dos Santos; Míriam Silva Rafael. "Insecticidal and Genotoxic Potential of Two Semi-Synthetic Derivatives of Dillapiole for the Control of Aedes (Stegomyia) Aegypti (*Diptera: culicidae)*." *Mutation Research/Genetic Toxicology and Environmental Mutagenesis* **2014,** *772*(1), 42–54.

Perumal Samy, Ramar; Ponnampalam Gopalakrishnakone. "Therapeutic Potential of Plants as Anti-Microbials for Drug Discovery." *Evidence-based Complementary and Alternative Medicine* **2010,** *7*(3), 283–294.

Pitchai Daisy, Rajamanickam P. O. N. Nivedha; and Rajamanickam Helen Bakiya. "In Silico Drug Designing Approach for Biotin Protein Ligase of *Mycobacterium tuberculosis*." *Asian Journal of Pharmaceutical and Clinical Research* **2013,** *6*(3), 4–8.

Premkumar Jayaraman; Meena K. Sakharkar; Chu Sing Lim; Thean Hock Tang; Kishore R. Sakharkar. "Activity and Interactions of Antibiotic and Phytochemical Combinations against *Pseudomonas aeruginosa* in Vitro." *International Journal of Biological Sciences* **2010,** *6*(6), 556–568.

Prerna N. Domadia; Anirban Bhunia, J. Sivaraman; Sanjay Swarup; Debjani Dasgupta. "Berberine Targets Assembly of *Escherichia coli* Cell Division Protein FtsZ." *Biochemistry* **2008,** *47*(10), 3225–3234.

Prerna N. Domadia; Sanjay Swarup; Anirban Bhunia, J. Sivaraman; Debjani Dasgupta. "Inhibition of Bacterial Cell Division Protein FtsZ by Cinnamaldehyde." *Biochemical Pharmacology* **2007,** *74*(6), 831–840.

Priyanka Gupta et al., "Citral, a Monoterpenoid Aldehyde Interacts Synergistically with Norfloxacin against Methicillin-Resistant *Staphylococcus aureus*." *Phytomedicine* **2017,** *34*(1), 85–96.

Queiroz, T. B. et al., "Cytotoxic and Genotoxic Potential of Geraniol in Peripheral Blood Mononuclear Cells and Human Hepatoma Cell Line (HepG2)." *Genetics and Molecular Research* **2017,** *16*(3), 1–7.

Rafie Hamidpour; Soheila Hamidpour; Mohsen Hamidpour; Mina Shahlari. "Camphor (*Cinnamomum camphora*), a Traditional Remedy with the History of Treating Several Diseases." *International Journal of Case Reports and Images* **2013,** *4*(2), 86–89.

Rajeswara Rao, B. R.; Kaul, P. N.; Syamasundar, K. V.; Ramesh, S. "Water Soluble Fractions of Rose-Scented Geranium (Pelargonium Species) Essential Oil." *Bioresource Technology* **2002,** *84*(3), 243–246.

Renata de F. Mendes et al., "The Essential Oil from the Fruits of the Brazilian Spice *Xylopia sericea* A. St.-Hil. Presents Expressive in-Vitro Antibacterial and Antioxidant Activity." *Journal of Pharmacy and Pharmacology* **2017,** *69*(3), 341–348.

Renée A. Street; Jasmeen Sidana; and Gerhard Prinsloo. "Cichorium Intybus: Traditional Uses, Phytochemistry, Pharmacology, and Toxicology." *Evidence-based Complementary and Alternative Medicine* **2013,** *6*(2), 1–13.

Richa Jaiswal; Tushar K. Beuria; Renu Mohan; Suresh K. Mahajan; and Dulal Panda. "Totarol Inhibits Bacterial Cytokinesis by Perturbing the Assembly Dynamics of FtsZ." *Biochemistry* **2007,** *46*(14), 4211–4220.

Rosa, L. S.; Silva, N. J. A.; Soares, N. C. P.; Monteiro, M. C.; Teodoro, A. J. "Nutrition & Food Sciences Anticancer Properties of Phenolic Acids in Colon Cancer–A Review." *Nutrition and Food Sciences* **2016,** *6*(2), 2–7.

Rowena L. Lock; Elizabeth J. Harry. "Cell-Division Inhibitors: New Insights for Future Antibiotics." *Nature Reviews Drug Discovery* **2008,** *7*(4), 324–38.

Ruan, H.; Zhan, Y. Y.; Hou, J.; Xu, B.; Chen, B.; Tian, Y. et al., "Berberine Binds RXRα to Suppress β-Catenin Signaling in Colon Cancer Cells." *Oncogene* **2017,** *36*(50), 6906–6918.

Ruixue Guo; Xinbo Guo; Tong Li; Xiong Fu; Rui Hai Liu. "Comparative Assessment of Phytochemical Profiles, Antioxidant and Antiproliferative Activities of Sea Buckthorn (*Hippophaë rhamnoides* L.) Berries." *Food Chemistry* **2017,** *221*(1), 997–1003.

Sainan Li et al., "Genistein Suppresses Aerobic Glycolysis and Induces Hepatocellular Carcinoma Cell Death." *British Journal of Cancer* **2017,** *117*(10), 1518–1528.

Saleh Al-Sohaibani; Kasi Murugan. "Anti-Biofilm Activity of *Salvadora persica* on Cariogenic Isolates of *Streptococcus Mutans*: *In Vitro* and Molecular Docking Studies." *Biofouling* **2012,** *28*(1), 29–38.

Saleh Hosseinzadeh; Azizollah Jafari Kukhdan; Ahmadreza Hosseini; Raham Armand. "The Application of *Thymus vulgaris* in Traditional and Modern Medicine : A Review." *Global Journal of Pharmacology* **2015,** *9*(3), 260–66.

Sanaa K. Bardaweel; Khaled A. Tawaha; Mohammad M. Hudaib. "Antioxidant, Antimicrobial and Antiproliferative Activities of *Anthemis palestina* Essential Oil." *BMC Complementary and Alternative Medicine* **2014,** *14*(297), 1–8.

Sandeep Sharma et al., "Piperine as an Inhibitor of Rv1258c, a Putative Multidrug Efflux Pump of *Mycobacterium tuberculosis*." *Journal of Antimicrobial Chemotherapy* **2010,** *65*(8), 1694–1701.

Sanjai Saxena. "*Glycyrrhiza glabra* : Medicine over the Millennium." *Natural Product Radiance* **2005,** *4*(5), 358–367.

Sara Edwirgens Costa Benício Vasconcelos et al., "*Plectranthus amboinicus* Essential Oil and Carvacrol Bioactive against Planktonic and Biofilm of Oxacillin and Vancomycin-Resistant *Staphylococcus aureus*." *BMC Complementary and Alternative Medicine* **2017,** *17*(1), 462–478.

Sara Rooney; Ryan, M. F. "Modes of Action of Alpha-Hederin and Thymoquinone, Active Constituents of *Nigella sativa*, against HEp-2 Cancer Cells." *Anticancer Research* **2005,** *25*(6 B), 4255–4259.

Sarac Zorica et al., "Biological Activity of Pinus Nigra Terpenes-Evaluation of FtsZ Inhibition by Selected Compounds as Contribution to Their Antimicrobial Activity." *Computers in Biology and Medicine* **2014,** *54*(1), 72–78.

Sarfaraj Hussain; Sheeba Fareed; Mohammad Ali. "Hyphenated Chromatographic Analysis of Bioactive Gallic Acid and Quercetin in *Hygrophila auriculata* (K. Schum) Heine Growing Wildly in Marshy Places in India by Validated HPTLC Method." *Asian Pacific Journal of Tropical Biomedicine* **2012,** *2*(2), 87–98.

Satish Kumar Bedada; Praveen Kumar Boga. "Effect of Piperine on CYP2E1 Enzyme Activity of Chlorzoxazone in Healthy Volunteers." *Xenobiotica* **2017,** *47*(12), 1035–41.

Satyal Prabodh; Prajwal Paudel; Ambika Poudel; William N. Setzer. "Chemical Composition and Biological Activities of Essential Oil from Leaf and Bark of *Nyctanthes arbortristis* L. from Nepal." *Journal of Medicinal and Aromatic Plants* **2011,** *3*(1), 1–4.

Se Ra Won et al., "Oleic Acid: An Efficient Inhibitor of Glucosyltransferase." *FEBS Letters* **2007,** *581*(25), 4999–5002.

Serge Ankri; David Mirelman. "Antimicrobial Properties of Allicin from Garlic." *Microbes and Infection* **1999,** *2*(2), 125–129.

Shankara S. Bhat, "Neem–A Green Treasure." *Electronic Journal of Biology* **2008,** *4*(3), 102–11.

Shasank S. Swain; Rabindra N. Padhy. "Isolation of ESBL-Producing Gram-Negative Bacteria and in Silico Inhibition of ESBLs by Flavonoids." *Journal of Taibah University Medical Sciences* **2016,** *11*(3), 217–229.

Sheema Bai; Leena Seasotiya; Anupma Malik; Pooja Bharti; Sunita Dalal. "Bioactive Compounds and Pharmacological Potential of *Rosa indica* L. and *Psidium guajava* L. Methanol Extracts as Antiurease and Anticollagenase Agents." *Der Pharmacia Lettre* **2015,** *7*(1), 179–184.

Silva, E. O.; Ruano-González, A.; dos Santos, R. A.; Sánchez-Maestre, R.; Furtado, N. A.; Collado, I. G.; Aleu, J. "Antifungal and Cytotoxic Assessment of Lapachol Derivatives Produced by Fungal Biotransformation." *Nat Prod Commun Commun* **2016,** *11*(1), 95–98.

Sinosh Skariyachan; Arpitha Badarinath Mahajanakatti; Narasimha Sharma; Murugan Sevanan. "Selection of Herbal Therapeutics against Deltatoxin Mediated Clostridial Infections." *Bioinformation* **2011,** *6*(10), 375–379.

Sinosh Skariyachan; Rao Shruti Krishnan et al., "Computer Aided Screening and Evaluation of Herbal Therapeutics against MRSA Infections." *Bioinformation* **2011,** *7*(5), 222–233.

Sospeter Ngoci Njeru; Josphat Matasyoh; Charles G. Mwaniki; Charles M. Mwendia; George Kimathi Kobia. "A Review of Some Phytochemicals Commonly Found in Medicinal Plants." *International Journal of Medicinal Plants* **2013,** *105*(6), 135–140.

Sridevi Duggirala; Rakesh P. Nankar; Selvakumar Rajendran; Mukesh Doble. "Phytochemicals as Inhibitors of Bacterial Cell Division Protein FtsZ: Coumarins Are Promising Candidates." *Applied Biochemistry and Biotechnology* **2014,** *174*(1), 283–296.

Subramani Parasuraman; Gan Siaw Thing; Sokkalingam Arumugam Dhanaraj. "Polyherbal Formulation: Concept of Ayurveda." *Pharmacognosy reviews* **2014,** *8*(16), 73–80.

Sudharsana Sundarrajan; Sajitha Lulu; Mohanapriya Arumugam. "Computational Evaluation of Phytocompounds for Combating Drug Resistant Tuberculosis by Multi-Targeted Therapy." *Journal of Molecular Modeling* **2015,** *21*(247), 3–16.

Sudipta Jena et al., "Chemical Composition and Antioxidant Activity of Essential Oil from Leaves and Rhizomes of *Curcuma angustifolia* Roxb." *Natural Product Research* **2017,** *31*(18), 2188–2191.

Suneel Kumar, A.; Venkatarathanamma, V.; Naga Saibabu V. "Phytochemical and Phytotherapeutic Properties of *Annona Squamosa*, *Annona Reticulata*, and *Annona Muricata:* A Review." *Asian Journal of Plant Science and Research* **2015,** *5*(8), 28–33.

Tambekar, D. H.; Dahikar, S. B. "Antibacterial Potential of Some Herbal Preparation: An Alternative Medicine in Treatment of Enteric Bacterial Infection." *International Journal of Pharmacy and Pharmaceutical Sciences* **2010,** *2*(4), 176–179.

Taro Nomura; Toshio Fukai; Toshiyuki Akiyama. Chemistry of Phenolic Compounds of Licorice (*Glycyrrhiza* Species) and Their Estrogenic and Cytotoxic Activities. *Pure Appl. Chem.* **2002,** *74*(7), 1199–1206.

Thirupathi, A.; Silveira, P. C.; Nesi, R. T.; Pinho, R. A. "And Anti-Apoptotic Effects on Dimethyl Nitrosamine–Induced Hepatic Fibrosis in Male Rats." *Human and Experimental Toxicology* **2016,** *11*(1), 1–10.

Tingting Cao; Yong Li; Ziyao Yang. "Synthesis and Biological Evaluation of 3,8-Dimethyl-5-Isopropylazulene Derivatives as Anti-Gastric Ulcer Agent." *Chem Biol Drug Des* **2016,** *8*(3), 1–8.

Tushar K. Beuria; Manas K. Santra; Dulal Panda. "Sanguinarine Blocks Cytokinesis in Bacteria by Inhibiting FtsZ Assembly a Bundling." *Biochemistry* **2005,** *44*(50), 16584–16593.

Verónica Ruiz-Torres et al., "An Updated Review on Marine Anticancer Compounds: The Use of Virtual Screening for the Discovery of Small-Molecule Cancer Drugs." *Molecules* **2017,***22*(7), 1–37.

Vinod, N. V.; Haridas, M.; Sadasivan, C. "Isolation of 1,4-Naphthalenedione, an Antibacterial Principle from the Leaves of *Holoptelea integrifolia* and Its Activity against Beta-Lactam Resistant *Staphylococcus aureus*." *Indian Journal of Biochemistry and Biophysics* **2010,** *47*(1), 53–55.

Vinod, N. V.; Shijina, R.; Dileep, K. V.; Sadasivan, C. "Inhibition of Beta-Lactamase by 1,4-Naphthalenedione from the Plant *Holoptelea integrifolia*." *Applied Biochemistry and Biotechnology* **2010,** *160*(6), 1752–1759.

Wei Shi et al., "Design, Synthesis and Cytotoxic Activities of Scopoletin-Isoxazole and Scopoletin-Pyrazole Hybrids." *Bioorganic & Medicinal Chemistry Letters* **2017,** *27*(2), 147–151.

Wen-Wan Chao; Chia-Chi Su; Hsin-Yi Peng; Su-Tze Chou. "*Melaleuca quinquenervia* Essential Oil Inhibits α-Melanocyte-Stimulating Hormone-Induced Melanin Production and Oxidative Stress in B16 Melanoma Cells." *Phytomedicine* **2017,** *14*(1), 1–29.

Xiao Guo et al., "Acaricidal Activities of the Essential Oil from Rhododendron Nivale Hook. F. and Its Main Compound, δ-Cadinene against *Psoroptes cuniculi*." *Veterinary Parasitology* **2017,** *236*(1), 51–54.

Xiaomu Yu et al., "β-Elemene Inhibits Tumor-Promoting Effect of M2 Macrophages in Lung Cancer." *Biochemical and Biophysical Research Communications* **2017,** *490*(2), 514–520.

Xiaoyu Ji et al., "Comparative Analysis of Methicillin-Sensitive and Resistant *Staphylococcus aureus* Exposed to Emodin Based on Proteomic Profiling." *Biochemical and Biophysical Research Communications* **2017,** *494*(1–2), 318–324.

Yeon Kim; Su Ji Jang; Hyeung Rak Kim; Seon Bong Kim. "Deodorizing, Antimicrobial and Glucosyltransferase Inhibitory Activities of Polyphenolics from Biosource." *Korean Journal of Chemical Engineering* **2017,** *34*(3), 1–5.

Yeonjoong Yong et al., "Dichamanetin Inhibits Cancer Cell Growth by Affecting ROS-Related Signaling Components Through Mitochondrial-Mediated Apoptosis." *Anticancer Research* **2013,** *33*(1), 5349–5356.

Yizhong Cai; Qiong Luo; Mei Sun; Harold Corke. "Antioxidant Activity and Phenolic Compounds of 112 Traditional Chinese Medicinal Plants Associated with Anticancer." *Life Sciences* **2004,***74*(1), 2157–2184.

Yong M. Zhang; Charles O. Rock. "Evaluation of Epigallocatechin Gallate and Related Plant Polyphenols as Inhibitors of the FabG and FabI Reductases of Bacterial Type II Fatty-Acid Synthase." *Journal of Biological Chemistry* **2004,** *279*(30), 394–401.

Yu-Jie Guo; Li-Qiong Sun; Bo-Yang Yu; Jin Qi. "An Integrated Antioxidant Activity Fingerprint for Commercial Teas Based on Their Capacities to Scavenge Reactive Oxygen Species." *Food Chemistry* **2017,** *237*(1), 645–653.

Yunfeng Xu et al., "Antimicrobial Activity of Punicalagin Against *Staphylococcus aureus* and Its Effect on Biofilm Formation." *Foodborne Pathogens and Disease* **2017,** *14*(5), 282–287.

Yuzhen Jia; Binger Xu; Jisen Xu. "Effects of Type 2 Diabetes Mellitus on the Pharmacokinetics of Berberine in Rats." *Pharmaceutical Biology* **2016,** *55*(1), 510–515.

Zahid Mehmood Mirza; Ashwani Kumar; Nitin Pal Kalia; Afzal Zargar; Inshad Ali Khan. Piperine as an Inhibitor of the MdeA Efflux Pump of *Staphylococcus aureus*." *Journal of Medical Microbiology* **2011,** *21*(1), 1472–1478.

Zeynep Aytac; Zehra Irem Yildiz; Fatma Kayaci-Senirmak; Turgay Tekinay; Tamer Uyar. "Electrospinning of Cyclodextrin/linalool-Inclusion Complex Nanofibers: Fast-Dissolving Nanofibrous Web with Prolonged Release and Antibacterial Activity." *Food Chemistry* **2017,** *231*(1), 192–201.

Zhonghong Wei et al., "Diallyl Trisulfides, a Natural Histone Deacetylase Inhibitor, Attenuate HIF-1α Synthesis, and Decreases Breast Cancer Metastasis." *Molecular Carcinogenesis* **2017,** *56*(10), 2317–2331.

Zulfiqar Ahmad; Thomas F. Laughlin; Ismail O. Kady. "Thymoquinone Inhibits *Escherichia coli* ATP Synthase and Cell Growth." *PLoS ONE* **2015,** *10*(5), 1–12.

CHAPTER 16

Ethnobotanical Study of Pharmacologically Important Medicinal Plants

S. C. SAHU

Department of Botany, North Orissa University, Baripada-757003, Odisha, India, E-mail: sudamsahu.bdk@gmail.com

ABSTRACT

Ethnobotanical study of Bisoi block, Mayurbhanj district of Odisha was carried out during 2016–17 for exploring the herbal therapy to cure different ailments. Traditional people were interviewed with predesigned questionnaires. A total of 32 plants belonging to 23 families and 22 genera were documented for curing different ailments. The dominant families were Verbenaceae, Asteraceae, Euphorbiaceae, and Amaranthaceae. The diseases such as skin disease, jaundice, asthma, snakebite, inflammation, etc. were cured by using these ethnomedicinal plants. Estimation of plant parts used revealed that leaves part represented the maximum percentage, i.e., 58.4% followed by flowers (16.6%), fruits (12.5%), roots (8.3%), and stems (4.2%). They were in different forms such as decoctions, extracts, paste, juices, and powders. Indigenous knowledge of tribal people regarding uses of medicinal plants needs to be preserved at an urgent basis prior to its loss.

16.1 INTRODUCTION

Since antiquity man has relied on nature for their fundamental needs such as cloth, food, shelter, transportation, and medicine as a mean for its survival. But the present day people face huge challenges for population explosion all over the world. To satisfy the fundamental needs of human beings such as food, people practice short-term crops and also apply chemical fertilizers to enhance crop yield which creates various diseases. So many pharmaceutical formulations are available which are used to treat the different ailments, but they are not economically viable, sometimes not effective and show numerous side effects. Hence, alternative use of medicine is an option which is naturally available and do not show any side effects. Medicinal plants are nature's gift to humanity, and a lot of herbal drugs have been discovered since the last few decades. By trial and error method, man learns that certain plants were found useful as food while others for treatment of different ailments. Adopting this method man was able to differentiate between harmful and useful plants. This utility gives rise to the creation of the word ethnobotany.

The interaction between plants and human, emphasizing on traditional tribal cultures is defined as ethnobotany. The term ethnobotany was first coined in1896 by John Hershberger (a botanist of America) as the knowledge on plants by original primitive people. An ethnobotanist usually documents the uses of plants available at his locality along with their local traditional culture of that place. According to Richard Evans Schulte: plants investigated in different geographical region of the world were used by primitive societies including their traditional culture. Ethnobotany has its roots in botany. Ethnobotany is the key approach to study the natural resource management of indigenous people. Acquiring indigenous knowledge on plants of medicinal importance and further developing herbal drugs in the recent time are not only for conservation of the present tradition and culture but also helps in conservation of biodiversity by the indigenous people.

Developing countries people are mostly dependent upon plants for their health (WHO). About 80% of Ethiopia's total populations depend on traditional medicine to treat different types of human diseases. Traditional medicine system was very common before 20th century. Most parts of the world traditional medicine men is practicing herbal therapy to cure different diseases of human and it is culturally and locally accepted. However, due to advance in research and modern techniques, allopathic drugs of chemical origin replaced the herbal medicines as the sources of most medicinal agents in industrialized countries. Further, the knowledge of ethnobotany is useful to develop the financial upliftment of the tribal people through proper utilizing of the forest products including medicinal plants.

Three major systems of medicines, namely, Ayurveda, Siddha, and Unani are adopted by Indian people. Ethnobotany has vital role in the conservation of nature, culture and the biological diversity of the world. In fact, conservation and biodiversity are interlinked with each other. Traditional knowledge systems are very old and involved not only about the medicinal plants and food but also for the sustainable utilization of plant resources. An ethnobotanist must be knowledgeable to find out the practical potential of native plants while studying the plants in a particular geographical region.

Folklore medicine is the practices through which different diseases/ailments can be cured using plants, animals, and minerals derived medicines along with spiritual therapies and exercises (WHO, 2003). Above 80% of human in the world are practicing and depending upon plants for their health problems (Gebremariam and Asress, 1998). Traditional medicine system has been neglected in the past by the modern societies. Nowadays, ethnobotanical studies are facilitated for collecting the primary data on the plants. The outcomes of these practices resulted in attitudinal changes of people. However, these studies are very less in context to the total natural resources of a country.

The plants of medicinal importance contain active ingredients and foundation compounds which can be utilized for formulation of drugs (Ibadan, 1993). Plants are known to have beneficial therapeutic effect documented in traditional Indian medicine system. There are so many works have been done on ethnomedicinal plants in India. A remarkable increase in the number of traditional natural products has been observed. The presences of bioactive compounds in the plants are major sources of drugs. From the past, the higher plants play a vital role in herbal drug development programme (Farombi, 2003). Study on antimicrobial activities of plants has great significance for solving therapeutic problem (Austin et al., 1999). Natural products extracted from plants may give a potential source of antimicrobial agents.

India feels a wide variation of climate and topography, possessing rich biodiversity. In recent years, Secondary metabolites of plant origin have been thoroughly investigated for medicinal properties. Phytochemicals which show better antibacterial activities can be used for any bacterial infections in near future (Weasth et al., 1985). The present day's resistance to antibiotic has been becoming a global issue as the multidrug-resistant pathogens are becoming antibiotic proof (Banbow et al., 2003). Mostly infectious diseases were treated with herbal medicines from the very beginning (Rojas et al., 2003). Hence, scientists are attracted to traditional herbal medicines and formulate herbal drugs for microbial infections (Benkeblia et al., 2003).

Of the 15,000 species of flowering plants found in India, more than 17% of plants are having medicinal value (Nadkarmi, 1954). Many species (1745) are from the Indian Himalayan region, and most of these are found in Uttarakhand (Bentley and Trimen, 1980; Kirtikar and Basu, 1993; Nadkarni, 1954; Chopra, 1956). Collection and commercialization of medicinal plants in India is a historical fact. Antimicrobials present in the plant is an incredible source for medicines and needs further exploration of plant therapeutic potential (Salau and Odeleye, 2007). Herbal medicines have been practiced to treat different ailments in traditional medicines system.

Mayurbhanj, a tribal dominated district, is very rich in ethnomedicinal plants. The landscape of the district is dominated by various tribal communities having unique tradition and culture. Similipal Biosphere Reserve (SBR) in Mayurbhanj district is an important Biosphere Reserve of India possessing more than 1076 plant species including 200 plants having medicinal value (Rout, 2004). The traditional vaidyas have acquired good knowledge of herbal therapy and practicing their healthcare services to local tribal people. A number of literatures are available on ethnobotany of Mayurbhanj district by various workers (Majumdar, 1971; Mudgal and Pal, 1980; Saxena et al., 1988; Pandey et al., 2002; Pandey and Rout, 2006; Rout and Pandey, 2007). However, ethnobotanical study in Bisoi block of Mayurbhanj district is lacking. Therefore, the present study focuses on exploration of ethnobotanical findings found in Bisoi block of Mayurbhanj district.

16.2 STUDY AREA

Mayurbhanj district is one of the tribal-dominated districts of Odisha extending over an area of 10,418 Sq.km. It lies in the tri-junction of three states-Odisha, West Bengal and Jharkhand. Baripada city is the headquarter of Mayurbhanj district. Forest cover is about 39% of the total geographical area (4049 sq. km.). There are about four Sub-divisions, 26 Blocks, 382 Gram Panchayats and 3945 villages in the district. Bisoi block is one of the tribal-dominated blocks among them (Figure 16.1). The total no of villages are 145. Dry deciduous forests are predominated in this block. More than 44% of the area of the district is occupied by forests. The total population of the block is 73, 899 and tribal population is 51, 299 (Census of India, 2011). The dominated scheduled tribes of Bisoi block are Santala, Bhuian, Bathudi, Bhumija, Munda. The climate of Bisoi block is tropical warm and humid. Three distinct seasons are felt during the year. Mostly depressions from the Bay of Bengal in monsoon season cause rain. The area gets an annual average rainfall of 165 cm from southwest monsoon (June-September) and retreating northeastern monsoon (November). Rainfall is irregular in distribution. The hot season starts from March and continues up to June last. May is the hottest month with temperature ranges between 40–45°C. Relative humidity is comparatively more throughout the year (about 70% to 90%).

The block is predominated by red-laterite soil. The soil is acidic in nature. In rainy season, deep standing water conditions prevail in the track. But in the off monsoon period of November to May, the soil profile is progressively recharged with salts through capillary rise to sub-soil salts due to desiccation. Practically the land remains fallow during the period, and hence the areas are monocropping. Exchangeable sodium percent ranges from 18 to 27% in the profile.

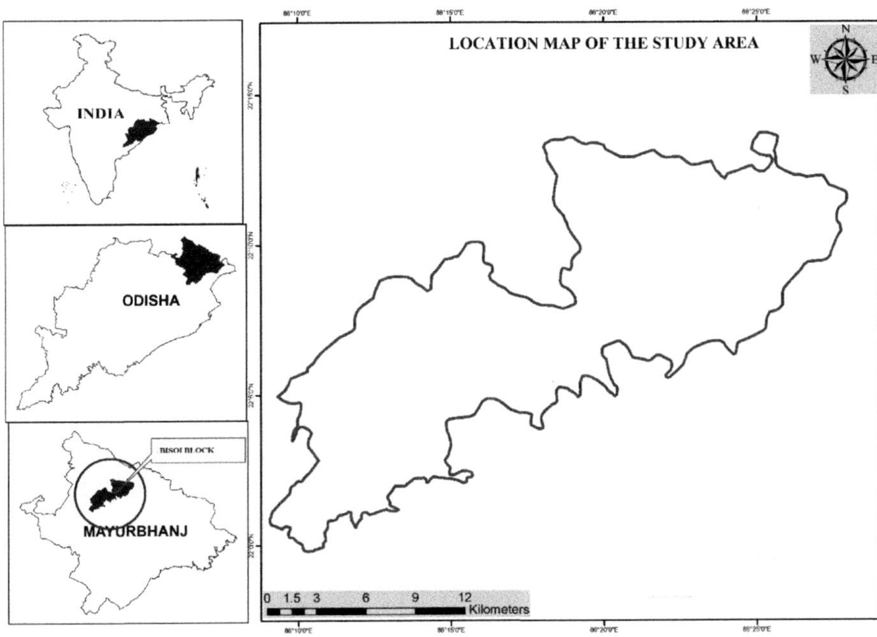

FIGURE 16.1 Map of study area.

16.3 METHODOLOGY ADOPTED

16.3.1 INTERACTION WITH LOCAL PEOPLE AND COLLECTION OF DATA

All the plant specimens were sampled at their mature stage by the local tribal vaidyas. Field observation on plant specimens was recorded in the field. Collected specimens were tagged with field number. Sample of plants was collected followed by identification and herbarium preparation. The taxa were critically identified using regional floras (Saxena and Brahmam, 1994–96; Haines, 1921–25; Gamble, 1915–36) and other available literatures.

Tribal healers were from different professions such as teachers, farmers, and daily laborers. Interestingly according to them, traditional healing procedure has higher efficiency than the modern synthetic medicines. It had been observed that many plants were taken for curing several diseases which may not be cured by the modern treatments. Some plants were specified for curing a particular disease. Out of several medicinal plants, some plants are selected which are mostly used by the villagers for their primary treatment.

16.3.2 IDENTIFICATION OF PLANTS

Some of the specimens were identified in the field. Unidentified specimens were brought to laboratory for their identification. They are identified in the Biosystematics laboratory cum herbarium at Department of Botany, North Orissa University, Baripada (Odisha) with the help of regional floras (Saxena and Brahmam, 1994–96; Haines, 1921–25; Gamble, 1915–36), monographs, revisions, and other taxonomic literatures. Preparation of herbarium specimens includes six phases: pressing, drying, poisoning, mounting, stitching, and labeling. Fleshy specimens, delicate flowers and large plant parts like tubers, rhizomes, corms, fruits, etc. which cannot be made into herbarium were preserved in the jars. We have consulted the Central National Herbarium (CNH), Howrah, Kolkata for problematic taxa. Voucher specimens were housed in the herbarium of Dept. of Botany, NOU, Baripada, Orissa.

16.3.3 CRIS-CROSS CHECKING

An effort had been made to cross check the ethnobotanical claims. It is a method for verifying the data authenticity. In this method, ethnobotanical claims were checked and rechecked by the data collector by asking that particular claim to the same tribe in different forest areas. It is essential for identifying authenticity of ethnobotanical claims of plant specimens.

16.3.4 ETHNOBOTANICAL NOTINGS

A brief but crisp note about the uses of the plants by the tribals was given at the end of the taxon. Every care has been taken to avoid ambiguity as regards to plant part, quantity, dosage, method of preparation, mode of administration, etc. The name of the place where the noting is recorded and the tribe from whom the data originated has been invariably given for future cross-checking.

The ethnobotanical information collected by the author included vernacular names of the plant species, their uses and preparation technique. Ethnobotanical information on different medicine plants was documented through semi questionnaire interviews with detail of their local names, locality, uses, mode preparation and parts used. A total of 32 taxa were recorded and documented for their medicinal value. The collected species were having various uses such as medicines, seed oil, vegetables, foods, and other uses. The quality and yield potential of these plants varies from one place to another due to differences in their climatic condition, soil, and topography. The collected species were widely taken by the local communities for curing different diseases. The enumeration details of 32 plant species along with correct botanical name, local name, family, short description and ethnobotanical uses were done below.

16.4 RESULTS

16.4.1 *Alternanthera Sessilis* (L.) R. Br. Ex DC.

Local names: Madaranga saga (Odia)

Description: Branched spreading prostrate herb. Leaves subsessile, oblong to lanceolate, entire margin, opposite, obtuse, glabrous. Flowers white, spineless spikes. Tepals 5, ovoid, ca. 2.5 mm long. Utricle obcordate compressed. Seeds minute, suborbicular, glabrous.

Ethnobotanical uses

- Leaves and shoots are edible as leafy vegetables.

16.4.2 *Achyranthes aspera L. (Amaranthaceae)*

Local name: Apamaranga

Description: Erect or subscandent herb. Leaves simple, opposite, elliptic-obovate, acute softly-finley, pubescent. Flowers greenish-white in axillary and terminal many flowered elongate spikes. Tepals 5, rigid, lanceolate, aristate, persistent. Utricle 2–3 mm long, oblong.

Ethnobotanical uses

- The plant decoction is orally taken against malaria.

16.4.3 *Aegle marmelos* L. (Rutaceae)

Local name: Bela

Description: Thorny trees up to10 m; bark greyish-white. Leaves 3-foliolate, ovate-elliptic, glabrous, gland-dotted, subcrenulate; petiole 2–5 cm long. Flowers greenish-white. Fruit globose, 6–10 cm diam. Seeds embedded in clear mucilage and yellow.

Ethnobotanical uses

- Leaf juice with black pepper is taken three times a day to cure from jaundice.
- Ripened fruit juice or syrup is useful in constipation.

16.4.4 *Aloe vera* (L.) Burm.f. (Liliaceae)

Local name: Geekuanri

Description: Leaves radical in a rosette, ensiform, margin spinous-dentate, apex tapering. Scape 40–60 cm long, raceme dense, 15–30 cm long. Perianth yellow or reddish-yellow and green, cylindric. 1.5–2.5 cm long.

Ethnobotanical uses

- The whole plant paste of this plant is applied to burns, abrasions, psoriasis and even bugs bites to relieve from pain.

16.4.5 *Azadirachta indica* L. (Meliaceae)

Local name: Neem

Description: Tall tree of height 20 meters or less. Leaves obliquely lanceolate, serrate, glabrous, base cuneate, oblique. Flowers white, fragrant. Fruit smooth, olive-like drupe which varies in shape from elongate oval to nearly roundish. The fruit exocarp is thin and pulpy. The white, hard inner shell (endocarp) encloses one, rarely two, or three, elongated seeds having a brown seed coat.

Ethnobotanical uses

- Leaf decoctions are massaged on the whole body, to prevent skin diseases.
- People use the leaves and flowers as food for balancing their blood sugar levels.
- Seed oil is rubbed on the hair for healthy black and strong hair.

*16.4.6 **Buchanania Lanzan** Spreng.*

Local names: Char (Odia)

Description: Small to medium sized tree. Leaves simple, alternate oblong or ovate-oblong, entire, obtuse, glabrous above and pubescent beneath. Flowers small, sessile, greenish-white, hermaphrodite in axillary and terminal panicles. Ovary 1-celled. Drupes globose, 1 cm diam.; stone hard, 2-valved.

Ethnobotanical uses

- The ripened fruits are eaten.
- The paste of leaf is used against swelling glands of neck.

16.4.7 Calotropis Procera W.T. Aiton (Apocynaceae)

Local name: Arakha

Description: Erect shrub with milky latex and ash-colored bark. Leaves subsessile, simple, opposite, obovate-oblong, entire, base cordate to semi-amplexicaul. Flowers bluish-purple in terminal and lateral umbellate-cymes. Follicles boat-shaped, cottony pubescent; seeds numerous, ovate.

Ethnobotanical uses

- A single dose of the aqueous suspension of the dried latex of this plant was effective to a significant level against the acute inflammatory response.

*16.4.8 **Clerodendrum inerme** (L.) Gaertn.*

Local names: Nutunga (Odia).

Description: Evergreen, glabrous straggling shrub 1–2 m tall. Leaves opposite obovate, broadly elliptic, entire, base narrowed; petioles up to 12 mm. long. Flowers 3-flowered axillary cymes. Drupes 2 cm long, pyriform, 4-grooved.

Ethnobotanical uses

- Leaf juice (warm) is prescribed as an ear drops to treat earache.

16.4.9 *Clerodendrum Viscosum* Vent.

Local names: Madhvi (Odia)

Description: Gregarious shrub. Leaf lamina ovate, cordate, dentate, acute, pubescent; petioles 5–12 cm long. Flowers whitish-pink, cymose or subcorymbose panicles; bracts ovate. Drupes 6–8 mm across, globose, bluish-black.

Ethnobotanical uses

- Paste prepared with 4 leaves and 8 black peppers (*Piper nigrum*) is prescribed for curing swollen legs. This herbal preparation can be applied externally for 7 days (twice a day).

16.4.10 *Curcuma longa* L. (Zingiberaceae)

Local name: Haldi

Description: Aromatic herb with cylindric, bright orange colored rhizomes. Leaves elliptic-oblong, entire, and glabrous. Spikes arise from middle of leafy tufts. Lip 1.4 cm long, pale yellow with deep yellow center. Anther reclinate with a small yellow lambella at its apex and 2 spurs in front. Capsules oblong.

Ethnobotanical uses

- The rhizome paste is applied to the wound, and it acts as antimicrobial paste.

16.4.11 *Cuscuta Reflexa* Roxb.

Local names: Nirmuli (Odia, Munda)).

Description: Twining parasitic leafless herb; stems slender, greenish-yellow. Flowers white, 6–7 mm long, solitary and in few to many flowered racemiform cymes, lobes 5, subequal. Ripe fruits are globose or ovoid.

Ethnobotanical uses

- The plant paste 2g grinded with 1g paste of long pepper and is applied on hydrocele to reduce its size and pain.

16.4.12 Cymbopogon flexuosus W.Watson (Poaceae)

Local name: Dhanantary

Description: Tall, aromatic grass; culms 0.9–2 m, terete, glabrous. Leaves linear, lanceolate, purple-tinged at base, 8–10 mm broad; ligules 3–4 mm long; sheaths terete, striate. Inflorescence congested with raceme pairs in masses. Sessile spikelets 4 x 1 mm.; lower glume flat, bidentate, 2-keeled; upper glume scabrid; lower lemma empty, ciliate in the upper half, upper lemma hermaphrodite, with 7–8 mm long awn in the sinus, epaleate. Pedicelled spikelet 3.5–4 mm long, awnless.

Ethnobotanical uses

- Decoctions prepared by boiling of these leaves with tulsi leaves and black pepper are taken orally to relief from cough and cold.

16.4.13 Cynodon dactylon (L.) Pers. (Poaceae)

Local name: Dooba

Description: Creeping grass by scaly rhizomes or stolons forming matted tufts. Culms 5–25 cm decumbent-ascending; nodes glabrous. Leaves linear-subulate, scabrid on the upper surface and margin; ligules minute, scarious, sheaths compressed, keeled, glabrous. Spikes 2–4, up to 3 x 2 cm. Spikelets 2–2.3 x 0.8–1 mm., lower glumes linear-lanceolate, acute, 1–2 mm long, 1-nerved, bidentate, ciliate on the keel.

Ethnobotanical uses

- Leaf juice added with honey is applied 2–3 times a day for the menstrual problem.

16.4.14 Datura metel L.

Local names: Kala dudura, Dudura (Odia).

Description: Erect shrub. Leaves broadly ovate, entire or obtusely toothed, acute, glabrous, petioles up to 8 cm long. Flowers erect; white or pale purple. Capsule 3–4 cm across, globose with deltoid spines, irregularly breaking.

Ethnobotanical uses

- Paste of one fruit mixed thoroughly with mustard oil applied for treatment of kibe.
- For worshiping Lord Shiva, the flowers are very demanded by the people.

16.4.15 *Erythrina variegata* L.

Local names: Paladhua (Odia); Sembed (Kondha).

Description: Medium sized tree armed with conical prickles. Leaves 3-foliate; petiole 10–15 cm long; leaflets 10–15 x 8–12 cm broadly rhomboid or ovate, entire, acute, glabrous. Flowers scarlet red, in dense puberulous 10–20 cm long racemes. Calyx tubular, 3 cm long. Corolla bright red, 5–6 cm long subequal. Stamens 10. Ovary stipitate; style curved; stigma capitate. Pods many, 15–25 cm long, stalked, subcylindric.

Ethnobotanical uses

- Buds are crushed, and the juice is administered twice a day in night time for 3 days against kurmi.

16.4.16 *Ipomoea carnea* Jacq.

Local names: Amari (Odia)

Description: Shrub up to 2 m, branches thick, hollow; with milky juice. Leaves broadly ovate, ovate-oblong, acuminate, glabrous above, thinly pubescent beneath, cordate at base, petiole 5–15 cm long. Flowers rose or purple-colored, 5–7 cm long; in umbellate cymes.

Ethnobotanical uses

- The wood of the plants are utilized for fuel purposes.

16.4.17 *Justicia adhatoda* L. (Acanthaceae)

Local name: Basanga

Description: Shrub 1–2 m tall. Leaves opposite, ovate-lanceolate or elliptic-oblong, entire, acuminate, glabrous; petioles 1–3 cm long. Flower white in colored with peduncles spikes. Capsule 2.5 cm long, clavate, stipitate.

Ethnobotanical uses

- The juice of leaf is taken orally to relieve from cough and cold.

16.4.18 Lantana camara L. Var. ***Aculeata*** (L.) Mold.

Local names: Naguari (Odia, Gonda).

Description: Bushy shrub with short recurved prickles. Leaves ovate, serrate, acute, scabrid above, subcordate at base. Flowers scented in short spikes, white, purple or orange. Drupes 4–6 mm across, globose, fleshy.

Ethnobotanical uses

- Ripe fruits are edible.
- Useful as fencing.
- The paste of leaf is applied on boils.

16.4.19 Lawsonia inermis L.(Lythraceae)

Local name: Manjuati

Description: Bushy shrub, 1.5–2 m tall. Leaves opposite, elliptic or obovate, entire, acute, glabrous. Flowers scented, creamy-white, large in corymbosely branched panicles. Capsules 6–8 mm. diam., globose.

Ethnobotanical uses

- The leaf and root juice are taken orally to cure from jaundice.

16.4.20 Moringa Oleifera Lam.

Local names: Sajana (Odia)

Description: Short heighted tree. Leaves alternate, 3-pinnate; leaflets 6–9 pairs, opposite elliptic, obovate, obtuse, dark green above, pale beneath. Flower color white in long panicles. Carpel syncarpous, 1-celled. Fruit an elongated loculicidally dehiscent, beaked woody capsule; seeds 10–20 winged.

Ethnobotanical uses

- Leaves and fruits are used popularly as vegetable throughout the district.

16.4.21 *Nyctanthes Arbor-Tristis* L. (Oleaceae)

Local name: Gangasiuli

Description: Small sized tree, branchlets quadrangular. Leaves glabrous; petiole 8–10 mm long. Flowers fragrant. Capsule elliptic or obovoid, glabrous. Seeds orbicular.

Ethnobotanical uses

- The fresh leaves juice with honey is administered orally for chronic fever.
- The powered dry leaf is added to betel leaf juice and applied for asthma and cough.
- Fresh juice (4–5 drops) is put in nostrils to check bleeding.

16.4.22 *Ocimum sanctum* L. (Lamiaceae)

Local name: Tulasi

Description: Erect, branched undershrub, 0.5–1.2 m tall, pubescent. Leaves elliptic, serrate, obtuse, and pubescent. Flowers very small, purplish, 3.5–4.5 mm long. Nutlets reddish or yellowish with small black markings, broadly ellipsoid.

Ethnobotanical uses

- The leaf juice mixed with little honey is usually taken in cough, fever and lungs disease.
- The leaf juice (3–4 drops) is put into nostrils for relief from headache.
- The paste of leaf is used on inflamed areas help to reduce pain and inflammation.

16.4.23 *Phyllanthus emblica* L. *(Euphorbiaceae)*

Local names: Anla (Odia); Meral (Gonda).

Description: Small deciduous tree, 4–8 m tall. Leaves linear or linear-oblong, close-set, distichous; glabrous. Flowers minute, white or yellowish, in axillary clusters. Fruit a globose, drupe, 1–2 cm in diameter; succulent; pale-yellow.

Ethnobotanical uses

- 'Triphala,' one of the important ayurvedic medicine prepared from fruits of this plant along with other ingredients.
- Fruits are made into pickles.

16.4.24 *Punica granatum* L. (Punicaceae)

Local name: Dalimba

Description: Small tree; often thorny. Leaves exstipulate, 2.5–6.5 cm long, obovate-oblong, entire obtuse, glabrous. Flowers bright red, in 3-flowered cymes. Fruit globose, reddish, crowned by the hardened persistent sepals; seeds numerous.

Ethnobotanical uses:

- Fruit juice usually taken to increase the blood and cool the stomach

16.4.25 *Rauvolfia serpentine* L. (Apocynaceae)

Local name: Patalagaruda

Description: Glabrous erect under shrub. Leaves up to 16 cm long, elliptic-lanceolate, entire, acute, glabrous. Flowers whitish pink, bracteate red cymes. Fruit drupaceous, black, obliquely ovoid, glabrous.

Ethnobotanical uses

- The juice of leaf is used to relief from stomach pain.

16.4.26 *Ricinus communis* L. (Euphorbiaceae)

Local name: Jada

Description: Shrub, 2–4 m tall. Leaves palmately 7-lobed, peltate; lobes elliptic-lanceolate, serrate; petiole 15–45 cm long, 1–2 glands at apex. Flowers yellow with racemose inflorescence. Perianth 3-lobed in males, 5-partite in females, membranous. Fruit a schizocarp, densely muricate; seeds oblong, carunculate.

Ethnobotanical uses

- Milky latex of this plant is applied to mouth once a day to cure from all kind of dental problems.

16.4.27 *Solanum virginianum* L. (Solanaceae)

Local name: Vegibaigana

Description: Diffuse prickly under shrub up to 40 cm long. Leaves elliptic, deeply sinuate, prickly glabrous. Flowers in supra-axillary cymes. Berries 1 x 1 cm diam., white with green veins turning yellow at maturity, many seeded.

Ethnobotanical uses:

- Root paste is locally applied for scabies.

16.4.28 *Syzygium cumini* L. (Myrtaceae)

Local name: Jamun

Description: Tree up to 10 m tall. Leaves simple, opposite, ovate-oblong, entire, acute to acuminate, glabrous; secondary nerves numerous, closely arranged. Flowers small, 3-chotomous panicles. Berries edible, one seeded.

Ethnobotanical uses

- Fruit juice is taken half glass twice a day for indigestion.

16.4.29 *Tagetes patula* L. *(Asteraceae)*

Local name: Gendu

Description: Annual herb. Leaves pinnately compound, serrate segments. Heads 2.5–4 cm diameter of various color. Involucre 1–2 cm long 5–7 bracts. Achenes ripen and are shed within two weeks of the start of bloom.

Ethnobotanical uses

- The leaf juice is applied to the wound site to heal the wound.

16.4.30 *Tamarindus indica* L. *(Caesalpiniaceae)*

Local name: Tentuli

Description: Large tree, bark dark-grey fissured. Leaves paripinnate; leaflets subsessile, 10–18 pairs, oblong, entire, obtuse, glabrous: Flowers pink, in lax few-flowered racemes. Pods curved, subcompressed with brittle epicarp, seeds 4–9, broadly oblong, compressed, dark brown, shining.

Ethnobotanical uses

- The dry leaf powder is taken by cattle for curing any stomach problem.

16.4.31 Tragia involucrata L.

Local names: Bichhuati (Odia); Sengelsing (Kondha).

Description: Trailing or straggling herb with stinging hairs. Leaves ovate-oblong, coarsely acuminate, hairy, 3-ribbed; petiole up to 3 cm long; stipules 2.5 mm long, lanceolate, deciduous. Flowers monoecious, usually in leaf-opposed subspicate racemes; males many, above few females. Capsule 7 mm diam., 3-lobed, hirsute.

Ethnobotanical uses

- The paste of root is used as an antidote to snake bite.
- Leaf juice (5 ml) is taken orally for whooping cough

16.4.32 Tridax procumbens L. **(Asteraceae)**

Local name: Bisalyakarani

Description: Procumbent hairy herb up to 30 cm tall. Leaves petioled, opposite, toothed, hairy. Heads solitary on 10–15 cm long peduncles. Achenes 2 mm long, oblong, hairy, black, faintly ribbed.

Ethnobotanical uses

- The leaf juice is directly applied on wounds. Its leaf extracts were externally applied for infectious skin diseases.

16.5 DISCUSSION

A total of 32 plants belonging to 23 families and 22 genera were observed to be used in curing different diseases and other purposes also (Figure 16.2, Figure 16.3). The dominant families were *Verbenaceae*, *Asteraceae, Euphorbiaceae,* and *Amaranthaceae. Ethnomedicinal* plants were used to cure skin disease, jaundice, cough, snake bite, inflammation, etc. Plants of various forms such as powder, juice, decoction, paste, etc. were preferred to use for curing of different ailments. The herbal medicines were combination of plants or other products such as saccharine, honey, milk, curd, etc. The data collected from the tribal people of Bisoi block in Mayurbhanj district in regard to the plant parts utilized in the medicinal preparation were whole plants, roots, bark, leaves, fruits, flowers, stem and seeds. The most frequently utilized plant parts percentage were leaves (58.4%), followed by the roots (8.3%), fruits (12.5%),

Stem (4.2%), flowers (16.6%) (Figure 16.2). The information of medicinal plants for curing different ailments such as asthma, diabetes, jaundice, paralysis, snakebite, fever, etc. was interesting. Leaves were the very commonly used parts of plants by the tribal people among other plant parts. The present study revealed that the people of Bisoi block preferred to take leaves orally.

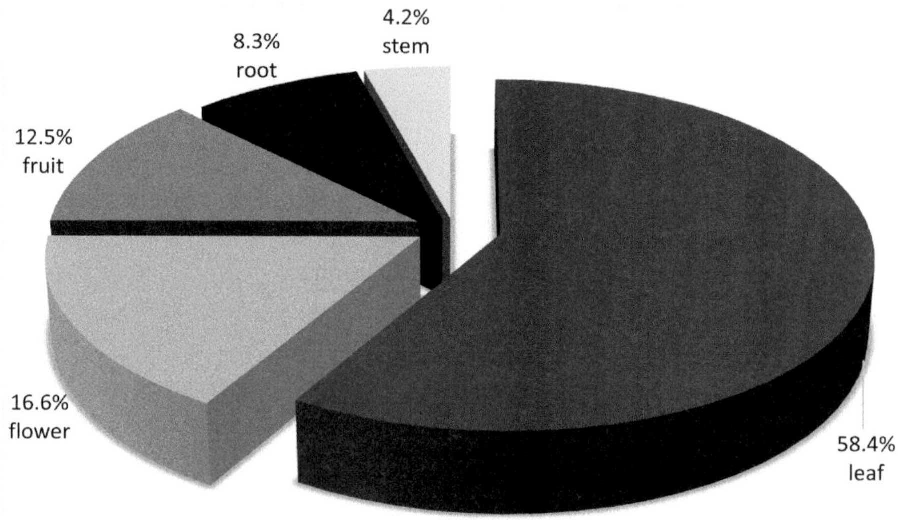

FIGURE 16.2 The pie chart showing percentage of ethnobotanical uses of different parts of plants.

This study observed that the indigenous knowledge on herbal therapy was mostly confined to the old village vaidyas and due to modernization this knowledge was not transferred to the next generation. The threat to extinction of this indigenous knowledge is very fast which cannot be revived again. Therefore, necessary steps would be implemented to conserve and document this valuable knowledge prior to its loss. Further, an unsustainable collection of medicinal plants leading to many medicinal plants fall under the threatened category.

A–*Azadirachta indica* L., B–*Ocimum sanctum* L., C–*Nyctanthes arbor-tristis* L. D–*Cynodon-dactylon* (L.) Pers., E–*Calotropis procera* W.T. Aiton F–*Tagetes patula* L., G–*Tridax procumbens* L., H–*Achyranthes aspera* L. I–*Aegle marmelos* L., J–*Tamarindus indica* L., K–*Ricinus communis* L., L–*Lawsonia inermis* L., M–*Cymbopogon flexuosus* W.Watson N–*Rauvolfia serpentina* L., O–*Phyllanthus emblica* L., P–*Punica granatum* L., Q–*Justicia adhatoda* L., R–*Syzygium cumini* L. S–*Aloe vera* (L.) Burm.f, T–*Curcuma longa* L.Most of the people of Bisoi block of Mayurbhanj district depend upon herbal medicines traditionally due to lesser side effects and cost-effective. This is an easy process to record the information on general uses of plants because a large number of people either consuming or sold in the nearby markets. But collecting indigenous information from tribal healers is a tough job for us. Tribal people are generally very shy in nature and don't want to spread their valuable knowledge to others. By creating a good rapport with these people, we can enable us to document this valuable information. The fast disappearance of indigenous knowledge and biodiversity due to population explosion, urbanization and industrialization result in the loss of valuable indigenous knowledge on medicinal plants forever. Hence, it is on a priority basis to document the valuable information and also conserve this knowledge for the next generation.

FIGURE 16.3 (See color insert.) Some important medicinal plants of Bisoi block, Mayurbhanj district.

16.6 CONCLUSION

This study revealed some preliminary information on ethnobotanical claims for different diseases. Although allopathic medicines are easily available, tribal people of Bisoi block still prefer to use fresh medicinal plants for different ailments such as skin diseases, headache stomach problem, snake bite, etc. Further higher investigation is imperative to find out the bioactive compounds present in these medicinal plants for herbal drug development programme. The recent generations of tribal vaidyas are in the mature/old stage. The next generations are not showing interest to learn or practice this traditional medicine system. Hence, it is necessary to document and conserve the knowledge prior to its loss.

KEYWORDS

- **Bisoi block**
- **ethnobotany**
- **Mayurbhanj district**
- **medicinal plants**
- **traditional knowledge**

REFERENCES

Austin, D. J.; Kristinsson, K. G.; Anderson, R. M. The relationship between the volume of antimicrobial consumption in human communities and the frequency of resistance. *Proceedings of National Academy of Science USA* **1999,** *96,* 152–156.

Awoyemi, O. K.; Ewa, E. E.; Abdulkarim, I. A.; Aduloju, A. R. Ethnobotanical assessment of herbal plants in southwestern Nigeria. *Academic Research International* **2012**, *2,* 50–57.

Balick, M. J.; Cox, P. A. Plants, People, and Culture: The Science of Ethnobotany. Scientific American Library, New York, **1996,** pp. 228.

Balik, M. J. Ethnobotanical screenings of medicinal plants most often yield higher hit rates than random screenings. Ethnobotany, Drug development and biodiversity conservation exploring the linkages. *Ciba Foundation Symposium* **1994**, *185,* 4–18.

Bekele, E. Study on Actual Situation of Medicinal Plants in Ethiopia. Prepared for Japan Association for International Collaboration of Agriculture and Forestry, *Addis Ababa,* **2007**.

Benkeblia, N. Antimicrobial activity of essential oil extracts of various onions (*Allium cepa*) and garlic (*Allium sativum*). Lebensmwiss u-Technol. **2004**, *37,* 263–268.

Bentley, R.; Trimen, H. Medicinal plants. Volos-IV (repr. Edn). International Book Distributor, Dehradun **1980**.

Chopra, R. N.; Chopra, I. C.; Handa, K. L.; Kapur, L. D. *Indigenous Drugs of India*: Second Edition (Reprinted), New Delhi, Academic Publishers, **1982**.

Chopra, R. N.; Nayar, S. L.; Chopra, I. C. In Glossary of Indian medicinal plants, vol. 1 *Council of Scientific and Industrial Research,* New Delhi **1956,** pp. 197.

Chopra, R. N.; Nayar, S. L.; Chopra, I. C. In Glossary of Indian medicinal plants Pub. And Info. Det., CSIR, New Delhi, **1956**.

Forombi, E. O. African indigenous Plants with chemotherapeutic potentials and biotechnological approach to the production of bioactive prophylactic agents. *African Journal of Biotechnology* **2003**, *2,* 662–671.

Gamble, J. S.; Fischer, C. E. C. *Flora of Madras Presidency:* Bishen Singh and Mahendrapal Singh. Dehradun, 1915–36.

Gebremariam, T.; Asress, K. Applied Research in Medicinal Plants. Programme and Abstract of National Workshop on "Biodiversity Conservation and Sustainable Use of Medicinal Plants in Ethiopia," April 28–May 1, Biodiversity Institute, *Addis Ababa,* 1998.

Haines, H. H. *The Botany of Bihar and Orissa 6 Parts*: Bishen Singh and Mahendrapal Singh. Dehradun, **1921–25**.

Hamil, F. A.; Apio, S.; Mubiru, N. K.; Bukenya-Ziruba, R.; Mosanyo, M.; Magangi, O. W. et al. Traditional herbal drugs of Southern Uganda, II. Literature analysis and antimicrobial assays. *Journal of Ethnopharmacol* **2003**, *84,* 57–78.

Jain, S. K.; Sinha, B. K.; Gupta, R. C. *Notable Plants in Ethnomedicine in India.* NewDelhi, Deep Publications, **1991**.

Jain, S. K. *A Manual of Ethnobotany*: Scientific Publishers, Jodhpur, **1995**.

Kirtikar, K. R.; Basu, B. D. Indian Medicinal Plants: Bishen singh & Mahendrapal singh. Dehradun, 1980,

Krithikar, K. R.; Basu, B. D. *Indian Medicinal Plants* In: Blatter, E., Caius, J. F. editors. Vol 1–5. Bishen Singh, Mahendra Pal Singh. Dehradun **1991,** pp. 1734.

Motsei, M. l.; Lindsey, K. L.; Van Staden, J.; Jaeger, A. K. Screening of traditionally used South African plants for antifungal activity against Candida albians. *Journal of Ethnopharmacol* 2003, *86,* 235–241.

Mudgal, V.; Pal, D. C. Medicinal Plants used by tribals of Mayurbhanj (Orissa). *Bull Bot Surv India* **1980**, *22,* 59–62.

Nandkarni, A. K. *Indian Materea Mediica.* Vol. 1 (3rd edn). Popular Book Depot, Bombay, 1954.

Pandey, A. K.; Rout, S. D. Ethnobotanical uses of Plants by tribals of Similipal Biosphere Reserve, Orissa. *Ethnobotany,* **2006**, *18,* 102–106.

Pandey, A. K.; Rout, S. D.; Pandit, N. *Medicinal Plants of Similipal Biosphere Reserve–I,* pp. 681–696. In: Das, A. P. (Ed.). *Perspectives of Plant Biodiversity*. Bishen Singh Mahendra Pal Singh, Dehra Dun, **2002**.

WHO: *Traditional Medicine*. Fact sheet No 134 **2003**.

Rojas, R.; Bustamante, B.; Bauer, J.; Antimicrobial activity of selected Peruvian medicinal plants. *Journal of Jthenopharmacuetical* **2003**, *88,* 199–204.

Rout, S. D.; Pandey, A. K. Ethnomedicobiology of Similipal Biosphere Reserve, Orissa. In: Das, A. P. & Pandey, A. K. (Eds.): *Advances in Ethnobotany:* Dehera Dun **2007,** pp. 247–252.

Saxena, H. O.; Brahmam, M. *The Flora of Orissa. Vols I-IV.* Orissa Forest Development Corporation Ltd, Bhubaneswar, India, 1996.

Saxena, H. O.; Brahmam, M.; Dutta, P. K. Ethnobotanical studies in Similipal Forests of Mayurbhanj District (Orissa). *Bull Bot Surv Indi.* **1988**, *10*(1–4), 83–89.

WHO: Traditional medicine, Fact Sheet No 134 **2003**.

World Health Organization. **2002**, WHO Traditional medicine strategy 2002–2005.

Zhang, X. *Traditional Medicine, Its Importance, and Protection*, In: Twarog, S., Kapoor, P. (eds). Protecting and promoting traditional knowledge system, National Experience, and International Dimensions Part-1. The Role of Traditional Knowledge in health care and Agriculture. New York; United Nations, **2004,** pp. 3–6.

CHAPTER 17

Garcinia xanthochymus Hook. f. ex T. Anderson: An Ethnobotanically Important Tree Species of the Similipal Biosphere Reserve, India

RAJKUMARI SUPRIYA DEVI,[1] SUBHENDU CHAKROBORTY,[2] SANJEET KUMAR,[3] and NABIN KUMAR DHAL[4]

[1]*Ambika Prasad Research Foundation, Regional Centre, Imphal–795001, Manipur, India, E-mail: supriyaaprf91@gmail.com*

[2]*Department of Biomedical Sciences and Engineering, National Central University, Jhongli City 320, Taiwan*

[3]*Ambika Prasad Research Foundation, Bhubaneswar–751006, Odisha, India*

[4]*CSIR–Institute of Mineral and Material Technology, Bhubaneswar, India*

ABSTRACT

Garcinia xanthochymus (Clusiaceae) is an evergreen tree species of tropical regions. The fruits are consumed by the rural and tribal communities of Similipal Biosphere Reserve, India. Fruits have sound nutritional and antibacterial activities. Less reports on leaf extracts are available in spite of sound anti-inflammatory activities and leaf paste is used against bacterial infections collected from the local communities. Keeping this in view, an attempt has been made to evaluate the antibacterial activities of aqueous extract of *Garcinia xanthochymus* leaves. The plant extract was prepared by maceration in water. The qualitative phytochemical analysis was done. The extract was used to evaluate the Minimum Inhibitory Concentration (MIC) using broth dilution method against selected five bacterial strains [*Vibrio cholerae* (MTCC 3909), *Salmonella typhi* (MTCC 1252), *Shiegella flexneri*

(MTCC 1457), *Streptococcus pyogenes* (MTCC 1926), *Streptococcus mutans* (MTCC497)]. Field survey revealed that leaves are used against inflammation and antibacterial infections. The MIC values showed that aqueous extract of experimental leaves are low against MTCC 1926 than other studied bacterial strains. Based on the validation of tribal claims, it is concluded that ethnobotanical values are the base of ethnopharmacological and pharmaceutical industries and could help to fight against AMR (Antimicrobial resistance).

17.1 INTRODUCTION

Plants contain many primary and secondary metabolites, which are commercially more important and used in pharmaceutical industries (Cowan, 1999). Secondary metabolites are "antibiotic" in a wide ambiance, protecting the plants against bacteria, animals, fungi and even other plants found in very smaller range from plants. Human beings have been dependent on plants for therapeutic purposes since the very beginning. In traditional systems, several medicinal plants are used as medicine by indigenous people for various ailments like stomachache, diarrhea, and dysentery, skin problems like wounds, eczema, urinary problems, hypertension, etc. The employ of plant extracts and secondary metabolites, both with bioactive properties can be of great magnitude in therapeutic treatments. The bioactive compounds and other plant pigments have therapeutic values in human life and could be refined to formulate the needed drugs. There are lots of reports are available in literature such as Quinine from the Cinchona, Morphine, and Codeine from the poppy, Digoxin from the Foxglove, Inulin from the roots of Dahlias, etc. (Meskin, 2002). In present era, most of our formulated known potential drugs are losing their curative properties due to resistance of pathogenic microorganisms. They have developed tolerance capacity against contemporary drugs due to natural mutation, overdoses and other wrong practices of treatments. This phenomenon is known as Antimicrobial resistance (AMR). Hence, it is very necessary to screen the new anti-microbial drugs from wild. Plants are the prime sources of antimicrobial substances represent a starting point for drug discovery due to disease precautionary properties (Nascimento et al., 2000). The foremost groups of antibacterial secondary metabolites include alkaloids, flavonoids, quinines, essential oils, lectins, phenolics, polyphenol tannins and terpenoid (Dreosti, 2000) are found in plant parts. The research of wild medicinal plants as antimicrobial agents is required for acquiring insight into medicinal flora and their rare value (Cong et al., 2004; Tiwari, 2004). The plants used in traditional medicine are still a large source of medicines that might serve as leads for the development of novel anti-microbial drugs (Lee et al., 2003). Among them, genus *Garcinia* belonging to family Clusiaceae consists of over 200 species distributed in the tropical area of the world. About 35 species exist in India, many of which are endemic and economically important with huge medicinal properties (Roberts et al., 1984). *Garcinia* species has demonstrated to be an interesting source of active compounds with a great biological versatility. The genus is also rich source of oxygenated and prenylated xanthones (Mbwambo et al., 2006; Chen et al., 2010). There are several species of the genus is reported throughout the all vegetations like *G. cambogia, G. kola, G. indica, G. atro-viridis, G. hombroniana, G. parvifolia, G. mangostana, G. cowa, G. mannii, G. prainiana, G. xanthochymus,* etc. During the survey in 2009 to 2015 on rare medicinal plants, wild tuber crops and Non-timber forest produces under different organizations (Ravenshaw University, Cuttack, India; National Bureau of Plant Genetic Resources, Cuttack, India; Regional Plant Resource Centre, Bhubaneswar, India), third author (Dr. Sanjeet Kumar) found that the tribal communities of Similipal Biosphere Reserve (SBR), Odisha, India were using the fruits of *Garcinia xanthochymus* for food and preparation of pickles. It was also observed that the leaves pastes were applying to cure skin infections. SBR has a unique assemblage of a number of ecosystems, such as wetlands, forests, grasslands and mountains that accumulate into an abutting patch with a range of diverse vegetation types. Its rich flora and fauna with many indicator species make the SBR a unique base for ethno-botanical studies. SBR laying between 21° 10′ to 22° 12′ N latitude and 85° 58′ to 86° 42′ E longitude, ranging between 300 m to 1180 m above sea level, the SBR is located in central part of Mayurbhanj district in Odisha (Kumar and Jena, 2017). The average rainfall of the region is 173 mm with a maximum of 225 mm. The average maximum temperature during May is about 44°C, and the average minimum temperature is 4°C during December. It exhibits a mixed type of vegetation such as tropical moist broadleaf forest, dry deciduous hill forest, tropical moist deciduous forest, Orissa semi-ever-green forest, high-level sal forest with grasslands and Savana. The densely forested area of SBR is the shelter of many tribal communities including two primitive (Hill-Kharia & Mankirdia). The major tribal communities are Ho, Kolha, Santhal, Bathudi, Bhumija, Saunti, Munda, Gonda, etc. (Kumar et al., 2017a). The above beautiful co-incidences with SBR and the skills of their tribal communities offered me an opportunity to collect the ethnobotan-ical values on *Garcinia xanthochymus* and make it a future source plants for the extraction of antimicrobial compounds to formulate a new drugs to fight against AMR. *Garcinia xanthochymus* is a evergreen tree species with horizontal dense branches. It grows up to about 17 m. trunk is straight. Leaves are dark green, oblong and about 10–25 cm long. Flowers are white with 5 sepals and 5 petals. Fruits are green and very similar to wild mango, turned orange to yellow when ripen. Many researchers reported the pharmacological values of fruits, but no or less reports are available in literature on leaves whereas authors found the ethnobotanical uses of leaves. Modern mind never believes the tribal claims before any scientific confirmation. Hence, keeping all in view, an attempt has been taken to collect the traditional therapeutic values of *Garcinia xanthochymus* leaves and to validate their claims through phytochemical screening and analysis of MIC (Minimum Inhibitory Concentration) values against some selected pathogenic bacterial strains (Figure 17.3). The present chapter provides a baseline data to the pharmacological industries and researchers who work on the formulation of new drugs to fight against AMR. The present chapter also highlights the importance of wild edible plants as a nutraceutical.

17.2 METHODOLOGY ADAPTED

During the survey, the ethnobotanical values on *Garcinia xanthochymus* (Figures 17.1b and 17.1d) were collected using standard passport data form (PDF). As per the collected therapeutic values, leaves of *Garcinia xanthocymus* were collected from Similipal Biosphere Reserve (Figure 17.1a) and identified by authors. Leaves extracts was obtained from aqueous extract by using Soxhlet apparatus (Figure 17.2a). 50 g of powdered tuber was fed into the apparatus for extraction. Extract was collected with water after 5–6 siphons. The extract was then dried over a period of 24 hours at room temperature. The dried sample was then weighed and stored in microfuge tubes under refrigeration (Trease and Evans, 1989).

Phytochemical analysis (Trease and Evans, 1989; Sofowora, 1993; Harborne, 1993).

Test for Tannin: 0.7 ml of the extract was taken in test tube with dissolution in 50ml of DI water and was heated for 10 minutes. After cooling few drops of 1%, ferric chloride was added. Color of sample changed from light to dark creamy precipitate.

Test for Saponin: 5 ml of extract was dried, and 1ml of Ethyl acetate was added to it. Then Ethyl acetate was removed, and DI water was added to it. Then the mixture was shaken vigorously and observed for persistent foam which lasts for at least 15 minutes.

FIGURE 17.1 (See color insert.) Ethnobotanical data and plant parts collections, (a) Panoramic view of Similipal Biosphere Reserve, (b) Leaves and fruits of experimental plant, (c) GPS data collection, (d) Fruits, (e–f) Ethnobotanical data collection.

Test for Flavonoids: Few amounts of extract was taken in a flask and dissolved in 10% NaOH. Few drops of HCl was added. Yellow color turned to colorless.

Test for Terpenoid: One ml of extract was thoroughly mixed with 400 μl $CHCl_3$. Then the mixture was treated with drop of H_2SO_4. A reddish brown interface indicates presence of terpenoid.

Test for Phenolic compounds: Extract was mixed with 3–4 drops of ferric chloride solution. Development of bluish black color shows the presence of phenolic compounds.

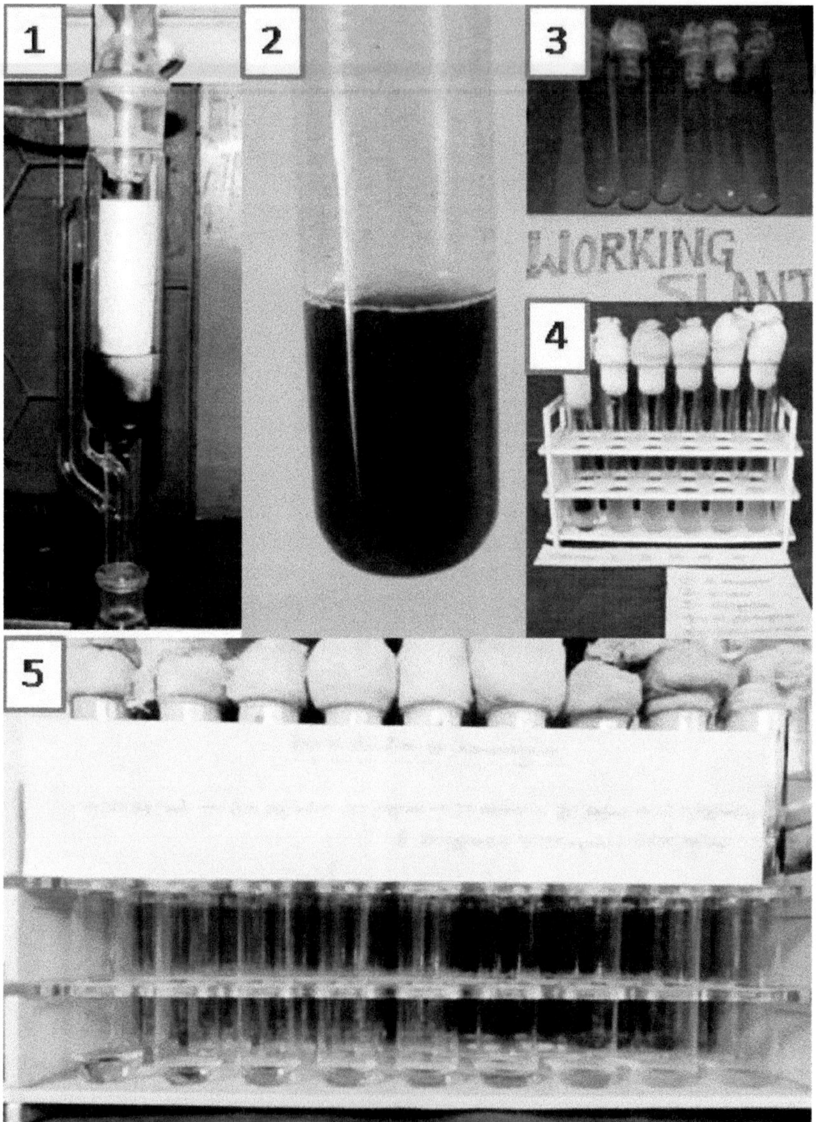

FIGURE 17.2 Phytochemical screening and antimicrobial activity of *G. xanthochymus* leaves extract, (a) Extraction, (b) Test of phenolic compounds, (c) Working slant, (d) Bacterial broth for MIC, (e) MIC values of Kanamycin.

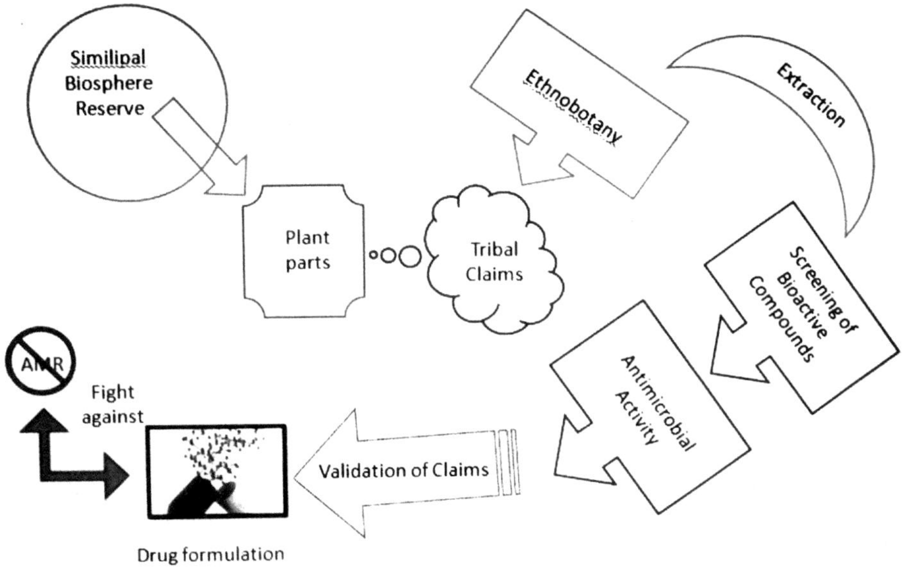

FIGURE 17.3 Pathway of formulation of new antimicrobial drugs from wild.

17.2.1 ANTIMICROBIAL ASSAY

The aqueous extract of *Garcinia xanthochymus* leaves were screened against two Gram-positive bacteria *Streptococcus mutans* (MTCC 497) and *Streptococcus pyogenes* (MTCC 1926); and three Gram-negative bacteria *Vibrio cholerae* (MTCC 3906), *Shigella flexneri* (MTCC 1457) and *Salmonella typhi* (MTCC 1252). All used MTCC (Microbial Type Culture Collection) bacterial strains (Figures 17.2c and 17.2d) were collected from Institute of Microbial Technology (IMTECH), Chandigarh. Antibacterial activity was done using slight modification of standard methods of Broth dilution assay (Rai et al., 2010; Kumar et al., 2017b)

17.3 OUTCOME AND DISCUSSION

The field survey indicated that the leaves of *Garcinia xanthochymus* are used against inflammation, skin infections and stomach pain by the tribal communities of Similipal Biosphere Reserve, India (Table 17.1). The ethnobotanical values indicated that the plant parts have rich bioactive compounds. Therefore, the qualitative phytochemical screening was done, and it was observed that aqueous extract of leaves are rich with saponin, tannin, flavonoids and phenolic compounds (Table 17.2, Figure 17.2b). The literature survey also indicated that the leaves are rich with diverse bioactive compounds (Table 17.3). The MIC values of leaves extract showed that aqueous extract is good against MTCC 1926 as compared to Kanamycin (Table 17.4, Figure 17.2e). The above results of phytochemical screening validate the tribal claims. The presence of tannin and saponin might be responsible for skin infections claimed by Kolho tribe. The presence of flavonoids and phenolic compounds in leaves extract might be responsible for curing inflammation and pain claimed by two primitive tribal communities (Mankirdia & Hill-Kharia) of SBR. The presence of terpenoids in leaves extract might be responsible to cure stomach problems claimed by Bathudi communities of study area (Figures 17.1e and 17.1f). The MIC values against MTCC 1926 (*Streptococcus pyrogens*-pathogenic bacteria responsible for various types of skin infections) validate the claimed by Kolha and Santhal communities that leaves paste is good against skin problems. Recently Misra et al. (2012) validate the tribal claims on *Celastrus paniculatus* (Intellect plant). In 2013, Kumar et al., did tribal claims validation on *Helicteres isora* and in 2014, Kumar et al., validate that *Dioscorea pentaphylla* has potential to cure various types of skin infections.

TABLE 17.1 Ethno-Botanical Values of *Garcinia xanthochymus* Collected from the Tribal Communities of Similipal Biosphere Reserve, India

Informants community	Collection site	Ethnomedicinal values	Source(s)
Mankirdia	Similipal Biosphere Reserve, India	Leaves are used against inflammation	Present work
Santal	Similipal Biosphere Reserve, India	Leaves are used to keep skin healthy	Present work
Kharia	Similipal Biosphere Reserve, India	Leaves are used against inflammation	Present work
Kolho	Similipal Biosphere Reserve, India	Leaves are used against infections (skin)	Present work
Bathudi	Similipal Biosphere Reserve, India	Leaves are used against stomach problems	Present work

TABLE 17.2 Most Common Secondary Metabolites Presence in *Garcinia xanthochymus*

Plant parts	Extract(s)	Bioactive Compounds	Detection	Responsible compound(s) for skin infections	Source(s)
Leaves	Aqueous	Saponin	+++	Saponin	Sofowora (1993)
		Tannin	+++		
		Flavonoids	+++		
		Terpenoids	+		
		Phenolic compounds	++++		

+: less, +++: rich. ++++: very rich.

TABLE 17.3 Structure Isolated from Leaves of *Garcinia xanthochymus* (Chen et al., 2010)

S. No.	Structure	Reference
1.	*Flavone group* 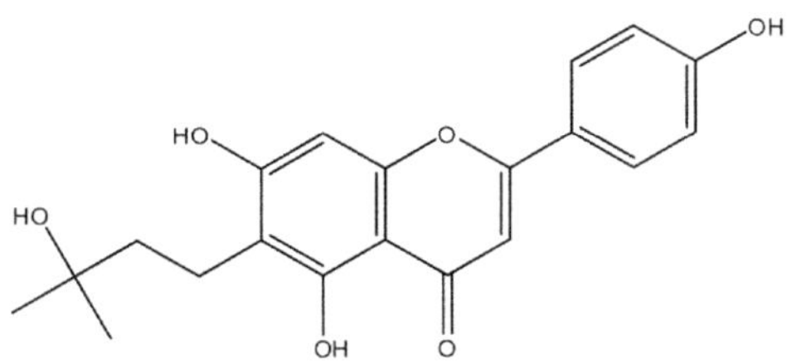	Chen et al., (2010)
2.	*Methoxyxanthone group* 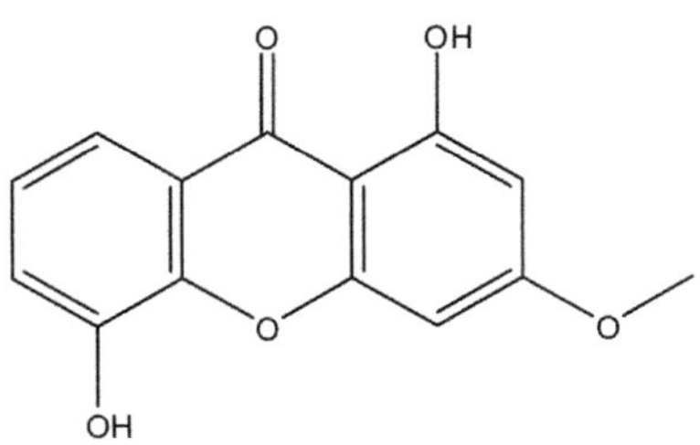	Chen et al., (2010)
3.	*Xanthone group* 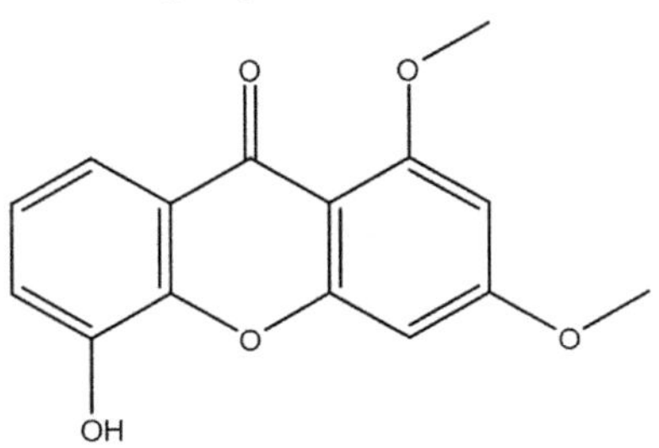	Chen et al., (2010)

TABLE 17.3 *(Continued)*

S. No.	Structure	Reference
4.	*Kaempferol group* 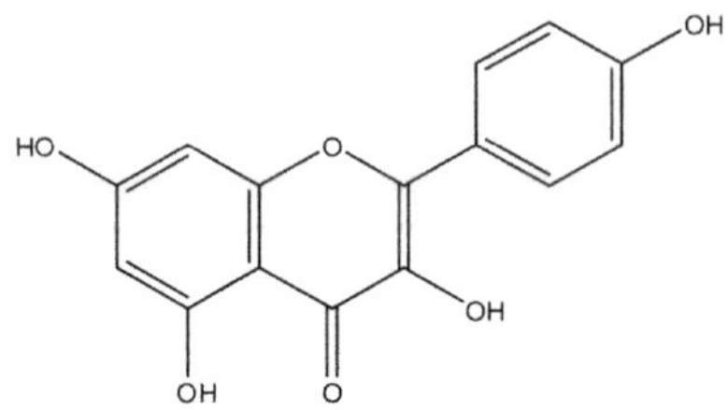	Chen et al., (2010)
5.	*Kaempferol group* 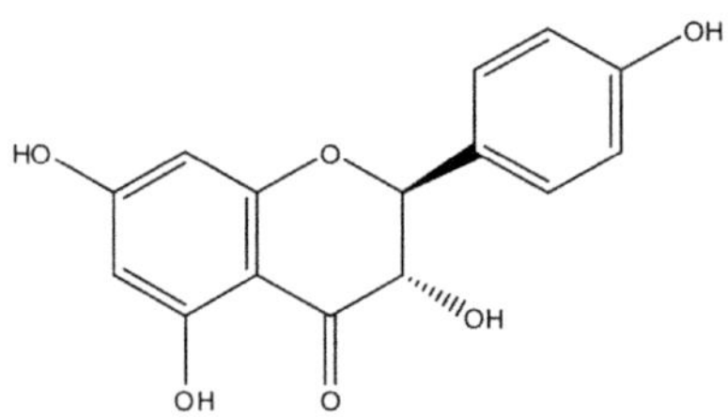	Chen et al., (2010)

TABLE 17.4 Antibacterial Activity (Minimum Inhibitory Concentration) to Validate the Tribal Claims on *Garcinia xanthochymus* Leaves

Agent	MTCC 3906	MTCC 1252	MTCC 1457	MTCC 1926	MTCC 497
Aqueous extract	500 µg/mL	500 µg/mL	500 µg/mL	450 µg/mL	500 µg/mL
Kanamycin	25 µg/mL	12.55 µg/mL	25 µg/mL	12.5 µg/mL	12.5 µg/mL
Inoculums control	Growth in all concentration	Growth in all concentration	Growth in all concentration	Growth in all concentration	Growth in all concentration
Broth control	No Growth	No Growth	No Growth	No Growth	No Growth
DMSO	Growth in all concentration	Growth in all concentration	Growth in all concentration	Growth in all concentration	Growth in all concentration

MTCC: Microbial type culture collection; MTCC 3906: *Vibrio cholerae;* MTCC 1252: *Salmonella typhi;* MTCC 1457: *Shigella flexneri;* MTCC 1926: *Streptococcus pyogenes;* MTCC 497: *Streptococcus mutans;* DMSO: Dimethyl sulfo-oxide.

17.4 CONCLUSION AND RECOMMENDATION

The present chapter highlights the medicinal importance of *Garcinia xanthochymus* available in Similipal Biosphere Reserve. The results obtained from the MIC analysis showed that aqueous extract is significant against MTCC

1926. Further research on the active extract is needed to isolate the active compounds to formulate new drugs from the said active extract, which will be helpful to fight against AMR. The chapter recommends that wild nutraceutical give ample scope to screen the new novel bioactive compounds. Therefore, more screening is needed under the sustainable extraction of wild valuable resources. The chapter also recommends that there is an urgent need to conserve such important bioresearches of the state for future generation.

ACKNOWLEDGMENTS

The authors are grateful to the Field Director, Similipal Biosphere Reserve, HOD, Department of Botany, Ravenshaw University, Cuttack for providing the facilities for the present study. The authors wish to thank the rural and tribal communities of the villages Hatibadi, Padampur, Gurguria and Kalika Prasad of Similipal Biosphere Reserve. Authors are also thankful to the Dr. R. C. Misra, National Bureau of Plant Genetic Resources, Base Centre, Cuttack and Chief Executive & Dr. P. C. Panda, Regional Plant Resource Centre, Bhubaneswar for their support.

KEYWORDS

- **antibacterial activity**
- **chemical constituents**
- **ethnobotany**
- ***Garcinia xanthochymus***
- **Similipal Biosphere Reserve**

REFERENCES

Chen, Y.; Fan, H.; Yang, G.; Jiang, Y.; Zhong, F.; He, H. Prenylated *xanthones* from the bark of *Garcinia xanthochymus* and their 1,1-Diphenyl-2-picrylhydrazyl (DPPH) radical scavenging activities. *Molecules*. **2010**, *15*, 7438–7449.

Cong, L.; Khan, S. I.; Jacob, M.; Walker, L. A.; Ferreira, D. Absolute configuration and biological activity of bioflavonoids from *Rheedia acuminata*. *Abstracts of Papers, 228th ACS National Meeting*, Philadelphia, PA, USA, **2004**, 22–26.

Cowan, M. M. Plant products as antimicrobial agents. *Clin. Microbiol. Rev.* **1999**, *12*, 564–582.

Dreosti, I. E., Recommended dietary intake levels for phytochemicals, Feasible or Fanciful. Asia Pac. J. Clin. Nutr. **2000**, *9*, 119–122.

Harborne, J. B. *Phytochemistry*. Academic Press, London. **1993**, 89–131.

Kumar, S.; Das G.; Shin, H. S.; Patra, J. K. *Dioscorea* spp. (A wild edible tuber): A study on its ethnopharmacological potential and traditional use by the local people of Similipal Biosphere Reserve, India. *Frontiers in Pharmacology.* **2017a**, DOI: 10.3389/fphar.2017.00052.

Kumar, S.; Jena, P. K. Tools from biodiversity: wild nutraceutical plants. Edt: Furze, J. N.; Swing, K.; Gupta, A. K.; McClatchey, R. H.; Reynolds, M. D. *Mathematical Advances Towards Sustainable Environmental Systems*. Springer. **2017**, 181–211.

Kumar, S.; Mahanti, P.; Singh, N. R.; Rath, S. K.; Jena, P. K.; Patra, J. K. Antioxidant activity, antibacterial potential and characterization of active fraction of *Dioscorea pentaphylla* L. tuber extract collected from Similipal Biosphere Reserve, Odisha, India. *Brazilian Journal of Pharmaceutical Sciences.* **2017b**, DOI: 10.1590/s2175–97902017000417006.

Lee, S. E.; Hyun, J. H.; Ha, J. S.; Jeong, H. S.; Kim, J. H. Screening of medicinal plant extracts for antioxidant activity. *Life Sci.* **2004**, *73*, 167–179.

Liu L.; Li, Y.; F, Gan, F.; Yang, G. Z.; Chen, Y. Chemical constituents from leaves of *Garcinia xanthochymus*. *China Journal of Chinese Material Medica*. **2016**, *41* (11), 2098–2104.

Mbwambo, Z. H.; Kapingu, M. C.; Moshi, M.; Machumi, J.; Apers, F.; Cos, S. P.; Ferreira, D.; Marais, J. P.; Berghe, D. V.; Maes, L.; Vietinck, L.; Pieters, L. Antiparasitic activity of some xanthones and bioflavonoids from the root bark of *Garcinia Livingstone*. J. *Nat. Prod.* **2006**, *69*, 369–372.

Meskin, M. S. *Phytochemicals in Nutrition and Health.* CRC Press, 2002.

Nascimento, F. G. G.; Locatelli, J.; Freitas, P. C.; Silva, G. L. Antibacterial activity of plant extracts and phytochemicals on antibiotic-resistant bacteria. *Braz. J. Microbiol.* **2002**, *31*, 247–256.

Rai, U. S.; Arun, M.; Isloor; Shetty, P.; Vijesh, A. M.; Prabhu, N.; Isloor, S.; Thiageeswaran, M. Novel Chromeno (2,3-b)-pyrimidine derivatives as potential anti-microbial agents. *Euro. J. Med. Chem.* **2010**, *45*, 2695–2699.

Roberts, E.; Sing, B.; Sing, M. P. *Vegetable Materia Medica of India and Ceylon*. Dehra Dun, India, **1984.**

Sofowora. *Medicinal Plants and Traditional Medicines in Africa*. Chichester: John Wiley and Sons: New York, **1993**.

Tiwari, A. K. Imbalance in antioxidant defense and human diseases: Multiple approaches of natural antioxidant therapy. *Curr. Sci.* **2004**, *8,* 1179–1187.

Trease, G. E.; Evans, W. C. *Pharmacognosy*, 14th Ed. W. B. Scanders Company, Ltd., London. **1989**, 89–300.

CHAPTER 18

Tribal Claims vs. Scientific Validation: A Case Study on Two Species of the Order Zingiberales *Curcuma longa* L. and *Costus speciosus* Koen.

R. S. DEVI,[1] ARCHITA BEHERA,[2] PUYAM DEVANDA SINGH,[1] MADHUSMITA MAHAPATRA,[3] BIKASH BHATTARAI,[4] and PADAN KUMAR JENA[2]

[1]*Ambika Prasad Research Foundation, Regional Centre, Imphal, Manipur, India, E-mail: supriyaaprf91@gmail.com*

[2]*Department of Botany, Ravenshaw University, Cuttack, Odisha, India*

[3]*Institute of Bioresources and Sustainable Development, Sikkim Centre, Tadong, Gangtok, Sikkim, India*

[4]*Regional Centre, Ambika Prasad Research Foundation, Khamdong, Singtam, Sikkim, India*

ABSTRACT

Medicinal plants play an important role to cure many diseases and disorders and also play a preeminent role in traditional medicines and modern day. Since ancient period, rural and tribal communities have been using different parts of the plant as remedies against various diseases. They claim particular plant parts for specific diseases, but modern era does not accept without scientific justification. Keeping the above point in view an attempt has been made to study the comparative understanding of the bioactive compounds, antibacterial activity and antioxidant potentials of two wild edible rhizome bearing plants [*Curcuma longa* L. (CL) and *Costus speciosus* Koen. (CS)] of Odisha. The plants have long been used as traditional medicines in herbal remedies and healthcare preparations Antibacterial activity was against five pathogenic bacterial strains namely *Salmonella typhii, Shigella flexneri, Vibrio cholerae, Streptococcus pyogenes* and *Streptococcus mutans* using Agar well diffusion assay. The experiment using ethanol extract showed the highest zone of inhibition against *Salmonella typhii* of CL whereas CS showed highest against *Streptococcus mutans*. The aqueous extract of CL was observed against *Vibrio cholera* whereas CS showed highest zone of inhibition against *Streptococcus mutans*. The ethnobotanical information was presented using standard methods. *In vitro*, antioxidant activity was assessed by reducing power and scavenging activity toward DPPH using spectrophotometric method. Ethanol extract of CL showed highest antioxidant scavenging activity with 77.37% reduction as compared to others. The rhizomes of CL and CS showed presence of diverse bioactive compounds. The correlation was made between the claims made by the tribal communities and the biological activities of the bioactive compounds present in the plant parts, which proves to be safe and effective for its uses as herbal medicine.

18.1 INTRODUCTION

The traditional practice of curing different diseases and disorders using floral parts is back to pre-historic times. These therapeutic skills of rural and tribal heelers from period to period flow into the prime life habit that has emerged as ethnobotanical systems. Aboriginals use the floral parts in single or multiple formulations as macerated paste, juice, decoction, powder paste, ash paste, etc. in raw form to treat diseases. These ethnobotanical values are the baseline of modern allopathic drugs, but the 21st century's people do not accept these healing properties of plants. Anyway, people are bending towards the green chemistry due to side effects of allopathic medicines. Pharmacological industries rapidly screening the new natured products due to the problems created by antimicrobial resistance (AMR). The properties of the plants to use as medicines are due to its bioactive compounds present in the plant. These bioactive compounds are not necessary for their normal growth and reproduction but are used as a defense mechanism. The aboriginals have developed different unique skills to use these constituents against the diseases. Therefore there is a need to study the ethnobotany, phytochemistry and biological activity of the plant and correlating with the previous literatures and aboriginals available. The present work highlights the comparative traditional uses of two rhizomes bearing wild plants, *Curcuma longa* L. and *Costus speciosus* Koen. There is a need for a scientific justification behind the logic of the claims made by the tribal communities for its safe and effective uses.

Costus speciosus (Koen.) Sm. (Plate 18.1:2) and *Curcuma longa* L. (Plate 18.1: 1) are important and commonly available rhizomatous plants possessing wide traditional uses among the common man of Odisha. Both the plant species belong to the same order Zingiberales. *C. longa* L. (CL) also known as turmeric belongs to the family Zingiberaceae which is characterized by the presence of volatile oils and oleoresins is a small rhizomatous, perennial aromatic plant distributed widely in South Asia and cultivated extensively throughout the warmer parts of the world. It has many rhizomes on its root system which are the source of its culinary spices and its medicinal extract called Curcumin (Chakraborty et al., 2011). The plant occupies a special position in India as it plays an important part in the rituals, ceremonies, and cuisine (Fischbash and Walsh, 1993). Because of its strong antiseptic properties, it has been used as a remedy for all kinds of skin infections, ulcers, and wounds (Hegde et al., 2012). It helps in adding to the complexion of the skin, and so it is applied on the face as a depilatory and facial tonic. It is also used as a blood purifier as it destroys pathogenic organisms (Kumar and Satapathy, 2011). A paste of turmeric alone, or combined with a paste of neem (*Azadirachta indica*) leaves, is used to cure ringworm, obstinate itching, eczema, and other parasitic skin diseases and in chicken pox and smallpox (Kumar et al., 2013). *C. speciosus* (Koen.) Sm. (Plate A-1) belongs to the family Costaceae, placed under family Zingiberaceae. It is a wild rhizomatous stout. It is rich with a saponin, diosgenin in its rhizome. It is generally used in the commercial production of steroidal hormones (Thomas et al., 1997). The rhizomes are also useful in vitiated conditions of kapha and pitta (Malabadi, 2005). It is also used against burning sensation, flatulence, constipation (Vasantharaj et al., 2009), helminthiasis, leprosy, skin diseases (Saraf, 2010), fever, hiccough (Vijayalakshm et al., 2009), asthma, bronchitis (Choi, 2009), inflammation and anemia (Gupta et al., 2015). Literatures revealed that both the plants are used as wild food among the rural and tribal communities and has a diverse traditional therapeutic importance. They contain diverse bioactive compounds that make the potent pharmacological agents which widely available in Odisha. Despite this, field survey indicated that both of valuable plants are disappearing from the wild population in India. The survey also indicated that very less information was available on the justification of claims. Therefore, keeping above all, an attempt was made to gather the ethnobotanical claims of said plants through field survey in tribal area of Khurda district, Odisha, India. Evaluation of antibacterial activity of the ethanol and aqueous extract of the plant showed zone of inhibition and IC_{50} values. It also showed a positive detection of bioactive compounds and antioxidant activity.

PLATE 18.1 (See color insert.) Studied plant species, (1) *Curcuma longa,* (2) *Costus speciosus*

18.2 EXPERIMENTAL SECTION

The literature was collected from the online journal from web and offprint from the Library, Regional Plant Resources Centre, Bhubaneswar, India. In the field survey taken during 2012–2014 in tribal areas of Khurda district and

Mayurbhaj district of Odisha, India, ethnobotanical claims were collected through semi-structured questioners (Hawkes, 1980; Christian and Brigitte, 2004) and rhizomes of CS and CL were collected from the tribal village of Khurda district near Chandaka Wildlife Sanctuary, and were authenticated at Botany Department of Ravenshaw University, Cuttack. The phytochemical analysis was done using standard methods (Harborne, 1973; Trease and Evans, 1989; Sofowara, 1993; Kumar, 2011; Misra et al., 2012). The ethanolic and aqueous extracts were prepared using soxhlet and rotary evaporator. The antibacterial activity carried out using agar well diffusion assay (Allen et al., 1991; Kumar et al., 2013; Tripathy et al., 2014) and disc diffusion assay (Scorzoni et al., 2007) against *Vibrio cholera* (MTCC 3909), *Salmonella typhii* (MTCC 1252), *Shigella flexneri* (MTCC 1457), *Streptococcus pyogenes* (MTCC 1926) and *Streptococcus mutans* (MTCC 1497) obtained from the microbial type culture collection (MTCC), IMTECH, Chandigarh. The data for the measure of zone of inhibition (mm) of the antibacterial activity was analyzed using standard deviation method. The antioxidant activity was evaluated using DPPH assay (Kumar et al., 2017).

18.3 IMPORTANT FINDINGS

Pathogenic bacteria have been a major caused by health complications and mortality in human. Although many pharmaceutical industries have developed many antibacterial drugs in the last few decades, resistance to such drugs is increasing which has now become a global concern. Among the potential sources of new agents, plants have long been investigated. Because they contain many bioactive compounds that can be of interest in therapeutic. Due to its less toxicity, the plant has been traditionally used in normal diet and in treatment of various diseases. Keeping the above context in view, the results of present study showed that the studied plants are suitable edible antibacterial agents to fight against resistance problems. The literature survey showed that both plants are used to cure skin infections, in cold and other diseases and disorders. The details are listed in Table 18.1. The ethnobotanical survey revealed that the rhizome of both plant is used against skin infections and some common diseases and disorders. The details are listed in Table 18.4. The qualitative detection of bioactive compounds showed plants possess diverse bioactive compounds, which make them strong pharmacological agents. Tannin was present in all extracts of both experimental plants whereas flavonoids, terpenoids, steroids, and saponin were detected in ethanolic extract of CS and tannin,

TABLE 18.1 Ethno-Medicinal Values of Selected Experimental Plant Species

Plant species	Parts used	Mode of use(s)	Medicinal Uses	Supporting Literature(s)
C. speciosus	Rhizome	Juice (Oral)	Skin infections	Vasantharaj et al., (2009); Kumar et al., (2012); Duraipandiyan & Ignacimuthu, (2011); Saraf, (2010)
			Antidiabetic	Duraipandiyan & Ignacimuthu, (2011); Rajesh et al., (2009); Vishalaxmi & Urooj, (2009)
			Antifertility	Choudhury et al., (2012)
		Dried powder	Hepatotoxicity	Verma and Khosa, (2009)
	Stem and Rhizome	Juice (Oral)	Diarrhea and Dysentery	Vasantharaj et al., (2009); Saraf, (2010)
	Leaves and Rhizome		Fever and Headache	Ariharan et al., (2012); Duraipandiyan & Ignacimuthu, (2011)
	Rhizome	Paste	Inflammation	Binny K. et al., (2010)
	Leaves	Juice (Oral)	Asthma and Bronchitis	Duraipandiyan & Ignacimuthu, (2011); Omakhua, (2011)
C. longa	Rhizome	Juice (Oral)	Skin infection	Duaa, (2013); Qaiser et al., (2013); Jha et al., (2013)
		Juice (Oral)	Antifertility	Ghosh et al., (2011)
		Paste	Hepatotoxicity	Kalantari et al., (2007)
		Juice (Oral)	Diarrhea and Dysentery	Jha et al., (2013); Arutselvi et al., (2012)
	Bruised leaves and rhizome		Fever and Headache	Arutselvi et al., (2012)
	Leaves		Asthma and Bronchitis	Linthoingambi et al., (2013); Qaiser et al., (2013)

terpenoid and reducing sugar were detected in aqueous extract of CS (Table 18.2).

TABLE 18.2 Bioactive Compounds Detected in *Costus speciosus* (Koen.) Sm. and *Curcuma longa* L.

Plant name	Parts	Extract	Detected compounds
C. speciosus	Rhizome	Ethanol	Tannin, Flavonoid, Saponin, Terpenoid, Steroid, Glycosides and Reducing Sugar
		Aqueous	Tannin, Terpenoid and Reducing Sugar
C. longa	Rhizome	Ethanol	Tannin, Flavonoid, Alkaloid, Phenolic compounds, Glycosides and Reducing sugar
		Aqueous	Tannin, Flavonoid, Alkaloid, Saponin, Terpenoid, Steroid, Glycosides and Reducing sugar

Tannin, flavonoid, alkaloid, phenolic compounds, glycosides were detected in ethanol extract of CL whereas tannin, steroid, flavonoid, saponin, and glycosides in aqueous extract of CL (Table 18.2). Gupta et al. (2015) reported the presence of alkaloids, tannin, flavonoids, and glycosides in MeOH extract of CL rhizome whereas Jagtap and Satpute (2014) showed the presence of alkaloids, saponin, flavonoid and glycosides in MeOH extract of CS rhizome. Sarf (2010) reported the presence of flavonoid, saponin, and tannin in the same extract of CS. Devi and Urooj (2010) also showed the presence of flavonoid, alkaloid, terpenoid, steroids, and tannin in MeOH extract of CS rhizome. The above bioactive compounds detected and reported revealed that the experimental plant has sound antibacterial and antioxidant activity. In this context, the highest zone of inhibition of the ethanol extract (rhizome) of CL and CS was shown against *Salmonella typhii.* The aqueous extract of CL showed its highest inhibitory activity against *Vibrio cholera* whereas the aqueous extract of CS showed its highest inhibitory activity against *Streptococcus mutans* (Table 18.3). R. B. Malabadi (2005) explained that the hexane, methanol and aqueous extracts of leaf and rhizomes of CS were used by Indian traditional healers for treating skin diseases and evaluated the antibacterial activity against *Shigella, Staphylococcus aureus, Escherichia coli, Klebsiella pneumoniae, Pseudomonas, Bacillus subtilis* and *Salmonella* while Sarfa (2010) documented the activity (15 mm) of aqueous extract against *S. aureus.* Gupta et al. (2015) also reported the activity of MeOH and aqueous extract of CL rhizome against *S. aureus* are 12 mm and 18 mm inhibition respectively at 50 mg/mL. In this context, there are very less reports or no reports available in literature against used strains in present study.

TABLE 18.3 Antibacterial Activity of CL and CS (rhizome) Using Agar Well Diffusion Assay

Plant species	Extracts	Agar well diffusion assay (Zone of inhibition, mm)					Concentration
		ST	**SP**	**SM**	**VC**	**SF**	
C. longa	Ethanol	15.0 ± 0.4	10.5 ± 0.00	14.6 ± 0.2	14.8 ± 0.2	13.6 ± 0.2	2000 μg/mL
C. speciosus	Ethanol	12.0 ± 0.4	7.80 ± 0.2	12.00 ± 0.4	11.00 ± 0.4	11.00 ± 0.4	
C. longa	Aqueous	13.0 ± 0.4	12.0 ± 0.00	10.8 ± 0.2	13.3 ± 0.2	13.00 ± 0.4	
C. speciosus	Aqueous	12.0 ± 0.4	10.5 ± 0.4	14.8 ± 0.2	12.5 ± 0.8	11.6 ± 0.6	
Standard		9.50 ± 0.00	12.00 ± 0.00	11.00 ± 0.00	9.00 ± 0.00	10.00 ± 0.00	100 μg/mL
Plant species	**Extract**	**Disc diffusion assay (Zone of inhibition, mm)**					**Concentration**
		ST	**SP**	**SM**	**VC**	**SF**	
C. longa	Ethanol	14.00 ± 0.00	13.33 ± 0.57	10.16 ± 0.28	11.00 ± 1.73	10.83 ± 1.04	50 μg /Disc
C. speciosus	Ethanol	9.66 ± 1.15	10.00 ± 0.00	9.00 ± 0.00	9.00 ± 0.00	9.00 ± 0.00	
C. longa	Aqueous	9.50 ± 0.86	10.83 ± 1.04	10.33 ± 0.57	10.00 ± 0.00	10.00 ± 0.00	
C. speciosus	Aqueous	8.50 ± 0.50	8.50 ± 1.32	8.00 ± 0.86	8.16 ± 0.76	8.00 ± 0.86	
Standard		17.33 ± 0.57	18.00± 0.00	15.67 ± 0.58	16.44 ± 1.53	14.00 ± 0.00	10 μg /Disc

ST–*Salmonella typhii*, SP–*Streptococcus pyogenes*, SM–*Streptococcus mutans*, VC–*Vibrio cholera*, SF–*Shigella flexneri*.

The results of IC_{50} revealed that ethanol extract of CL showed the lowest IC_{50} value (77.37 µg/mL) as compared to CS. The details are documented in Table 18.4. Vijayalakshm and Urooj (2009) studied the different parts of CS for their polyphenol content and antioxidant activity. Chakraborty (2009) revealed that the antioxidant activity of chloroform extract of CS leaves for its free radical scavenging activity. Jagtap and Satpute (2014) reported the antioxidant activity (78.03% scavenging) of MeOH extract (rhizome) of CS whereas Choi (2009) reported the IC_{50} values of MeOH extract (rhizome) of CL is 58.17 µg/mL using DPPH assay.

TABLE 18.4 IC_{50} Values of CL and CS Using DPPH Assay

Plant name/standard	Extract	IC_{50} values
C. speciosus	Ethanol	83.46
	Aqueous	95.23
C. longa	Ethanol	77.37
	Aqueous	97.41
Standard (BHT)		32.14

The above literature survey and experimental results concluded that the studied plants have sound pharmacological potentials. The traditional medicinal uses of the rhizome of CL and CS indicate the wide healing potential, but the modern scientific ways do not accept the claims without justification. And therefore the justification was carried out with experimental results and validation was done with the previous studies reported. The justification and correlation revealed that the traditional therapeutic uses are interlinked with the bioactive compounds present in the rhizome of experimental plants and their antibacterial and antioxidant activities. The details are presented in Table 18.5.

18.4 CONCLUSION

The present investigation indicated that there are number of rhizome bearing plants that are available in biodiversity having sound traditional therapeutic values and natural pharmacological agents. The screening of such plants provides a baseline to fight against AMR and formulation of new pharmacological agents. The results revealed that the experimental rhizome possesses diverse bioactive compounds as the rhizomes are variously used to cure different diseases and disorders, thus selected rhizome might be a source

TABLE 18.5 Validation of Tribal Claims and Their Correlation with Bioactive Compounds Present in CL and CS

Plant species	Tribal Claims (Present study)	Correlation with bioactive compounds	Correlation with Antibacterial and antioxidant activity	Supporting literature(s)
CS	Juice is taken orally in case of diarrhea and dysentery.	Presence of Tannin, Terpenoid and Saponin may be responsible for antidiarrhoeal activity.	Activity against *Shigella flexneri*	Saraf, (2009)
	The paste is effective against scabies, ringworms, and eczema.	Presence of saponins in aqueous extract might be responsible for antifungal/antimicrobial activity/effective against skin diseases.	Activity against *Streptococcus pyogenes*	Vasantharaj et al., (2009)
	The juice is applied along with brushed leaves of *Costus speciosus* in case of fever/headache	Presence of terpenoids in aqueous and ethanol extract showing analgesic effect.		Ariharan et al., (2012)
	The rhizome juice is consumed as an antifertility agent	Presence of steroid in ethanol extract may be responsible for antifertility agent.		Choudhury et al., (2012)
CL	Decoction of stem and rhizome in case of diarrhea	Maybe due to presence of saponin and tannin in aqueous extract.	Activity against *Shigella flexneri*	Jha et al., (2013)
	The juice of rhizome is applied externally on burns, cuts, inflammation and other skin diseases.	Presence of saponin and flavonoids in water extract might be curing infectious diseases.	Activity against *Streptococcus pyogenes*	Qaiser et al., (2013)
	The dried powder of the rhizome is taken daily for liver disorders.	Phenolic compounds in the ethanol extract may be responsible for hepatotoxicity.		Kalantari et al., (2007)
	Paste of rhizome is used to cure muscular pain	Presence of Tannin and phenolic compounds might be responsible		Capasso et al., (2003)
	Aqueous paste is used as anti-aging agent	Presence of phenolic compounds might be responsible	Antioxidant activity of the rhizome	Ma et al., (2015)

of high pharmacological importance making them suitable nutraceuticals. The experimental results of pharmacological values are interlinked with the medicinal claims. The study also highlights the importance of plants for conserving the biodiversity for sustainable development.

ACKNOWLEDGMENTS

Authors are grateful to the Chief Executive, Regional Plant Resources Centre, Bhubaneswar; HOD and Staff members of the Department of Botany, Ravenshaw University, Cuttack for their kind support.

KEYWORDS

- **antibacterial activity**
- **antioxidant activity**
- ***Costus speciosus***
- ***Curcuma longa***

REFERENCES

Allen, K. L.; Molan, P. C.; Reid, G. M. A survey of the antibacterial activity of some New Zealand honeys. *The Journal of Pharmacy and Pharmacology* **1991,** *43,* 817–822.

Ariharan, V. N.; Meena, D. V. N.; Rajakokila, M.; Nagendra, P. P. Antibacterial activity of *Costus speciosus* rhizome extract on some pathogenic bacteria. *International Journal of Advanced Life Sciences.* **2012,** *4,* 24–27.

Arutselvi, R.; Balasaravanan, T.; Ponmurugan, P.; Saranji, N. M.; Suresh, P. Phytochemical screening and comparative study of antimicrobial activity of leaves and rhizomes of Turmeric varieties. *Asian Journal of Plant Science and Research.* **2012,** *2*(2), 212–219.

Binny, K.; Sunil, K. G.; Dennis T. Anti-inflammatory and antipyretic properties of the rhizome of *Costus speciosus* (Koen.) Sm. *Journal of Basic and Clinical Pharmacy.* **2010,** *1*(3), 177–181.

Capasso, F.; Gaginella, T. S.; Grandolini, G.; Izzo, A. A. Active principles. *Phytotherapy* **2003,** 31–44.

Chakraborty, P. S.; Ali, S. A.; Kaushik, S.; Ray, R. K.; Yadav, R. P.; Rai, M. K.; Singh, D.; Bhakat, A. K.; Singh, V. K.; John, M. D.; Das, K. C.; Prasad, V. G.; Nain, S. S.; Singh, M.; Chandra, P. K.; Singh, D. K.; Rai, Y.; Singh, P.; Singh, O.; Singh, A. K. N.; Shah, M.; Pradhan, P. K.; Bavaskar, R. L. D.; Nayak, C.; Singh, V.; Singh, K. *Curcuma longa*: A multicentric clinical verification study. *Indian Journal of Research in Homoeopathy.* **2011,** *5* (1), 19–27.

Choi, H. Y. Antioxidant activity of *Curcuma longa* L., Novel Foodstuff. *Molecular Cell Toxicology* **2009,** *5*(3), 237–242.

Choudhury, N.; Chandra, K. J.; Ansarul, H. Effect of *Costus speciosus* (Koen.) Sm. On reproductive organs of female Albino mice. *International Research Journal of Pharmacy.* **2012,** *3*(4), 200–202.

Christian, R. V.; Brigitte, V. L. Tools and methods for data collection in ethnobotanical studies of home gardens. *Field Method.* **2004,** *16* (3), 285–306.

Devi, V. D.; Urooj, A. Nutrient profile and antioxidant components of Costus speciosus Sm. And Costus igneus Nak. *Indian Journal of Natural Products and Resources.* **2010,** *1*(1), 116–118.

Duaa, S. S. Screening the Antibacterial potency of *Curcuma longa* L. Essential oil extract against boils causing *Staphylococcus* species. *International Journal of Advanced Biological Research.* **2013,** *3*(4), 490–500.

Duraipandiyan, V.; Ignacimuthu, S. Antifungal activity of traditional medicinal plants from Tamil Nadu, India. Asian Pacific *Journal of Tropical Biomedicine.* **2011,** 204–215.

Fischbash, M. A.; Walsh, C. T. Antibiotics for emerging pathogens. *Science.* **2009,** *325,* 1089–1093.

Ghosh, A. K.; Das, A.K; Patra, K. K.; Studies on antifertility effect of rhizome of *Curcuma longa* L. *Asian Journal of Pharmacy and Life Science.* **2011,** *1*(4), 349–353.

Gupta, A.; Mahajan, S.; Sharma, R. Evaluation of antimicrobial activity of *Curcuma longa* rhizome extract against *Staphylococcus aureus. Biotechnology Reports.* **2015,** *6,* 51–55.

Harborne, J. B. Phytochemicals methods. London. Chapman and Hall, Ltd. **1973,** 49–188.

Hawkes, J. G. Crop genetic resources: field collection manual. International Board of Plant Genetic Resources, Rome, Italy and EUCARPIA, University of Birmingham, England. **1980.**

Hegde, M. N.; Shetty, S.; Yelapure, M.; Patil, A. Evaluation of Antimicrobial activity of Aqueous and Hydro-alcoholic *Curcuma longa* extracts against endodontic pathogens. IOSR *Journal of Pharmacy.* **2012,** *2*(2), 192–198.

Jagtap, S.; Satpute, R. Phytochemical screening and antioxidant activity of rhizome extract of *Costus speciosus* (Koen.) J. E. Smith. *Journal of Academia and Industrial Research.* **2014,** *3*(1), 40–47.

Jha, H.; Barapatre, A.; Prajapati, M.; Keshaw, R. A.; Senapati, S. Antimicrobial activity of rhizome of selected *Curcuma* variety. *International Journal of Life Sciences, Biotechnology and Pharma Research.* **2013,** *2*(3), 183–189.

Kalantari, H.; Khorsandi, L. S.; Taherimbarakeh, M. The protective effect of the *Curcuma longa* extract on acetaminophen-induced hepatotoxicity in mice. *Jundishapur Journal of Natural Pharmaceutical Products.* **2007,** *2*(1), 7–12.

Kumar, S. Qualitative studies of bioactive compounds in leaf of *Tylophora indica* (Burm.f.) Merr. *International Journal of Research in Pharmaceutical and Biomedical Sciences.* **2011,** *2* (3), 1188–1192.

Kumar, S.; Behera, S. P.; Jena, P. K. Validation of tribal claims on *Dioscorea pentaphyplla* L. through phytochemical screening and evaluation of antibacterial activity. *Plant Science Research.* **2013,** *35*(1, 2), 55–61.

Kumar, S.; Jena, P. K.; Tripathy, P. K. Study of wild edible plants among tribal groups of Similipal Biosphere Reserve Forest, Odisha, India; with special reference to *Dioscorea* species. *International Journal of Biological Technology.* **2012,** *3*(1), 11–19.

Kumar, S.; Mahanti, P.; Singh, N. R.; Rath, S. K.; Jena, P. K.; Patra, J. K. Antioxidant activity, antibacterial potential and characterization of active fraction of *Dioscorea pentaphylla* L. tuber extract collected from Similipal Biosphere Reserve, Odisha, India. Brazilian Journal of Pharmaceutical Sciences. **2017,** *53*(4), e17006,

Kumar, S.; Satapathy, M. K. Medicinal plants in an urban environment; herbaceous medicinal flora from the campus of Regional Institute of Education, Bhubaneswar, Odisha. *International Journal of Life Science and Pharma Research.* **2011,** *2*(11), 1206–1210.

Linthoingambi, W.; Satyavama, D. A.; Singh, M. S.; Laitonjam, W. S. Antioxidant and antimicrobial activities of different solvent extracts of the rhizomes of *Curcuma leucorrhiza* Roxb. *Indian Journal of Natural Products and Resources.* **2013,** *4*(4), 375–379.

Ma, Z.; Cui, F.; Gao, X.; Zhang, J.; Zheng, L.; Jia, Le. Purification, characterization, antioxidant activity and antiaging of exopolysaccharides by Flammulina velutipes SF-06. *Antonie Van Leeuwenhoek.* **2015,** *107*(1), 73–82.

Malabadi R. B. *Journal of Phytological Research.* **2005,** *18*(1), 83–85.

Misra, R. C.; Kumar, S.; Pani, D. R.; Bhandari, D. C. Empirical tribal claims and correlation with bioactive compounds: a study on *Celastrus paniculata* Willd., a vulnerable medicinal plant of Odisha. *Indian Journal of Traditional Knowledge.* **2012,** *11*(4), 615–622

Omukhua, G. E. Medicinal and socio-cultural importance of *Costus afer* (Ker Grawl) in Nigeria. *International Multidisciplinary Journal.* **2011,** *5*(5), 282–287.

Qaiser, J.; Shahzad, M.; Sherwani, S. K.; Mohammad, S.; Uzma, J.; Malik, M. S.; Hussain, M. Antibacterial activity of two medicinal plants: *Withania somnifera* and *Curcuma longa. European Academic Research.* **2013,** *1*(6), 1335–1345.

Rajesh, M. S.; Harish, M. S.; Sathyaprakash, R. J.; Shetty, A. R.; Shivananda, T. N. Antihyperglycemic activity of the various extracts of *Costus speciosus* rhizomes. *Journal of Natural Remedies.* **2009,** *9*(2), 235–241.

Saraf, A. Phytochemical and antimicrobial studies of medicinal plant *Costus speciosus* (Koen.). *E-Journal of Chemistry.* **2010,** *7*(S1): S405-S413.

Scorzoni, L.; Benaducci, T.; Almeida, A. M. F.; Silva, D. H. S.; Bolzani, V. S.; Mendes, M. J. S. Comparative study of disc diffusion and microdilution methods for evaluation of antifungal activity of natural compounds against medical yeasts *Candida* spp and *Cryptococcus* sp. *Revista de Ciencias Farmaceuticas Basica et Aplicada.* **2007,** *28*(1), 25–34.

Sofowora, A. Medicinal plants and traditional medicine in Africa. Spectrum Books limited. Ibadan. **1993**.

Thomas, J.; Joy, P. P.; Mathew, S. S.; Skaria, B. P. Indigenous less known essential oils perspective. *Pafai Journal.* **1997,** *20*(1), 13–20.

Trease, G. E.; Evans, W. C. Pharmacognosy. WB Scanders Company, Ltd. London. **1989,** 89–300.

Tripathy, P. K.; Kumar, S.; Jena, P. K. Assessment of food, ethnobotanical and antibacterial activity of *Trichosanthes cucumerina* L. *International Journal of Pharmaceutical Sciences and Research.* **2014,** *5*(7), 2919–2926.

Vasantharaj, S.; Sathiyavimal, S.; Hemashenpagam, N. Antimicrobial potential, and screening of antimicrobial compounds of *Costus igneus* (N. E. Br). *International Journal of Natural Products and Resources.* **2009,** *1*(1), 116–118.

Verma, N.; Khosa, R. L. Evaluation of protective effects of Ethanolic extract of *Costus speciosus* (Koenig) Sm. Rhizomes on carbon tetrachloride-induced hepatotoxicity in rats. *Natural Product Radiance.* **2009,** *8*(2), 123–126.

Vijayalakshmi A.; Sarada, N. C. Screening of *Costus speciosus* extracts for antioxidant activity. *Fitoterapia,* **2008,** *79* (3), 197–198.

Vishalakshi, D. D.; Urooj, A. Nutrient profile and antioxidant components of *Costus speciosus* (Koen.) Sm. and *Costus igneus* (N. E. Br). *International Journal of Natural Products and Resources.* **2009,** *1*(1), 116–118.

Index

B

F

J

K

L

M

T

X

Y

Z